Agroforestry

Subhabrata Panda

Agroforestry

Nature Based Solution for Climate Change and Food Security

Springer

Subhabrata Panda
Department of Soil and Water Conservation and AICRP on Agroforestry
Bidhan Chandra Krishi Viswavidyalaya (BCKV)
Mohanpur, West Bengal, India

ISBN 978-981-96-6854-0 ISBN 978-981-96-6855-7 (eBook)
https://doi.org/10.1007/978-981-96-6855-7

This Springer imprint is published by the registered company Springer Nature Singapore Pte Ltd.
The registered company address is: 152 Beach Road, #21-01/04 Gateway East, Singapore 189721, Singapore

Preface

Agroforestry is a useful farming system to protect human health in today's busy era burdened with continuously limited availability of land resources for cultivation. This farming system can equally cater to the need for multipurpose food and neutralize the effects of global warming through carbon sequestration and also create carbon credits through agroforestry, a form of carbon farming. In this way, agroforestry will also be another assured source of income for farmers from their agroforestry-related carbon credit in the near future. This book covers the versatile nature of agroforestry systems by modulating farming practices through continuous provisioning for production of food, feed, and other materials in a sustainable way related to livelihoods of the community. The unique nature of agroforestry facilitates the supply of healthy foods, as well as animal and fish feeds, by immobilizing contaminants in the soil–water–plant–animal continuum.

Topics in this book are concentrated on plant-origin food and feed sources made available from the cultivation of alley crops in the interspaces of trees in agroforestry. Incorporating fruit trees within an agroforestry system creates an extra support for human nutritional sources of various fruits. Food security of the community, thus availed through agroforestry farming practices, can be ensured by provisioning improvement in livelihood security, and this can be achieved by value addition of agroforestry products right from woods to foods produced from such farming and from associated animal husbandry and aquacultural practices. The growth of selected trees for agroforestry is substantially rapid, and thus, growing these trees can be helpful in generating more financially lucrative livelihoods from farming on the same agricultural plot. This is achieved by producing timber for common uses, as well as reducing pressure on existing forest resources. In this way, agroforestry farming helps to conserve forests, protect the environment, and mitigate climate change through carbon sequestration.

This book envisions that agroforestry can contribute to achieving the Sustainable Development Goals (SDGs) of the United Nations. Students, interns, and entrepreneurs in agroforestry will be attracted to go through the numerical problems and solutions and the way of financially examining the validity of an agroforestry project at its formation. Likewise, this book would be helpful as a brief commentary for students and researchers in various soil, water, sustainable agriculture, natural resource conservation, and natural science related courses in the universities. It is a ready reference for students for appearing in competitive exams, field practitioners, interns, and individual farmers, concerned about awareness generation for sustaining food production as well as healthy livelihoods among farming community while initiating agroforestry as a nature-based solution in the prevailing challenges of climate change for better soil and water conservation in each farming plot.

The author is deeply grateful to the Hon'ble Vice-Chancellor, BCKV, and all the members of BCKV authority, as well as ICAR-CAFRI, for allowing association with regular routine work of the AICRP on Agroforestry and with the Department of Soil and Water Conservation at BCKV. This has helped the author to form such a book on agroforestry addressing toward the present needs of the intelligentsia as well as students, interns, and entrepreneurs.

The author is grateful to all the members associated with the publication of this book from Springer Nature, especially Fumiko Yamaguchi, William Achauer, Anil Chandy, Naren Aggarwal, Momoko Asawa, Muthuneela Muthukumar, and Priyanka Meganathan.

The author expresses his sincere thanks to all his colleagues at BCKV, especially Dr. R. Nath, Dr. D. Mukherjee, ADR, RRS, Jhargram, Dr. B. Biswas, OIC AICRP on AF, Joydev Rana, Pradip Kumar Nayek, ex-colleagues with AICRP on AF, Prof. S. K. De, Prof. R. Ray, Prof. N. C. Das, and Prof. P. K. Dhara of the SWC Department for their continuous encouragement for such works.

I would like to express my gratitude to Subhankar Roy of Connectica Management Service for creating a smooth computer system and to Binapani, Bodhi, Bappa, Piu, Dada, Boudi, Saheb, Bulbuli, and Abir for creating a cool atmosphere at home during manuscript preparation.

I am grateful to Smritirekha Panda, Abir Panda, and Dr. Bappaditya Mishra for sharing some photographs from their personal archives.

Jhargram, India
March 2025

Subhabrata Panda

Competing Interests The author has no competing interests to declare that are relevant to the content of this manuscript.

Contents

About the Author

Subhabrata Panda is an Assistant Professor in Soil and Water Conservation at Bidhan Chandra Krishi Viswavidyalaya (BCKV), the State Agricultural University in West Bengal, India, and is associated with the "All India Coordinated Research Project on Agroforestry" of the Indian Council of Agricultural Research with the ICAR-Central Agroforestry Research Institute (CAFRI). He has more than 30 years of experience in research, extension, and teaching from his present and previous associations as Assistant Agricultural Chemist at the State Agricultural Research Institute and additionally with the Soil Survey Laboratory and the State Nodal Cell of the Soil Health Management (SHM) under the National Mission for Sustainable Agriculture (NMSA), Government of West Bengal, Kolkata, India. Additionally, he has served as a Research Associate on the ICAR-MNES All India Network Project on Solar Photovoltaic Pumping Systems for crop water management and on the ICAR-sponsored inter-institutional Arsenic Project at BCKV, which focused on studying mobilization and management of arsenic from groundwater in Nadia district in West Bengal. He also worked as a Post-Graduate Apprentice on the National Watershed Development Project for Rainfed Areas (NWDPRA) at Arapanch Microwatershed, South 24 Parganas Zilla Parishad, West Bengal. Dr. Panda has authored a book titled, *Soil and Water Conservation for Sustainable Food Production* (SpringerBriefs). His research contributions include developing technologies for the reclamation of coastal salt-affected soils, land shaping for rainwater harvesting, and sustainable management of contaminated groundwater. His work has earned recognition, including his selection as an Expert Reviewer for the IPCC Sixth Assessment Report (2022) and his contribution to identifying unresolved problems in hydrology (2019). He has also received awards, such as the National Scholarship (1978) and the Certificate of Honour from a workshop on river hydraulics (1995). With significant experience in external collaborations and seminar organizations, Dr. Panda's contributions to the field of soil and water conservation, including saline, sodic, acid soil amelioration, watershed management, coastal agricultural water management, water quality management, rainwater harvesting, agroforestry, continue to shape sustainable agricultural practices.

1 Introductory Remarks: Agroforestry - The Sustainable Food Source

Abstract

Agroforestry, a land use system that integrates growing of trees and agriculture on the same piece of land. This land use is potential to serve as a promising sustainable food source to address the global challenges of exponentially growing population and so does the demand for food. This increasing pressure on food production has led to the exploitation of natural resources, causing environmental degradation and unsustainable land use practices. The role of agroforestry is established in promoting sustainable food production, ensuring food security, and extra benefit for farmers through carbon credit and maintaining environmental stability.

Keywords

Agroforestry · Agriculture · Tree · Food security · Natural resources · Environmental degradation · Land use practices · Sustainable · Environmental stability · Pulses · Oilseeds · Silvopasture · Silvoculture · Agrosilvopasture · Aquaculture · Livestock · Soil carbon sequestration · Greenhouse gas · Agricultural landscape · Root exudates · Food production · Carbon storage · Carbon credit · Carbon emission · Fossil fuel · Climate change · Value addition · Economic return · Economic resilience · Environmental conservation · Ecological balance · Ecosystem services · Ecotourism · Multipurpose food · Multipurpose tree · Multipurpose crop · Numerical problems · Problems and solutions · UN SDG 2030

Abbreviations

AF	Agroforestry
AFS	Agroforestry systems
CO_2	Carbon dioxide
D & D	Diagnosis and design (survey)

S. Panda, *Agroforestry*, https://doi.org/10.1007/978-981-96-6855-7_1

ES	Ecosystem services
GHG	Greenhouse gas
MPF	Multipurpose food
MPT	Multipurpose tree
SOC	Soil organic carbon
SOM	Soil organic matter
SDG	Sustainable development goal(s)
nbs	Nature-based solution
UN	United Nations

1.1 Agroforestry: A Brief Overview

Agroforestry combines agricultural and forestry practices on the same land, creating a harmonious blend of trees, crops, and livestock. It encompasses a wide range of systems, including alley cropping, silvopasture, and forest gardens, among others. By integrating trees with crops or livestock or aquaculture or fishery, agroforestry systems can enhance soil fertility, reduce erosion, increase biodiversity, and improve overall ecosystem health. Moreover, these systems can provide multiple products and services, including food, timber, fuelwood, fodder, and ecotourism, and environmental benefits making those farming systems more resilient and profitable than monoculture systems.

1.1.1 Sustainable Food Source

Agroforestry plays a vital role in promoting sustainable food source by addressing some of the critical challenges faced by conventional agricultural systems. These challenges include soil degradation, nutrient depletion, and the overuse of chemical fertilizers and pesticides. Agroforestry systems can help overcome these challenges by harnessing the following six ecosystem services (ES) of agroforestry to achieve UN SDG 2030 (UN SDG 2030 2015, van Noordwijk 2021) as nature-based interventions (Figs. 1.1 and 1.2).

- "Agroforestry — the treasure of multipurpose foods," the provisioning ES.
- "Agroforestry influencing production of cereals, pulses, oilseeds, and fruits," the regulating ES.
- "Agroforestry — helping soil organic carbon built-up," the provisioning ES.
- "Agroforestry — improving soil health and healthy food production," the regulating and provisioning ES,
- "Agroforestry — mitigating climate change through carbon sequestration," the regulating ES,
- "Agroforestry for sustainable and improved livelihoods through value addition and animal husbandry," the cultural and supporting ES.

Fig. 1.1 Ecosystem services of agroforestry systems

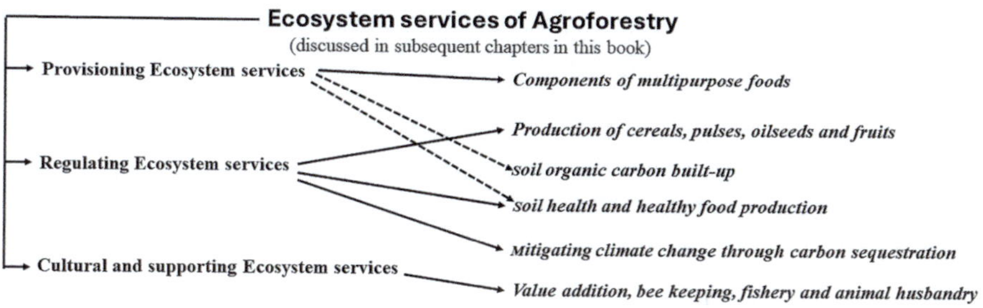

Fig. 1.2 Ecosystem services of agroforestry system (flow chart)

1.2 Multipurpose Food Components from Agroforestry

In the face of climate change as well as managing good health, it now requires attention, as described in Chap. 2, to the fascinating concept of multipurpose foods procured from agroforestry systems. Agroforestry (AF), a sustainable land use practice that combines agriculture and forestry, offers numerous benefits including food security, income generation, and environmental conservation. One of its most valuable aspects of AF is the production of multipurpose foods, which can be utilized in various ways, contributing to a diverse and nutritious diets. Multipurpose foods from agroforestry systems include fruits, nuts, vegetables, and even edible leaves, which can be consumed fresh, processed, or cooked in various recipes. These foods often have high nutritional value, providing essential vitamins, minerals, and dietary fiber. They also possess medicinal properties and can be used as natural remedies for minor ailments. Multipurpose food (MPF) is seemingly a "wonder food" source and has the potentiality to solve the problem of "Meals for Million," with the protein, minerals and vitamin-rich food components (Simonow 2021). As the single food source has the limitation to serve all those components, agroforestry (AF) can accommodate different types of food production to serve as the source of various components of MPF. Because, AF can serve as the source of production of staple foods of cereals and pulses; and oilseeds, vegetables, and fruits. With regard to agroforestry, being the source of multipurpose foods, AF should be the combination of multipurpose trees as well as multipurpose crops to enable the AF system as a sustainable source of multipurpose foods, i.e., the growing of different food materials from the same piece land, various combinations of which are viable to provide multipurpose food preparations from an AF system. As multipurpose food is the preparation of balanced food by mixing different food items, so an AF system could be a source of various food items required to prepare a multipurpose food. Again, as an AF system being combination of growing crops and trees on same piece of land, so an AF system, being the source of multipurpose foods, should be the combination of both the multipurpose trees (MPTs), for example, various multipurpose trees to provide fruit, therapeutic leaves or bark, and wood for construction like moringa, mango, cashew, and many more; and multipurpose crops, for example, rice, sorghum, potato, pepper, cabbage, eggplant, soybean, groundnut, carrot, cauliflower, and many more; also the provisions for procuring chitinous foods from crustaceans, molluscs, fungi like mushroom, and many more; and combination of animal husbandry and aquaculture for obtaining animal proteins as a component of multipurpose food. Combinations of some parts of plants and trees serve the purpose of environmentally friendly food serving and packaging. Presently, agricultural and tree products could serve for uses beyond food only, because agriculture and tree growing are the backbone of bioeconomy. Climate change, pollution, and finite fossil resources have driven the need for biodegradable alternatives. This has led to the emergence of a bioeconomy, where agriculture and tree growing provide raw materials for products (such as alternative of plastics, polymer composites, and paints) those were previously oil-based. For example, traditional crops like rice provides food but also plant material for housing, utensils, paper, and more like uses as home remedies; sorghum is cultivated for grain, sweet stems, forage, and broomcorn; and it also serves as a source of fuel, bioethanol, alcoholic beverages, and building materials; hemp is a valuable multipurpose crop as it provides raw materials for both food and non-food industries; above all trees in AF systems can produce therapeutic leaves or bark, wood for construction and fruit trees provide fruit. All such food and non-food AF products are also equally important for more environmentally friendly and economically viable future (Bastos Lima 2018; Ananda et al. 2020; Anna 2023).

1.3 Production of Cereals, Pulses, Oilseeds, and Fruits from Agroforestry

Agroforestry, a sustainable land use practice that integrates trees and shrubs with crops and/or livestock, as elaborated in Chap. 3, has been gaining prominence in recent years due to its numerous ecological, social, and economic benefits. AF influences the production of cereals, pulses, oilseeds, fruits, and vegetables. AF systems can enhance cereal production by improving soil fertility, providing shade to reduce soil moisture loss and acting as windbreaks. Trees and shrubs in these systems can fix atmospheric nitrogen which improves soil fertility and promotes cereal growth. Furthermore, the diversified AF systems can offer a more stable food supply as they provide alternative food sources during years with poor cereal harvests. The integration of trees and crops in agroforestry systems can positively impact pulse production. Trees can help control pests and diseases, reduce soil erosion, and provide organic matter that improves soil fertility. Additionally, certain tree species can act as cover crops or green manures, which can increase nitrogen availability in the soil and improve pulse yields. AF systems can improve oilseed production by offering better soil fertility, reduced soil erosion, and improved pest and disease control. Trees can also provide shade and wind protection, which can benefit oilseed crops that are sensitive to these environmental factors. Moreover, some tree species can fix atmospheric nitrogen, which can enhance the growth and yield of oilseed crops. AF systems can significantly contribute to fruit production by providing a suitable microclimate, improving soil fertility, and offering biological control of pests and diseases. The presence of trees and shrubs can create a favorable environment for fruit trees by reducing temperature fluctuations and by increasing humidity and can provide shelter against strong winds. Furthermore, the diverse species composition in agroforestry systems can lead to better pollination and increased fruit yields. The integration of trees and vegetables in AF systems can result in improved vegetable production.

1.4 Impact of Agroforestry Practices on Enhancing Soil Organic Carbon Content

Appraising the significant potential of agroforestry in increasing soil organic carbon (SOC), as detailed in Chap. 4, is crucial for maintaining soil fertility and overall ecosystem health. Agroforestry combines the cultivation of trees and crops on the same land, creating a sustainable and productive agricultural system.

Soil organic carbon refers to the carbon content as organic matter present in soil, primarily derived from plant and animal residues. It plays a vital role in soil structure, water retention, nutrient cycling, and overall soil fertility.

1.4.1 Benefits of Increasing Soil Organic Carbon (SOC)

The benefits of increasing soil organic carbon content can be listed as follows:

(i) Improved soil structure: Higher SOC content leads to better aggregation and stability of soil particles, enhancing water infiltration and this is helpful for reducing erosion.

(ii) Water retention and drought resilience: Soils with higher SOC can hold more water, reducing the risk of drought which requires improving overall water-use efficiency.
(iii) Nutrient cycling and fertility: SOC serves as a reservoir for essential nutrients, promoting plant growth and overall soil fertility.
(iv) Greenhouse gas mitigation: Soils with high SOC act as carbon sinks, helping to mitigate climate change by sequestering atmospheric carbon dioxide.

1.4.2 Natural Processes of Increasing SOC Through Agroforestry Practices

(i) Tree canopies and root systems contribute to the accumulation of organic matter by shading the soil surface. This helps in reducing soil erosion, and trapping organic debris.
(ii) Trees and crop plants release root exudates that stimulate microbial activity which act in promoting decomposition of organic matter and nutrient cycling.
(iii) Agroforestry systems often involve the incorporation of leguminous crops or trees which can fix atmospheric nitrogen, enhancing soil fertility and promoting the growth of other plants.
(iv) Diversified agroforestry systems can provide a continuous supply of organic matter through crop residues, pruned and fallen leaves, and small twigs, which thus help further increasing SOC levels.

1.4.3 Implementing Agroforestry Practices for SOC Enhancement

Implementation of agroforestry practices for SOC enhancement necessarily require considerations for area or region specific (i) designing appropriate AF systems and (ii) suitable diverse AF practices. Suitably created AF practices would yield expected benefits from created AF systems following their conspicuous implementing steps. These AFS implementation steps necessarily require due consideration for challenges and opportunities in practicing agroforestry in enhancing SOC.

1.4.3.1 Designing AF System

This implies establishment of locally suitable designing diverse agroforestry systems that combine trees, crops, and sometimes livestock husbandry. As the world grapples with climate change, soil organic carbon (SOC) has emerged as a critical component in the global effort to mitigate its effects. Soil is a vital carbon sink, and its ability to store carbon can play a significant role in reducing greenhouse gas (GHG) emissions. Agroforestry practices have been identified as a promising approach to enhance SOC levels, while also promoting sustainable agriculture and forestry. This requires exploring the concept of agroforestry, its benefits, and how it can be implemented to increase SOC.

1.4.3.2 Diverse AF Practices

Agroforestry, thus, is an agricultural practice that involves integrating trees into farming systems, alongside crops and livestock. This approach combines the benefits of both agriculture and forestry. This helps in creating a more diverse and resilient agroecosystem. Agroforestry system can be practiced in various ways, including:

(a) Silvopasture: Combining trees with livestock grazing.
(b) Silvoculture: Planting trees among crops.
(c) Agrosilvopasture: Combining trees with crops, pasture/livestock.
(d) Silvihorticulture: Combining trees with horticultural plants and trees and other various combinations of trees, crops, pasture/livestock, and aquaculture.

1.4.3.3 Benefits of Agroforestry Practices for SOC Enhancement

(a) Tree roots help retain soil: Tree roots hold soil in place, preventing soil erosion and soil degradation and thus, resisting SOC loss.
(b) Soil organic matter (SOM) accumulation: Leaf litter and self-pruning branches of agroforestry trees add organic matter to the soil, and thereby increasing soil carbon content.
(c) Microbial activity stimulation: Trees provide habitat for microorganisms, which break down organic matter and release nutrients. In that way, soil microbes help increasing SOC under agroforestry.
(d) Soil structure improvement: SOM and tree roots help improve soil structure, allowing for better water infiltration and aeration, and thus, agroforestry can enhance SOC retention.

1.4.3.4 Implementing Steps of Agroforestry Practices for SOC Enhancement

This can be grouped into three categories (i) D & D survey (a, b, c as hereunder), (ii) establishing AFS (b as hereunder), and (iii) Monitoring and maintenance of AFS (e as hereunder).

(a) Conducting a site assessment: Evaluation of topography of land, climate, soil type, and existing vegetation to determine the best agroforestry system.
(b) Choosing suitable tree species: Selecting tree species that are well-suited to the region and compatible with the farming or forestry goals.
(c) Designing agroforestry system: Planning the layout of agroforestry system, considering factors such as tree spacing, crop rotation, and livestock grazing.
(d) Establishing agroforestry system: Planting trees and establishing agroforestry system according to design plan.
(e) Monitoring and maintaining agroforestry system: Monitoring agroforestry system regularly for its performance and making adjustments as needed to optimize SOC enhancement. For example:

- Forest Garden Approach: In Rwanda, farmers have adopted forest gardens, integrating trees with crops and livestock, resulting in increased SOC levels (Kay et al. 2019; Nkurikiye et al. 2024),
- Silvopasture systems: In Brazil, silvopasture systems have increased SOC by up to 300% compared to traditional pasture systems (Freitas et al. 2020),
- Agroforestry farms: In humid and sub-humid tropics, agroforestry farms have increased SOC by 21% compared to conventional farming practices (Muchane et al. 2020).

1.4.3.5 Challenges and Opportunities in Practicing Agroforestry in Enhancing SOC

While agroforestry practices offer numerous benefits for SOC enhancement, there are also following challenges and opportunities in practicing agroforestry in enhancing SOC.

(a) Initial investment costs: Establishing an agroforestry system requires significant upfront investment in tree planting and infrastructure.
(b) Land use conflicts: Conflicts may arise between agricultural and forestry interests when implementing agroforestry practices (FAO and ICRAF 2019).
(c) Scalability: Agroforestry practices can be scaled up or down depending on local conditions and resources.

Thus, it is required to note that implementing agroforestry practices can significantly enhance soil organic carbon levels while promoting sustainable agriculture and forestry (Lorenz and Lal 2014). By understanding the benefits of agroforestry and following best practices for implementation, farmers, foresters, and policymakers can work together to mitigate climate change through soil carbon sequestration. As the world continues to grapple with climate change, the adoption of agroforestry practices will play a critical role in reducing greenhouse gas (GHG) emissions and promoting sustainable development.

1.5 Agroforestry in Improving Soil Health and Healthy Food Production

Agroforestry, being a land use management system, integrating trees and shrubs into agricultural landscapes, has been gaining attention for its numerous benefits on soil health and food production. This practice, as laid out in Chap. 5, not only promotes ecological balance and biodiversity but also contributes significantly to the improvement of soil quality and the provision of nutritious food. In that way, agroforestry can enhance soil health, and thereby improves healthy food production.

1.5.1 Soil Health Enhancement

Agroforestry systems contribute to the improvement of soil health in several ways. Trees and shrubs in these systems help in improving soil health in the following ways:

(a) Soil erosion control: The presence of perennial vegetation and annual crop covers in agroforestry systems reduce the impact of raindrops on the soil surface, which helps in preventing soil erosion and maintaining soil structure.
(b) Soil organic matter (SOM) accumulation: As trees and shrubs shed leaves, branches, and through root decomposition, they contribute to the organic matter content in the soil. This organic matter improves soil fertility, soil water retention, and soil aeration.
(c) Soil nutrient cycling: Agroforestry systems encourage the cycling of nutrients within the soil ecosystem. For example, trees and crops in these systems help fix atmospheric nitrogen, making it available for plant uptake, and also release essential nutrients like phosphorus and potassium.

1.5.2 Healthy Food Production

Agroforestry systems, being capable of enhancing soil health, can also promote the production of diverse and nutritious food by following three farming objectives:

(a) Crop diversification: The integration of trees and crops in agroforestry systems allows for the cultivation of a variety of food crops. This diversification reduces the risk of crop failure and ensures a continuous supply of nutritious food throughout the year.
(b) Improved soil fertility: As agroforestry systems contribute to soil fertility by adding organic matter and promoting nutrient cycling, this farming system results in healthier crops with higher nutritional value.
(c) Pest and disease management: The presence of trees and shrubs in agroforestry systems can help suppress pests and diseases by providing habitats for beneficial insects and microorganisms. These natural pest controllers help maintain a healthy ecosystem and reduce the need for chemical pesticides.

Thus, agroforestry plays a vital role in enhancing soil health and promoting sustainable food production. By integrating trees and crops, the resilient agricultural systems, thus created, is enable for continuous soil health enhancement and healthy food production.

1.6 Agroforestry Serves Toward Mitigating Climate Change Through Carbon Sequestration

Agroforestry, a sustainable land use practice, combining the roles of both agriculture and forestry, has emerged as an effective nature-based solution against climate change for environmental protection. This innovative approach, as detailed in Chap. 6, not only enhances agricultural productivity but also plays a crucial role in mitigating climate change by sequestering carbon from the atmosphere.

Carbon sequestration is the process of capturing and storing atmospheric carbon dioxide (CO_2) to prevent it from contributing to the greenhouse effect. Agroforestry systems, which integrate trees and shrubs with crops and livestock, provide an excellent opportunity for carbon sequestration. The trees and plants in these systems absorb CO_2 during photosynthesis, converting it into organic matter which is then stored in the soil biomass (i.e., SOM) and products.

The benefits of agroforestry in carbon sequestration can be attributed to following several factors:

(i) Trees and shrubs: The presence of trees and shrubs in agroforestry systems significantly increases the amount of carbon stored in the ecosystem. These woody plants have larger root systems that help retain soil organic matter (SOM), further enhancing carbon sequestration.
(ii) Soil improvement: Agroforestry practices improve soil quality by increasing organic matter content, promoting soil structure, and reducing soil erosion. Healthy soils are better at retaining carbon, thus contributing to the overall carbon sequestration capacity of the system.
(iii) Diversified cropping systems: Agroforestry systems often involve diverse crop and tree species, which can lead to increased carbon sequestration rates. Different species have varying carbon storage capacities, and their combined effect can be more substantial than that of a single species.
(iv) Reduced emissions: Agroforestry practices can also reduce greenhouse gas emissions from agricultural activities. For instance, incorporating trees in farming systems can lead to reduced reliance on fossil fuel-based inputs, such as fertilizers and pesticides, thereby lowering carbon emissions associated with their production and application.

Thus, agroforestry serves as a protection for the environment by promoting carbon sequestration, which is crucial for mitigating climate change. This sustainable land use practice not only offers environmental benefits but also enhances agricultural productivity and supports rural livelihoods. Embracing agroforestry as a key strategy in global efforts to combat climate change can contribute significantly to a more sustainable and resilient future for our planet.

1.7 Agroforestry for Sustainable Improved Livelihoods Through Value Addition and Animal Husbandry

For assuring sustainable development and economic growth, it is crucial to explore innovative ways to improve livelihoods while preserving the environment. One such approach is by focussing on the value addition of agroforestry products. Chap. 7 deals at a length on how this strategy can significantly contribute to enhancing the livelihoods of rural communities and promote sustainable agroforestry practices. Agroforestry, being the integration of trees and shrubs into agricultural systems, offers numerous ecological, social, and economic benefits. Value addition to the products derived from these systems can further maximize these benefits and create opportunities for rural communities due to following ecosystem services of AF system:

(a) Diversification of income sources: Value addition in agroforestry products allows farmers to diversify their income sources, reducing their dependency on a single crop or product. This diversification helps in stabilizing their income and coping with market fluctuations.
(b) Enhanced market value: Market value of raw agroforestry products can be increased by processing those raw AF products into more refined or processed forms. This, in turn, leads to higher profits for the farmers and encourages them to invest more in their agroforestry systems.
(c) Job creation and skill development: The value addition process often requires additional labor and specialized skills. This creates job opportunities, particularly for rural communities, and promotes skill development, ultimately contributing to poverty alleviation.
(d) Sustainable use of resources: Adding value to agroforestry products encourages the sustainable use of natural resources. Farmers are more likely to conserve and protect their trees and other resources when they see the potential for economic returns from these resources.
(e) Environmental conservation: Agroforestry practices contribute to environmental conservation by promoting soil and water conservation, carbon sequestration, and biodiversity preservation. The value addition process further supports these environmental benefits by creating economic incentives for sustainable land management.
(f) Improved food security: Agroforestry systems often provide multiple products, including food, fodder, fuelwood, and timber. Value addition of those products can further improve the availability and accessibility of those resources, contributing to healthy food.

The nature-based solutions for livelihood opportunities through agroforestry are presented with real numerical problems and solutions. These will be helpful for agroforesters and researchers on conducting diagnosis & design (D & D) survey and invigorate all farming communities to satisfy their inquisitiveness about reaping the benefits of agroforestry and possible way of extra income through carbon credit as per government approval. These will also be helpful for students in understanding and practicing agroforestry for solutions of their assignments in internships and formulating various kinds of agroforestry projects.

1.8 Agroforestry as Nature-Based Solution for Food Security

Nature-based solution (nbs) for food security through agroforestry can be assured because agroforestry is an innovative and sustainable approach to farming that combines agricultural and forestry practices on the same piece of land. AF involves the intentional integration of trees and shrubs into farming systems and land use plans. This nature-based solution, as extended in concluding Chap. 8, offers numerous benefits, including improved food security, environmental conservation, and economic resilience as follows.

(a) Enhanced food security: Agroforestry systems provide a diverse range of food sources, including grains, fruits, nuts, and vegetables. The integration of trees and crops can lead to assured crop yields, as trees often provide shade and windbreaks that protect crops from harsh weather conditions. Additionally, trees can improve soil fertility by adding organic matter and fixing nitrogen, which, in turn, benefits the growth of other crops.
(b) Environmental conservation: Agroforestry systems contribute to the conservation of natural resources by reducing soil erosion, maintaining soil fertility, and improving soil water retention (Bates et al. 2008). Trees help to stabilize the soil, preventing it from being washed away during heavy rains. This, in turn, prevents the loss of topsoil, which is essential for agriculture. Furthermore, agroforestry systems can help combat climate change by sequestering carbon in the trees and soil, and thereby, reducing greenhouse gas (GHG) emissions.
(c) Economic resilience: By diversifying the types of crops grown on a piece of land, agroforestry systems can help farmers reduce their economic risks. Farmers can benefit from multiple income sources, as they can sell various products, such as grains, fruits, nuts, and timber. Additionally, agroforestry systems can reduce the need for external inputs, such as fertilizers and pesticides, which can be costly. This can lead to significant cost savings for farmers and contribute to their overall economic resilience.
(d) Biodiversity enhancement: Agroforestry systems promote biodiversity by providing habitats for various plant and animal species. The presence of trees and shrubs in these systems, including animal husbandry, and aquaculture creates a diverse and complex ecosystem, which can support a wider range of species than monoculture farming systems. This can contribute to the conservation of local flora and fauna and maintain the ecological balance in the area.

1.9 Summary

Agroforestry is a sustainable land use practice that combines agriculture and forestry, offering numerous benefits including food security, income generation, and environmental conservation. By integrating trees with crops or livestock or fishery, agroforestry systems can enhance soil fertility, biodiversity, and reduce soil erosion. This helps in providing multiple products and services such as food, timber, and fuelwood, etc. This practice also promotes sustainable food production by addressing challenges like soil degradation, nutrient depletion, and chemical fertilizer overuse. Agroforestry systems can help mitigate climate change through carbon sequestration and provide a diverse range of multipurpose foods that can be consumed fresh, processed or cooked in various recipes, contributing to a nutritious diet. Multipurpose foods, rich in nutrients and medicinal properties, can be produced through agroforestry systems by combining trees and crops to provide a sustainable source of food. Agroforestry systems can accommodate various food production, including cereals, pulses and legumes, oilseeds, vegetables, and fruits, offering a stable food supply and alternative food sources during poor harvests. The integration of trees and crops also enhances soil fertility, reduces soil erosion, and provides pest and disease control, benefiting pulse and oilseed production. Additionally, trees can provide shade and wind protection for sensitive crops, and fix atmospheric nitrogen (i.e., legume trees) to improve soil fertility and enhance fruit yields. This approach can also lead to better pollination and increased fruit yields making agroforestry a viable solution for producing multipurpose foods while promoting environmentally friendly practices. Agroforestry practices can significantly enhance soil organic carbon (SOC) levels, leading to improved soil fertility, soil water retention, and greenhouse gas (GHG) mitigation. By integrating trees and crops on the same land, agroforestry systems can create a sustainable and productive agricultural system. Tree roots, leaf litter, and self-pruning branches contribute to the accumulation of soil organic matter (SOC), while root exudates stimulate microbial activity and promote decomposition of organic matter under agroforestry. Diversified agroforestry systems can provide a continuous supply of SOM increasing SOC levels. Implementing agroforestry practices requires careful planning, including site assessment, tree species selection, system design, establishment, and monitoring of AFS. While there are challenges to implementing agroforestry practices, such as initial investment costs and land use conflicts, the benefits of increased SOC levels make it a promising approach to mitigate climate change. Agroforestry, a land use management system that integrates trees and crops, plays a crucial role in improving soil health and promoting healthy food production. By integrating trees and shrubs into agricultural landscapes, agroforestry systems can enhance soil quality, fertility, and biodiversity, leading to the production of nutritious and diverse food. Additionally, agroforestry systems can also contribute to carbon sequestration by storing atmospheric carbon dioxide in the soil, biomass, and products. This sustainable land use practice offers numerous benefits, including environmental protection, enhanced agricultural productivity, and support for rural livelihoods. Agroforestry is a promising approach to improve rural livelihoods and promote sustainable development. By integrating trees and shrubs into agricultural systems, agroforestry offers numerous benefits including diversified income sources, enhanced market value, job creation, sustainable use of resources, and environmental conservation. Value addition to agroforestry products can further maximize these benefits by creating opportunities for rural communities, reducing poverty, and promoting sustainable land use management.

References

Ananda GKS, Myrans H, Norton SL, Gleadow R, Furtado A, Henry RJ (2020) Wild sorghum as a promising resource for crop improvement. Front Plant Sci 11:1 I 08. https://doi.org/10.3389/fpls.2020.01108

Anna T (2023) Multi-purpose crops: hemp, a nature based solution towards sustainable agriculture and water productivity. National Research Council of Italy, Institute of Biosciences and Bioresources, Research Division Portici (CNR-IBBR), via Portici, Italy—24 Apr 2023. Blog-sustainable land & water management for food & agriculture inter-Regional Technical Platform on Water Scarcity (iRTP-WS), FAO

Bastos Lima MG (2018) Toward multipurpose agriculture: food, fuels, flex crops, and prospects for a bioeconomy. Glob Environ Polit 18:2. https://doi.org/10.1162/glep_a_00452

Bates B, Kundzewicz ZW, Wu S, Arnell N, Burkett V, Doll P, Gwary D et al (2008) Climate change and water. Intergovernmental Panel on Climate Change (IPCC) Working Group 11 Technical Support Unit., chaired by Dr. R. K. Pachauri, (Finalized at the 37th Session of the IPCC Bureau). IPCC Secretariat, c/o WMO, Geneva 2

FAO and ICRAF (2019) Agroforestry and tenure. Forestry Working Paper no. 8. Rome. 40 pp. Licence: CC BY-NCSA 3.0 IGO

Freitas ICD, Ribeiro JM, Araújo NCA et al. (2020) Agrosilvopastoral systems and well-managed pastures increase soil carbon stocks in the Brazilian Cerrado. Rangel Ecol Manag 73(2020):776–785. https://doi.org/10.1016/j.rama.2020.08.001. https://www.researchgate.net/publication/343866782_Agrosilvopastoral_Systems_and_Well-Managed_Pastures_Increase_Soil_Carbon_Stocks_in_the_Brazilian_Cerrado

Kay S, Rega C, Moreno G, den Herder M et al. (2019) Agroforestry creates carbon sinks whilst enhancing the environment in agricultural landscapes in Europe. Land Use Policy 83:581–593. https://doi.org/10.1016/j.landusepol.2019.02.025. ISSN0264-8377. https://www.sciencedirect.com/science/article/pii/S0264837718310470

Lorenz K, Lal R (2014) Soil organic carbon sequestration in agroforestry systems. A review. Agron Sustain Dev 34(2):443–454. https://doi.org/10.1007/s13593-014-0212-y. hal-01234833

Muchane MN, Sileshi GW, Gripenberg S et al. (2020) Agroforestry boosts soil health in the humid and sub-humid tropics: a meta-analysis. Agri Ecosyst Environ 295:106899. https://doi.org/10.1016/j.agee.2020. ISSN 0167-8809. https://www.sciencedirect.com/science/article/pii/S0167880920300840

Nkurikiye JB, Uwizeyimana V, Van Ruymbeke K et al. (2024) Farmers' preferences for adopting agroforestry in the Eastern Province of Rwanda: a choice experiment. Trees Forests People 16(2024):100592. https://doi.org/10.1016/j.tfp.2024.100592. https://www.sciencedirect.com/science/article/pii/S2666719324000992?via%3Dihub

Simonow J (2021) The rise and demise of multi-purpose food in India: food technology, population control and nutritional development in the post war era, c. 1944-66. South Asia: J South Asian Stud 44(I):I 67–I184. https://doi.org/10.1080/00856401.2021.1859080

UN SDG 2030 (2015) Transforming our world: the 2030 Agenda for Sustainable Development. 70/1. Resolution adopted by the General Assembly on 25 September 2015 [without reference to a Main Committee (A/70/L.1)]. United Nations General Assembly. Seventieth session, Agenda items 15 and 116. 21 October 2015

van Noordwijk M (2021) Agroforestry-based ecosystem services: reconciling values of humans and nature in sustainable development. Land 10:699. https://doi.org/10.3390/land10070699

Websites

Climate-smart crop production practices and technologies. https://www.fao.org/climate-smart-agriculture-sourcebook/production-resources/module-b1-crops/chapter-b1-2/en/

Multi-purpose Crops: Hemp, a Nature Based Solution Towards Sustainable. https://www.fao.org/platforms/water-scarcity/Outreach/blog-on-water-scarcity/blog-detail/sustainable-land-water-management-for-food-agriculture/2023/04/25/multi-purpose-crops-hemp-a-nature-based-solution-towards-sustainable-agriculture-and-water-productivity/en

Toward Multipurpose Agriculture: Food, Fuels, Flex Crops, and Prospects. https://mgbastoslima.com/wp-content/uploads/2020/04/Bastos-Lima-2018-Towards-multipurpose-agriculture.pdf

Wild Sorghum as a Promising Resource for Crop Improvement. https://www.frontiersin.org/journals/plant-science/articles/10.3389/fpls.2020.01108/full

Agroforestry—The Treasure of Multipurpose Foods

2

Abstract

Agroforestry systems, which combine agriculture and forestry, offer a promising solution to address the twofold task of climate change linked public health problems. By integrating trees into agricultural landscapes, farmers can improve food security, income generation, and environmental conservation, while also providing a more resilient and adaptable food system, as a sustainable source of components of multipurpose foods toward aspiring one health in the society. Agroforestry systems, being naturally endowed with multipurpose qualities, can promote food security by sustaining yields, diversifying income streams, and improving nutritional quality, while also generating income through value-added products like timber and non-timber forest products. Additionally, agroforestry systems can mitigate the impacts of climate change by sequestering carbon dioxide, reducing soil erosion, and promoting biodiversity. To achieve this vision, policymakers must provide support for research and development initiatives, while civil society organizations must raise awareness about the benefits of agroforestry systems.

Keywords

Agroforestry · Agriculture · Forestry · Multipurpose foods · Climate change · Public health · Agricultural landscapes · Farmer · Food processing industries · Food security · Fruits · Leafy greens · Legume · Bean · Nuts · Seeds · Grain · Gluten · Fat · Protein · Amino acid · Fiber · Micronutrient · Vitamin · Natural remedies · Minor ailments · Digestive issues · Respiratory issues · Skin issues · Minor infections · Zoonotic diseases · Risk assessments · Epidemiology · Laboratory diagnostic capacity · Bioeconomy · Electricity · Biofuel · Biosafety and biosecurity · Antimicrobial resistance · Food safety · Medicinal properties · Wound healing · Antimicrobial properties · Tourism · Ecotourism · Anti-inflammatory properties · Antioxidant properties · Water quality · One health · Income generation · Environmental conservation · Incinerator plant · Resilient · Sustainable · Sdgs · Organization · Meals for million

S. Panda, *Agroforestry*, https://doi.org/10.1007/978-981-96-6855-7_2

Abbreviations

A, C, D, E, K and B-complex	Vitamins
AFS	Agroforestry system
CBD	Convention on Biological Diversity
COPD	Chronic obstructive pulmonary disease
FAO	Food and Agriculture Organization
GEF	Global Environment Facility
GLEWS+	Global Early Warning System for health threats and emerging risks at the human-animal ecosystems interface
IBS	Irritable bowel syndrome
IFAD	International Fund for Agricultural Development
IYRP	International Year of Rangelands and Pastoralists
LEAP	Livestock Environmental Assessment and Performance
MPF	Multipurpose foods
MPT	Multipurpose trees
nbs	Nature-based solution
NTFPs	Non-timber forest products
SDGs	Sustainable Development Goals
TBO	Tree-borne oilseeds
UNDFF	United Nations Decade of Family Farming
UNEP	United Nations Environment Programme
WHO	World Health Organization
WOAH	World Organization for Animal Health

2.1 Agroforestry: Treasure of Multipurpose Foods for Solution of Dual Challenges of Climate Change and Maintaining Good Health

The pressing concerns of climate change and maintaining good health have led to a growing recognition of the importance of sustainable food systems. Agroforestry, an approach that integrates trees into agricultural landscapes, has emerged as a promising solution. Among its numerous benefits, agroforestry systems can provide multipurpose foods that offer a wide range of nutritional and health benefits. This requires exploring the concept of multipurpose foods from agroforestry systems and their potential to address the dual challenges of climate change and public health.

As global agricultural practices face increasing pressures from climate change, soil degradation, and the need for enhanced food security, agroforestry systems are gaining recognition as a viable solution for sustainable agricultural intensification. These innovative systems, which integrate crops, animals, and trees, provide a holistic approach that balances economic viability with environmental stewardship. By exploring the nuances of agrosilvopastoral and agrosilvicultural systems, we can better understand their potential to transform agriculture into a more sustainable and resilient practice as per UN SDG 2030 (2015).

2.1.1 Multipurpose Foods and Their Importance

Multipurpose food (MPF) sources are reaped from crops or plants that provide multiple benefits, including nutrition, income, and environmental services. These foods can be found in various forms, such as fruits, vegetables, grains, nuts, and seeds. In the context of agroforestry systems, multipurpose foods are often produced through integrated farming practices that combine trees with crops or livestock or fishes. Thus, agroforestry can be normal source of food components required for production of multipurpose foods. Those food components are produced from farming or from natural sources of plants, trees, animals, fishes, fungi, algae, etc. The globe is rich in genetic diversity, with species and ecosystems ranging from ants to elephants, moss to enormous trees, and deserts to rainforests. In actuality, there are more than ten million distinct species of fungi, plants, animals, and microbes on Earth. The survival of humans depends on this biological diversity. In addition to provide a wide range of nutrients for people's diets, it also helps people be resilient to unforeseen crop failures, give food and money for those who might not otherwise be able to afford it, produce clean water, control flooding, rich soils, pollinate plants, and much more. Food and biodiversity are interlinked and react to each other's changes. In recent decades, monocrop culture that focus on a small portion of the available food source and eradicate species deemed weeds or pests have replaced thriving ecosystems, giving intensive food production precedence over biodiverse landscapes. 98% of the world's food is thought to come from just 12 plant crops and 14 animal species, and many of their wild cousins have already vanished (CIFOR 2024). In this regard, agroforestry is the hands-on single-window know-how toward ensuring food security as well as protecting biodiversity, i.e., the nature-based sustainable solution.

An agroforestry farming system-based national policy can ensure a normal resource of multipurpose food production effort throughout the year. Agroforestry system (AFS) can be varied unlimited sources of different food components required to prepare multipurpose foods. AFS, being desired combination of various trees and crops, can include fruit trees and other useful crops necessarily opted for preparation of multipurpose foods. In this way, multipurpose agroforestry can necessarily be targeted for multipurpose food production. Thus, it requires inclusion of both multipurpose tree (including fruit trees, oil-producing trees, i.e., TBOs) culture and multipurpose agriculture.

2.1.2 Types of Multipurpose Foods from Agroforestry

Agroforestry systems can produce a wide range of multipurpose foods that cater to diverse nutritional needs and cultural preferences. Some examples include:

(a) Fruits: Fruits like mangoes, avocados, and citrus trees provide essential vitamins, minerals, and antioxidants. They also serve as a source of income for smallholder farmers.

(b) Leafy greens: Leafy greens like lettuce, spinach, and kale are rich in vitamins A, C, and K. They also require low-maintenance and can be grown under shade, making them an excellent choice for agroforestry systems.
(c) Legumes: Legumes like beans, lentils, and peas are rich in protein, fiber, and micronutrients. They also have the ability to fix atmospheric nitrogen, reducing the need for synthetic fertilizers.
(d) Nuts and seeds: Nuts like almonds, walnuts, and pistachios provide healthy fats, protein, and fiber. Seeds like sunflower seeds and pumpkin seeds are rich in vitamins and minerals.
(e) Grains: Grains like quinoa, amaranth, and sorghum are rich in protein, fiber, and essential amino acids. They are also gluten-free or have low gluten content.

Those multipurpose foods can be procured from multipurpose agriculture in combination with fruit-based agroforestry.

2.1.3 Health Wellness Values of Multipurpose Foods

Multipurpose food (MPF) has twofold qualities of both the nutritional and medicinal properties. These properties of MPF make it a valuable nature-based therapeutic solution for minor illness through food. For this unique nature, MPF is able to act as a potential solution to the problem of providing adequate nutrition for large populations. For this reason, MPF has the potential to provide "Meals for Million."

2.1.3.1 Medicinal Properties

(a) Antioxidant properties: MPF contains high levels of antioxidants, which help protect the body from free radicals and oxidative stress, reducing the risk of chronic diseases such as cancer, cardiovascular disease, and neurodegenerative disorders.
(b) Anti-inflammatory properties: The food components in MPF have anti-inflammatory properties, which can help alleviate symptoms of conditions like arthritis, allergies, and respiratory problems.
(c) Antimicrobial properties: MPF has been shown to possess antimicrobial properties, making it effective against bacteria, viruses, and fungi, which can help combat infections and diseases.
(d) Wound healing properties: The nutrients and bioactive compounds in MPF have been found to promote wound healing by enhancing tissue repair, reducing inflammation, and improving collagen synthesis.

2.1.3.2 Natural Remedies for Minor Ailments

(a) Digestive issues: MPF can be used as a natural remedy for digestive issues such as constipation, diarrhea, and irritable bowel syndrome (IBS).
(b) Respiratory issues: The antimicrobial and anti-inflammatory properties of MPF make it a potential treatment for respiratory conditions like bronchitis, asthma, and chronic obstructive pulmonary disease (COPD).

(c) Skin issues: MPF's antioxidant and antimicrobial properties can help soothe skin irritations, acne, and other skin conditions.
(d) Minor infections: The antimicrobial properties of MPF can be used to treat minor infections such as ear infections, sore throats, and eye infections.

2.1.3.3 Nutritional Value

(a) Protein-rich: MPF is an excellent source of protein, making it an ideal food for individuals seeking to increase their protein intake.
(b) Mineral-rich: MPF contains high levels of essential minerals such as iron, zinc, calcium, and potassium.
(c) Vitamin-rich: MPF is a rich source of vitamins A, C, D, E, K, and 8-complex vitamins.

Thus, multipurpose food (MPF) is a nutrient-rich food source that offers a wide range of medicinal and nutritional benefits. Its antioxidant, anti-inflammatory, antimicrobial, and wound healing properties make it an effective natural remedy for minor ailments. With its high protein content and mineral-rich profile, MPF has the potential to solve the problem of providing adequate nutrition for large populations. Further research is needed to fully explore the potential benefits of MPF and its applications in various industries.

2.1.4 Multipurpose Agriculture with Fruit Trees: Fruit-Based Agroforestry

For millennia, many plants have served multiple purposes and farmers naturally used to grow those plants in combination with various other trees. For the time being, such farming practices were disrupted due to extension of intensive cropping sequences innovated among farmers. Such changes among farming community can be categorically pointed out as follows.

2.1.4.1 Tree Culture Conventionally Followed in Farming Systems

Trees could provide fruit, therapeutic leaves or bark, and wood for packaging, construction, etc. For multipurpose agroforestry, it requires desirable selections of trees and crops for food production as well as for production of bio-materials (like fibers and wood) for food packaging. Because multipurpose agroforestry should be the total nature-based solution (nbs) not only for satisfying multipurpose food production but also for its packaging, and other purposes.

For selection of crops plants, traditional crops like rice, providing food but also plant material for housing, utensils, paper, and more, should be given importance. This is necessary for conserving traditional biodiverse ecosystem reliance in the farming area as well as sustaining food production and food security in future also.

In recent times, the scope of agriculture has broadened significantly due to developments in biotechnology. Agriculture is no longer solely about food; it now integrates into various industries, like food processing industries for production of packaged multipurpose foods.

2.2 Multipurpose Crop Culture

Multipurpose crops are capable of delivering improved biomass usability as well as being a source of non-food oils, bio-materials, nutraceutical products, and other valuable bioproducts beneficial for the society as a whole. In a general concern, the growing world population is driving up demand for food and housing. This puts environmental sustainability in jeopardy and widens the gap between the amount of resources available and the amount to meet the human needs. This is where multipurpose crops come into play as essential components of a larger agricultural system. In order to increase crop yields, diversify farm products, and enhance farm sustainability over time, multipurpose crops could be effectively combined with farming and tree culture, i.e., agroforestry. Natural resources are being used up and, thereby depleted by humans in order to meet their requirements. Nowadays, finding crop species that can support animal and human life needs without harming the environment has grown to be a significant concern. Therefore, plant experts are giving very much attention in researching underutilized crops. Because of their rapid growth, high nutritional content, and numerous uses in animal feed, medicines, textiles, and environmental preservation for which plants with a long history are the subject of current research. Those may be grown in a variety of climates of temperate, tropical, and subtropical areas.

For thousands of years, farmers have been growing crops for a variety of uses. The majority of plants regularly cultivated are cereal crops, like rice, barley, wheat, maize, and millets; pulses, oilseeds, nuts; and fibre crops, like jute, hemp, *mesta* (roselle, kenaf), etc. Multipurpose crop plants are grown for multiple important uses, such as medicinal, fuelwood, shade, fiber, fuel, and conservation. Multipurpose crops are grown by farmers in a variety of combinations with other crops for a variety of uses. Farmers are used to alter the crops in cultivation to adapt to the shifting environmental conditions, and also to satisfy consumer demands.

2.2.1 Objectives of Production of Multipurpose Crops

The objectives of production of multipurpose crops are multifaceted and aimed at achieving a range of benefits. Some of the key objectives include:

(i) Food security: To provide a reliable source of food, particularly for small-scale farmers and rural communities, to ensure their food security and nutritional well-being.
(ii) Income generation: To generate income for farmers and rural communities through the sale of crops, thereby contributing to their economic well-being.

(iii) Diversification of farming systems: To promote diversification of farming systems by providing an alternative to monoculture farming practices, which can lead to soil degradation, pest and disease buildup, and reduced biodiversity.
(iv) Improved soil fertility: To improve soil fertility through the use of legume-based cropping systems, which can fix nitrogen in the soil and reduce the need for synthetic fertilizers.
(v) Pest and disease management: To promote integrated pest management (IPM) practices by using multipurpose crops that can help to reduce the reliance on chemical pesticides and fungicides.
(vi) Biodiversity conservation: To promote biodiversity conservation by providing a habitat for beneficial insects, such as pollinators and predators, that can help to maintain ecological balance.
(vii) Climate change mitigation: To contribute to climate change mitigation by promoting sustainable agriculture practices that sequester carbon in soils, reduce greenhouse gas emissions, and promote climate-resilient agricultural systems.
(viii) Nutritional security: To provide a source of essential nutrients, such as protein, fiber, and micronutrients, for humans and animals.
(ix) Water conservation: To promote water conservation by using drought-tolerant crops that require less water for irrigation, reducing the pressure on water resources.
(x) Rural development: To promote rural development by providing employment opportunities, improving livelihoods and enhancing the overall quality of life for rural communities.
(xi) Ecosystem services: To provide ecosystem services, such as soil erosion control, windbreaks, and shelterbelts, that can help to maintain ecological balance and promote sustainable agriculture practices.

Thus, the objectives of production of multipurpose crops are to promote sustainable agriculture practices that are environmentally friendly, socially decent, and economically viable.

2.2.2 Innovations for Beneficial Uses of Multipurpose Crops

Various multi-crop cultivation methods are usually followed in tropical agriculture, e.g., mixed cropping, *paira* cropping, intercropping, and many more. Farmers are generally interested to grow multiple crops on the same cultivated land to increase their whole farm income, e.g., they are used to grow crops like cowpea, pearl millet, sorghum, paddy, and many more, useful for dual purposes of food and fodder. Innovations for using cultivated crops for multipurpose uses also would be certainly helpful for procuring more income from cultivation. Some of the following examples of multipurpose crops would attract the farmers to innovate more in such purposes.

2.2.3 Food Crop—Innovative Use as an Energy Crop

In the present concern of satisfying world energy needs, renewable energy is the answer for its quickest nature of growth. Under such circumstances, biomass is regarded as a good source of renewable energy globally, both in industrialized and developing nations. In this scenario, some crops like sugarcane have attracted attention for its large share in bioethanol, heat, and electricity production. Thirty percent of the electricity needed for households in most emerging countries comes from thermal power plants. After crushing the sugarcane and extracting the juice, it is heated to dry and then crushed. The dehydrated residues are burned in a combustion chamber at the incinerator plant. The heat generated is used to bring water to a boil, and the steam created is used to power the turbine of the generator to provide electricity.

It can also be noted that sorghum is cultivated for grain, sweet stems, forage, and broomcorn. It also serves as a source of fuel, bioethanol, alcoholic beverages, and building materials. Some other crops can also be taken into consideration like production of palm oil, corn (maize) oil, soybean oil, canola (rapeseed) oil, sunflower oil, and many more.

2.2.4 Food Crop and Trees—Innovative Source of Fiber

Hemp is a valuable multipurpose crop. It provides raw materials for both food and non-food (fiber) industries. *Kenaf* plants of east Africa is used as food, feed, and fibers. Some other examples are cotton, flax (linseed), coir (coconut), jute, and many more.

2.2.5 Food Trees and Crops—Innovative Use as Sources of Components of Multipurpose Foods

There are ample examples of innovative uses of fruit trees and food crops as sources of components of multipurpose foods. Some of these are given here:

Moringa is a fast-growing tree native to the sub-Himalayan regions of India, Pakistan, and Afghanistan. It is the source of nutrient-rich leaves, seeds, and pods. The moringa leaves are rich in vitamins A, C, and E, calcium, and protein, while the seeds can be used as a condiment, ground into flour, or pressed for oil.

Mango, a popular fruit worldwide, is not only delicious but also packed with vitamins A and C, antioxidants, and dietary fiber. This fruit can be consumed fresh, made into juices, or used in various culinary preparations.

Cashew tree produces edible seeds (cashew nuts) and fruits (cashew apples) that can be consumed fresh or processed into juices, vinegar, and even liquor.

Groundnut, a legume, is rich in protein, iron, and calcium. It can be consumed as a snack, added to stews, or ground into flour for making porridge and bread.

In such ways, multipurpose agriculture is a dynamic trend that aligns with the need for sustainable and versatile crop production. It not only addresses food security but also contributes to a more environmentally friendly and economically viable future. This can be aligned with tree culture to create the multipurpose agroforestry systems.

2.3 Agroforestry Systems—Sources of Multipurpose Food Components

Agroforestry systems offer a wide range of multipurpose food components that promote ecological balance, biodiversity, and sustainable food production. These foods are not only nutritious and source of energy from foods but also offer medicinal and environmental benefits. By incorporating these foods into diets and food systems, agroforestry systems can aid in promoting healthy eating habits while supporting sustainable agriculture practices. The following are the examples of multipurpose food components produced from agroforestry and highlight their significance in modem food systems.

2.3.1 Multipurpose Fruits—Some Examples

Mangifera indica (Mango): Rich in vitamins A and C, potassium, and antioxidants, mangoes are a major source of nutrition, used as a natural sweetener, juice, or dried fruit.

Citrus sinensis (Orange): High in vitamin C, flavonoids, and antioxidants, oranges are an excellent source of nutrition, consumed as juice, fresh fruit, or dried peel.

Psidium guajava (Guava): Rich in vitamin C, potassium, and antioxidants, guavas are a versatile fruit used as fresh fruit, jam, or juice.

2.3.2 Multipurpose Grains—Some Examples

Zea mays (Maize): High in protein, fiber, and nutrients like iron and zinc, maize is a staple food in many societies and can be consumed as whole grains, cornmeal, or tortillas (a very thin round Mexican bread).

Oryza sativa (Rice): Rich in carbohydrates, rice is a staple food in many human cultures consumed as white rice, brown rice, or rice flour.

Triticum aestivum (Wheat): High in carbohydrates, fiber, and nutrients like iron and selenium, wheat is a staple food in many cultures consumed as whole grain bread, pasta, or baked goods.

2.3.3 Multipurpose Legumes—Some Examples

Phaseolus vulgaris (Bean): Rich in protein, fiber, and nutrients like iron and zinc, beans are an excellent source of nutrition and consumed as whole beans, canned beans, or bean flour.

Vigna unguiculata (Cowpea): High in protein, fiber, and nutrients like iron and potassium, cowpeas are a versatile legume consumed as fresh pods, cooked peas, or dried beans.

Lens culinaris (Lentil): Rich in protein, fiber, and nutrients like iron and potassium, lentils are an excellent source of nutrition and consumed as whole lentils or lentil flour.

2.3.4 Multipurpose Nuts—Some Examples

Juglans regia (Walnut): High in protein, fiber, and nutrients like omega-3 fatty acids and antioxidants, walnuts are an excellent source of nutrition and consumed as fresh nuts or oil.

Carya illinoensis (Pecan): Rich in protein, fiber, and nutrients like magnesium and potassium, pecans are a versatile nut consumed as fresh nuts or oil.

Prunus dulcis (Almond): High in protein, fiber, and nutrients like vitamin E and magnesium, almonds are an excellent source of nutrition and consumed as fresh nuts or oil.

These are some examples as the sources of components of multipurpose foods affordable from agroforestry systems.

2.4 Challenges and Opportunities of Multipurpose Foods from Agroforestry

While agroforestry systems offer numerous benefits, there are also challenges to be addressed:

(a) Scaling up production: To meet the demands of a growing global population, agroforestry production needs to scale up while maintaining environmental sustainability.
(b) Market access: Smallholder farmers often struggle to access markets for their products due to lack of infrastructure or market information.
(c) Climate-smart agriculture: Agroforestry systems need to be designed with climate change mitigation and adaptation in mind to ensure long-term sustainability.
(d) Policy support: Governments and policymakers need to provide support for agroforestry initiatives by promoting policies that encourage sustainable agriculture practices.

2.4.1 Benefits of Multipurpose Foods from Agroforestry Systems

The integration of multipurpose foods into agroforestry systems offers several benefits:

(a) Improved nutrition: Agroforestry systems provide a diverse range of nutritious foods that cater to different dietary needs and preferences.
(b) Increased income: Multipurpose foods can generate income for smallholder farmers through sales at local markets or exports.
(c) Environmental services: Agroforestry systems help maintain biodiversity by providing habitat for beneficial insects, pollinators, and wildlife.
(d) Climate resilience: Agroforestry systems can mitigate climate change by sequestering carbon dioxide, reducing soil erosion, and regulating water cycles.
(e) Food security: Agroforestry systems can improve food security by providing a reliable source of nutritious food year-round.

2.4.2 Multipurpose Agroforestry and Bioeconomy

Multipurpose agroforestry, being aimed at practicing nbs, should accommodate production of biofuel plants for obtaining biofuels to replace conventional fuels to some extent targeting to reduce pollution for processing agricultural products. In this way, multipurpose agriculture would serve usefulness of tree and crop production beyond food. This has become important in the context of climate change, and the associated increase in pollution, and finite fossil resources have driven the need for biodegradable alternatives. This has led to the emergence of a bioeconomy, where agriculture provides raw materials for products (such as alternative of plastics, polymer composites, and paints) that were previously petroleum oil-based. Sustaining bioeconomy through agroforestry can be achieved by adopting agroforestry policy, assuring (i) food security, (ii) income generation, (iii) environmental conservation, and rational solution of (iv) challenges and opportunities of multipurpose agroforestry systems.

In 2019, the Food and Agriculture Organization (FAO) released a report on "The State of the World's Biodiversity for Food and Agriculture," which was the first global assessment of biodiversity for food and agriculture under the guidance of its Commission. In response to the findings from that report, the Commission requested FAO to develop a global plan of action, which was later endorsed as the Framework for Action on Biodiversity for Food and Agriculture in 2021. The framework aims to promote sustainable use and conservation of biodiversity and has been integrated into FAO's operations through the revised Environmental and Social Management Guidelines and Framework. Additionally, FAO has mainstreamed biodiversity conservation into its project cycle through its Climate Risk Toolbox, which has been applied in over 80 Global Environment Facility (GEF) projects since 2020. FAO is also promoting sustainable bioeconomy practices through its Priority Area on

Bioeconomy for Sustainable Food and Agriculture, which aims to ensure that the dimensions of sustainability are observed within the bioeconomy.

2.4.2.1 Food Security Through Multipurpose Agroforestry

Agroforestry systems can improve food security by sustaining crop yields and diversifying the types of crops grown. Trees provide shade, which can reduce soil temperature and retain moisture, leading to improved soil health and increased crop yields. Additionally, the diverse canopy of an agroforestry system can provide habitat for beneficial insects and pollinators, which can increase crop pollination rates. This can be particularly important for crops that are dependent on pollinators, such as coffee and cocoa. Agroforestry systems can also provide a more stable food supply by reducing the reliance on external inputs such as fertilizers and pesticides. The natural processes of decomposition and nutrient cycling within the agroforestry system can provide a steady supply of nutrients to crops, reducing the need for external inputs. This not only benefits the farmer but also reduces the environmental impacts associated with synthetic fertilizer production and application.

2.4.2.2 Income Generation Through Multipurpose Agroforestry

Agroforestry systems can also provide additional income streams for farmers through the sale of forest products such as timber, fuelwood, and non-timber forest products (NTFPs) like fruits, nuts, and medicinal plants. The integration of trees into agricultural landscapes can also provide habitat for livestock, which can increase their productivity and value. This can be particularly important for small-scale farmers who may not have access to large areas of land or resources. Agroforestry systems can also provide opportunities for ecotourism and sustainable recreation, which can generate additional income for local communities. For example, agroforestry farms with scenic views or unique products like fruit orchards or botanical gardens can attract tourists and generate revenue through tourism activities like guided tours or workshops.

2.4.2.3 Environmental Conservation Through Multipurpose Agroforestry

Agroforestry systems can play a critical role in environmental conservation by protecting biodiversity, improving water quality, and sequestering carbon. The integration of trees into agricultural landscapes can provide habitat for a wide range of species, including threatened and endangered species. Agroforestry systems can also help to maintain ecosystem services like pollination, pest control, and climate regulation. Agroforestry systems can also improve water quality by reducing soil erosion and sedimentation in waterways. The roots of trees help to stabilize soil and prevent erosion, while the leaf litter from trees helps to filter out pollutants from runoff water. This is particularly important in areas where water scarcity is a concern. Agroforestry systems can also sequester carbon through the storage of carbon in tree biomass and soils. Trees absorb carbon dioxide from the atmosphere during photosynthesis and store it in their biomass and soils. This can help to mitigate climate change by reducing greenhouse gas emissions.

2.4.2.4 Challenges and Opportunities Through Multipurpose Agroforestry

Despite the numerous benefits of agroforestry systems, there are several challenges that need to be addressed. One of the main challenges is the lack of awareness about agroforestry practices among farmers and policymakers. There is a need for education and training programs to raise awareness about the benefits of agroforestry and to equip farmers with the skills they need to establish and manage agroforestry systems. Another challenge is the need for policy support to promote agroforestry practices. Governments can play a critical role in promoting agroforestry by providing incentives such as subsidies for tree planting or certification programs that recognize sustainable agroforestry practices. Despite these challenges, there are many opportunities for agroforestry development. With growing concerns about climate change, biodiversity loss, and sustainable development, there is increasing interest in agroforestry as a way to address these challenges.

2.5 UN Actions Toward Challenges of Biodiversity Conservation in Agriculture

The Food and Agriculture Organization (FAO) uses the term "agricultural sectors" to encompass crop, livestock, forestry, fisheries, and aquaculture. These sectors intersect with several thematic programs and cross-cutting issues under the Convention on Biological Diversity (CBD). FAO is implementing the United Nations Decade of Family Farming (UNDFF 2019–2028) with the International Fund for Agricultural Development (IFAD), aiming to support family farming and achieve the Sustainable Development Goals (SDGs). The organization has published a Global Action Plan to mobilize the coordinated efforts to address challenges faced by family farmers. The Mongolian government proposed the International Year of Rangelands and Pastoralists (IYRP) in 2026, which was endorsed by FAO and adopted by the United Nations General Assembly. The IYRP aims to promote sustainable land management, improved ecosystems, and increased investment in the pastoral livestock sector. FAO also coordinates the Livestock Environmental Assessment and Performance (LEAP) partnership, which works to improve the environmental performance of livestock supply chains while ensuring economic and social viability. LEAP provides guidance and methodology for assessing the environmental impact of livestock production on biodiversity. In addition, FAO collaborates with other organizations, including the World Health Organization (WHO), World Organization for Animal Health (WOAH), and United Nations Environment Programme (UNEP), to address health challenges at the animal-human-ecosystems interface through the One Health approach. This includes building capacities and providing technical assistance for zoonotic diseases, risk assessments, epidemiology, laboratory diagnostic capacity, biosafety and biosecurity, antimicrobial resistance, food safety, and emergency outbreak responses. FAO also manages the EMPRES Global Animal Disease Information System (EMPRES-i) and comanages the Global Early Warning System for health threats and emerging risks at the human-animal ecosystems interface (GLEWS+). FAO participates in national One Health assessments through the Joint External Evaluation process and supports various programs, projects, and initiatives that require national "One Health" implementation.

2.6 Concluding Remarks

Multipurpose foods from agroforestry systems offer a promising solution to address the dual challenges of climate change and public health. By promoting the integration of trees into agricultural landscapes, we can improve nutrition, income generation, environmental services, food security, and climate resilience. To realize these benefits, it is essential to address the challenges facing agroforestry production while providing policy support for smallholder farmers. By working together toward sustainable food systems that prioritize multipurpose foods from agroforestry systems, we can create a healthier planet for future generations.

Agroforestry is a sustainable land use practice that offers numerous benefits including improved food security, increased income generation, and enhanced environmental conservation. By integrating trees into agricultural landscapes, farmers can sustain crop yields, reduce external inputs, and generate additional income streams. Agroforestry systems can also play a critical role in environmental conservation by protecting biodiversity, improving water quality, and sequestering carbon. While there are challenges associated with agroforestry development, there are many opportunities for growth and development in this area. With education, policy support, and investment in research and development, agroforestry can prove its potential to become a key strategy for achieving sustainable agriculture and rural development goals.

As the world grapples with the far-reaching consequences of climate change, it has become increasingly evident that a multifaceted approach is needed to ensure the long-term sustainability of our planet. Two of the most pressing concerns that require immediate attention are food security and public health. Fortunately, agroforestry systems, which combine agriculture and forestry, offer a fascinating solution that can help address both these issues while promoting environmental conservation. Agroforestry systems are designed to mimic the natural ecosystem by integrating trees into agricultural landscapes. This approach not only provides numerous benefits for farmers and local communities but also offers a vital solution to the global challenges posed by climate change, food security, and public health. One of the primary advantages of agroforestry systems is their ability to promote food security. By combining crops with trees, farmers can increase their yields, diversify their income streams, and improve the overall nutritional quality of their produce. Agroforestry systems also provide a more resilient and adaptable food system, better equipped to withstand the impacts of climate change, such as droughts, floods, and heat stress or temperature fluctuations. For instance, agroforestry systems can be designed to incorporate fruit trees, nut trees, and other perennial crops that provide a steady supply of fresh produce throughout the year. This not only ensures a consistent food supply but also reduces the reliance on monoculture farming practices that can lead to soil degradation and erosion. In addition to promoting food security, agroforestry systems can also generate significant income for farmers. By selling value-added products such as timber, non-timber forest products (NTFPs), and organic fertilizers, farmers can diversify their income streams and reduce their reliance on a single crop or market. This increased economic resilience can help farmers adapt to climate-related shocks and improve their overall well-being. Furthermore, agroforestry systems offer numerous benefits for environmental conservation. By integrating trees into agricultural landscapes, farmers can reduce soil erosion, improve soil health, and increase biodiversity. Trees also provide habitat for wildlife, help regulate the water cycle and sequester carbon dioxide from the atmosphere. From a public health perspective, agroforestry systems can provide access to fresh fruits, vegetables, and nuts that are rich in essential nutrients. The integration of trees into agricultural landscapes can also reduce exposure to pesticides and other chemicals used in monoculture farming practices. Moreover, agroforestry systems can promote social cohesion and community building by providing opportunities for communal management and decision-making.

Another significant benefit of agroforestry systems is their ability to mitigate the impacts of climate change. By sequestering carbon dioxide from the atmosphere, agroforestry systems can help reduce greenhouse gas emissions and slow down global warming. Trees also provide shade for crops during hot temperatures, reducing evapotranspiration and improving water use efficiency. In conclusion, agroforestry systems offer a fascinating concept that addresses multiple challenges facing humanity today. By combining agriculture and forestry in a sustainable manner, we can promote food security, income generation, environmental conservation, and public health. As we move forward in this era of climate change, it is essential that we prioritize agroforestry systems as a vital component of our global strategy for sustainable development.

To achieve this vision, policymakers must provide support for research and development initiatives that promote agroforestry practices. This includes investing in training programs for farmers, providing access to credit and insurance services for small-scale farmers, and creating markets for value-added products derived from agroforestry systems. Additionally, civil society organizations must play a crucial role in raising awareness about the benefits of agroforestry systems among farmers, policymakers, and consumers. This can be achieved through educational campaigns, community outreach programs, and advocacy efforts. Ultimately, the adoption of agroforestry systems will require a collective effort from governments, civil society organizations, farmers, and consumers alike. By working together toward this goal, we can create a more sustainable food system that promotes food security, income generation, environmental conservation, and public health while mitigating the impacts of climate change.

Further Readings

Al var-Beltran J, Elbaroudi I, Gialletti A, Heureux A, Neretin L, Soldan R (2021) Climate Resilient Practices: typology and guiding material for climate risk screening. FAO, Rome. https://www.fao.org/3/cb3991en/cb3991en.pdf

CBD (2022) Report by the food and agriculture organization of the united nations. Conference of the parties to the convention on biological diversity, Fifteenth meeting - Part II, Montreal, Canada, 7–19 Dec 2022. Agenda items 15, 16 and 24. https://www.cbd.int/doc/c/3119/8611/72d7962a5b058cc30e2299c7/cop-15-17-en.pdf

CIFOR (2024) Food and biodiversity. https://www2.cifor.org/food-biodiversity/

FAO (1999) Women: the key to food security. Women and Population Division, Sustainable Development Department, Food and Agriculture Organization of the United Nations. https://www.fao.org/4/x0171e/x0171e03.htm

FAO (2006) Building on gender, agrobiodiversity and local knowledge—a training manual. Chief, Publishing management Service, information Division, FAORome. (https://www.fao.org/4/y5956e/Y5956E03.htm)

FAO (2015). Environmental and social management Guidelines. Rome. https://www.fao.org/3/i4413e/i4413e.pdf

FAO (2019a) The State of the World's Biodiversity for Food and Agriculture. In: Belanger J, Pilling D (eds) FAO Commission on Genetic Resources for Food and Agriculture Assessments, Rome. https://doi.org/10.4060/CA3129EN

FAO (2019b) The state of the world's aquatic genetic resources for food and agriculture. In: FAO Commission on Genetic Resources for Food and Agriculture Assessments, Rome. https://doi.org/10.4060/CA5256EN

FAO (2019c) Developing sustainable value chains for small-scale livestock producers. In: Leroy G, Fernando M (eds) FAO Animal Production and Health Guidelines No. 21. Rome. https://doi.org/10.4060/CA5717EN

FAO (2019d) Measuring and modelling soil carbon stocks and stock changes in livestock production systems: Guidelines for assessment (Version 1). Rome. https://www.fao.org/documents/card/en/c/CA2934EN/

FAO (2019e) Report of the expert meeting on other effective area-based conservation measures in the marine capture fishery sector, Rome, 7–10 May 2019. FAO Fisheries and Aquaculture Report No. 1301. Rome.

FAO (2020) Biodiversity and the livestock sector—Guidelines for quantitative assessment—Version 1. Rome, Livestock Environmental Assessment and Performance Partnership (FAO LEAP). https://doi.org/10.4060/ca9295en

FAO (2021a) Strategic framework 2022–31, Rome. https://www.fao.org/3/cb7099en/cb7099en.pdf

FAO (2021b) Aspirational principles and criteria for a sustainable bioeconomy. Rome, 16 pp. https://www.fao.org/documents/card/en/c/cb3706en/

FAO (2022a) Framework for environmental and social management. Rome. https://doi.org/10.4060/cb9870en

FAO (2022b) Framework for action on biodiversity for food and agriculture. FAO Commission on Genetic Resources for Food and Agriculture, Rome. https://doi.org/10.4060/cb8338en

FAO (2022c) The state of world fisheries and aquaculture 2022. Towards blue transformation. FAO, Rome. https://doi.org/10.4060/cc0461en

FAO (2022d) Global plan of action for the conservation, sustainable use and development of aquatic genetic resources for food and agriculture. Commission on Genetic Resources for Food and Agriculture, Rome. https://doi.org/10.4060/cb9905en

FAO (2022e) The State of World Fisheries and Aquaculture 2022. Towards Blue Transformation. Rome. doi:https://doi.org/10.4060/cc0461en

FAO (2022f) Voluntary guidelines on the responsible governance of tenure of land, fisheries and forests in the context of national food security. First revision. Rome. https://www.fao.org/3/i2801e/i2801e.pdf

FAO (2024) The future of food: the challenge of feeding the world sustainably in the 21st century. https://www.fao.org/global-perspectives-studies/fofa/en/

FAO, UNEP, WHO and WOAH (2022) One health joint plan of action (2022–2026). Working together for the health of humans, animals, plants and the environment., Rome. http://www.fao.org/3/cc2289en/cc2289en.pdf

Friedman KJ, Bartley DM, Rodríguez-Ezpeleta N, Mair GC, Ban N, Beveridge M, Carolsfeld J et al. (2022) Current and future genetic technologies for fisheries and aquaculture: implications for the work of FAO. FAO Fisheries and Aquaculture Circular. No. 1387. FAO, Rome. https://doi.org/10.4060/cc1236en

Górska-Warsewicz H, Rejman K, Laskowski W, Czeczotko M (2019) Milk and dairy products and their nutritional contribution to the average Polish diet. Nutrients 11(8):1771. https://doi.org/10.3390/nu11081771. PMID: 31374893; PMCID: PMC6723869. https://pmc.ncbi.nlm.nih.gov/articles/PMC6723869/

IPES-Food (2016) From uniformity to diversity: a paradigm shift from industrial agriculture to diversified agroecological systems. International Panel of Experts on Sustainable Food systems. www.ipes-food.org

UN SDG 2030 (2015) Transforming our world: the 2030 Agenda for Sustainable Development. 70/1. Resolution adopted by the General Assembly on 25 September 2015 [without reference to a Main Committee (A/70/L.1)]. United Nations General Assembly. Seventieth session, Agenda items 15 and 116. 21 October 2015.

WHO, FAO & OIE (2019) Taking a multisectoral, one health approach: a tripartite guide to addressing zoonotic diseases in countries, Rome. http://www.fao.org/3/ca2942en/ca2942en

Agroforestry: Influencing Production of Cereals, Pulses, Legumes, Oilseeds, and Fruits

3

Abstract

Agroforestry is a sustainable approach to cereal-based cropping systems, promoting ecological balance, improved crop yields, and farm resilience. This holistic approach combines trees with other crops or livestock, creating a more diverse and resilient ecosystem. Benefits include soil conservation, improved crop yields, biodiversity conservation, climate change mitigation, farm resilience, and economic benefits. To design agroforestry systems for sustainable cereal-based cropping, farmers should consider tree species selection, crop selection, system design, soil management, and irrigation management.

Keywords

Agroforestry systems · Cereals · Pulses · Legumes · Fruits · Cereal-based cropping · Agricultural landscape · Four dimensions of food security · Aquaforestry · Tree species · Shade · Support trees · Indo-Gangetic Plains · Benefit: Cost Ratio · Wood · Spacing · Profitable · Light · Foot-candles · Lux · Electromagnetic radiation · Sunshine · High-intensity discharge · Plant metabolism · Photosynthesis · Photosynthetically active radiation · Daily Light Integral · Photosynthetic photon flux density · Energy · Wavelength · Moles · High-intensity discharge

Abbreviations

AF	Agroforestry
BCR	Benefit:Cost Ratio
DAP	Diammonium phosphate
DLI	Daily light integral
FS	Full sun
HID	High-intensity discharge
IGPs	Indo-Gangetic Plains

S. Panda, *Agroforestry*, https://doi.org/10.1007/978-981-96-6855-7_3

IRR	Internal Rate of Return
LER	Land equivalent ratio
MPTs	Multipurpose tree species
NPV	Net present value
PAR	Photosynthetically active radiation
PPFD	Photosynthetic photon flux density
ROI	Return on investment
w/m^2	Watt per square meter

3.1 Agroforestry for Sustainable Cereal-Based Cropping-Integrating Trees into Cereal Production Systems

Agroforestry, a farming practice that involves integrating trees into agricultural landscapes. This has gained significant attention in recent years as a promising approach to sustainable agriculture. Abundance of cereals is the foundation of food security in any community at any time scale. In this regard, searching for the effect of combination of cereals and tree in cultivation is of very much importance. The integration of trees into cereal-based cropping systems offers numerous benefits for both farmers and the environment. Thus, it is now important to explore the concept of agroforestry for sustainable cereal-based cropping, highlighting its potential to promote ecological balance, improve crop yields, and enhance farm resilience.

3.2 The Need for Sustainable Cereal-Based Cropping

Cereal-based cropping is a crucial component of global food security, providing staple foods for millions of people around the world. However, the widespread adoption of monoculture practices in cereal production has led to concerns about soil degradation, water pollution, and biodiversity loss. Conventional cereal-based cropping systems often rely on intensive use of synthetic fertilizers, pesticides, and irrigation, which can have negative environmental impacts.

3.3 Agroforestry: A Holistic Approach to Sustainable Agriculture

Agroforestry offers a holistic approach to sustainable agriculture by combining trees with other crops or livestock. By integrating trees into cereal-based cropping systems, farmers can create a more diverse and resilient ecosystem that benefits from the synergies between tree species and crops. Agroforestry systems can be designed to mimic natural ecosystems, promoting ecological balance and minimizing external inputs (Soemarwoto 1987; Nair 1993; Sharma 1996; Nguyen and Catacutan 2012; Viswanath et al. 2018).

3.4 Benefits of Agroforestry in Cereal-Based Cropping

(i) **Soil conservation and improvement:** Tree roots help stabilize soil and reduce erosion, while leaf litter and other organic matter contribute to soil fertility and structure.
(ii) **Improved crop yields:** Shaded crops can experience reduced water loss and improved soil moisture, leading to increased yields and reduced water consumption (Box 3.1)
(iii) **Biodiversity conservation:** Agroforestry systems provide habitat for beneficial insects, pollinators, and other wildlife, contributing to biodiversity conservation.
(iv) **Climate change mitigation:** Trees sequester carbon dioxide from atmosphere, reducing greenhouse gas emissions and contributing to climate change mitigation.
(v) **Farm resilience:** Agroforestry systems can help farmers adapt to climate change by providing a diversified income stream and reducing dependence on external inputs.
(vi) **Economic benefits:** Agroforestry systems can generate additional income through the sale of forest products, such as timber or non-timber forest products.

3.5 Designing Agroforestry Systems for Sustainable Cereal-Based Cropping

To effectively design agroforestry systems for sustainable cereal-based cropping, farmers should consider the following factors:

(i) **Tree species selection:** Choose tree species that are well-suited to the local climate and soil conditions, and that provide shade or other benefits for crops (Box 3.1).

Box 3.1 Effect of Tree Canopy Shade on Crop Growth[a]
Light is an electromagnetic radiation, whether from natural sunshine or high-intensity discharge (HID) lamps, differs in quality (colour or wavelength), and quantity at each colour or wavelength, and duration (energy over time). People can see light in the 400–700nm wavelength range, which is known as photosynthetically active radiation (PAR), and is best suited for crop production. An increase in energy in the PAR range is required to boost plant metabolism and photosynthesis. There are common ranges and rates of light for crop production, although each crop species has an ideal light intensity at which photosynthesis will be maximised. The goal of every farm is to assure enough, i.e., optimum light for crop production. This can be achieved by minimising plant

(continued)

Box 3.1 (continued)

light stress, wasted energy, and electricity costs. Foot-candles and lux, i.e., "photometric" units that primarily consist of green light commonly used to measure light based on visible light that is visible to the human eye. A single measurement is not appropriate for measuring PAR because light levels vary throughout the day and cannot accurately reflect the amount of light that a plant has received in a day. Rather, the majority of horticulture researchers use instantaneous readings of PAR, or micromoles per square meter per second, or "$\mu mol.m^{-2}.s^{-1}$" to measure light. The quantity of photons, which are discrete energy particles used in photosynthesis, that fall on a square metre (10.8 square feet) every second is measured using this "quantum" unit'. One single measurement with this unit is quantifiable and suitable for monitoring photosynthesis. The amount of PAR a plant receives daily, based on the time and intensity of light, is known as the daily light integral (DLI). It is expressed in $mol.m^{-2}.d^{-1}$ (moles per day), or moles of light (mol) per square metre (m^{-2}) per day (d^{-1}). This photosynthetic photon flux density (PPFD) is converted to Lux unit by multiplying the Lux by the conversion factor. For example, full sunlight is 108,000 Lux or 2000 $\mu mol.m^{-2}.s^{-1}$ (108,000 * 0.0185).

- For outdoor natural sunlight: 1 W/m^2 = 122 ± 1 lux.

Classes of plants based on photometric or quantum values:

Low light plants: DLI of 5–10 $mol.m^{-2}.day^{-1}$: 217–435 lux.
Moderate light plants: DLI of 10–20 $mol.m^{-2}.day^{-1}$: 435–870 lux.
High light plants: DLI of 20–40 $mol.m^{-2}.day^{-1}$ or more: 870–1739 lux to maximum 2217 lux.

Thus, it is interesting to note that photometric value for crop production ranges from 217 to 2217 lux, which is relatively very very low compared to the 108,000 lux of full sunlight.

So, planting crops under the canopy of trees does not pose a disadvantage to crop growth *w.r.t.* light illumination, although it can actually provide beneficial effects.

[a]Mõttus et al. (2012), G2V Optics Inc. (2020), Michael et al. (2020), https://www.nexsel.tech/how-to-measure-light-intensity-for-plants.php.

(ii) **Crop selection:** Select crop species that are compatible with the chosen tree species and can benefit from the shade or other benefits provided (Box 3.1).
(iii) **System design:** Design the agroforestry system to ensure adequate spacing between trees and crops, and to promote ecological balance.
(iv) **Soil management:** Implement soil conservation practices, such as contour farming or mulching, to reduce soil erosion and improve soil fertility.
(v) **Irrigation management:** Implement efficient irrigation practices that minimize water waste and reduce the risk of overwatering.

3.6 Effects of Agroforestry on Grain Yield—Few Examples from India, a Treasure of Agroforestry

As India is regarded as the treasure of agroforestry due to its age-old AF practices for which this country has evolved various successful agroforestry models throughout the country and could set exemplary evidence on multifarious aspects of AF systems under widely varying ecological regions from snow clad mountainous regions to hot desert, from plain lands to rugged slopy hilly areas, from coastal lands to high altitudes, etc. of geographical settings. So, few examples from some year-long experimental results from those various AF models in India are given here to show the effects of AF on grain yields in the different agro-ecological regions.

3.6.1 Indo-Gangetic Plains (IGPs)

Indo-Gangetic Plains (IGPs) occupies the Indian states of Uttarakhand, Uttar Pradesh, Bihar, Jharkhand, West Bengal, Punjab, and Haryana. Cereals are the main crop in this area. More than 70% of the gross cultivated area is under the rice-wheat cropping sequence. About 7% of the gross cropped area is made up of pulses. All the cropping sequences (Box 3.2) are followed under the Poplar-based AF.

The Indus-Ganga plains (IGPs), also known as the Great Indian Plains, are large floodplains of the Indus, Ganges, and Brahmaputra river systems, extending parallel to the Himalaya mountains and draining most of Northern and Eastern India.

Box 3.2 Cropping Sequences Followed Under Poplar-Based Agroforestry in Indo Gangetic Plains in India[a]

(i) Rice (*Oryza sativa*)—Wheat (*Triticum aestivum*),
(ii) Rice (*Oryza sativa*)—Rai (*Brassica nigra*),
(iii) Rice (*Oryza sativa*)—Sweet Potato (*lpomoea batatas*),
(iv) Rice (*Oryza sativa*)—Maize (*Zea mays*)-Wheat (*Triticum aestivum*),
(v) Rice (*Oryza sativa*)—Lentil (*Lense culinaris*),
(vi) Rice (*Oryza sativa*)—Linseed (Linum usitatissimum),
(vii) Rice (Oryza sativa)—Gram (*Cicer arietinum*),
(viii) Rice (*Oryza sativa*)—Wheat (*Triticum aestivum*)—Moong (*Vigna radiata*),
(ix) Rice (*Oryza sativa*)—Toria (*Brassica campestris* L. *var. toria*),
(x) Maize (Zea mays)—Sweet Potato (*lpomoea batatas*),
(xi) Maize (*Zea mays*)—Rai (*Brassica nigra*),
(xii) Jute (*Corchorus capsularis*)—Wheat (*Triticum aestivum*),
(xiii) Jute (*Corchorus capsularis*)—Potato (*Solanum tuberosum*),
(xiv) Jute (*Corchorus capsularis*)—Kalai (Vigna mungo),
(xv) Jute (*Corchorus capsularis*)—Mustard (*Brassica campestris*)

[a]ICFRE (2020).

Poplar tree (*Populus deltoides*) is highly preferred by farmers, and thus, a dominant agroforestry tree species in the western parts of the Indian Gangetic Plains (IGPs) region because of its short rotation, compatibility with agricultural crops, strong demand in wood-based industries, deciduous nature, and high profitability. In IGPs, encompassing the Indian states of Punjab, Haryana, and Uttar Pradesh, poplar is cultivated as block plantations, with alley cropping, and border plantings. The rotation age is set at 6–9 years depending on the demand for wood size. This wood is thought to be a source of raw materials for the paper, pulpwood, matchbox, veneer, packaging; and plywood industries. Although 5 × 4 m spacing is preferred by farmers in order to accommodate more trees, 5 × 5 m spacing is the most suitable for cultivating poplar in AF system. Wheat is the most favored crop under poplar since it is a winter crop, and poplar loses its leaves from October through February, when new leaves begin to emerge around the end of March. The production of the wheat crop does not decrease in the first year, but as trees get older, a yield loss of 10–46% has been noted. As trees get older, the competition for resources both above and below ground typically shifts, but as poplar rotation ages, the system yield as a whole rises. Any winter or monsoon crop grown in the IGPs region (except rice) may be grown under poplar for the first 3 years. After that, crops such as barley (*Hordeum vulgare*), berseem (*Trifolium alexandrinum*), cabbage (*Brassica oleracea var. capitata*), chilly (*Capsicum acuminatum, C. annuum*), coriander (*Coriandrum sativum*), ginger (*Zingiber officinale*), mustard (*Brassica juncea*, *B. oleracea*, and *B. nigra*), tomato (*Lycopersicon esculentum*), turmeric (*Curcuma domestica*), colocasia, potato, spinach, and garlic, as well as fruit crops (citrus, guava, and mango), wheat (*Triticum aestivum*), etc., can be grown under poplar until the poplar reaches its rotation age. It has also been discovered that sugarcane (*Saccharum officinarum*) as intercrops for the first 2 years is more profitable. After 8 years of rotation, poplar has been shown to yield 150–200 $m^3\ ha^{-1}$ of wood under block plantations and 12–20 $m^3\ ha^{-1}$ under border plantations. It has been advised to space poplar trees five by four meters apart for optimal poplar tree performance in agri-silviculture systems. In the Indo-Gangetic plains, the total above-ground biomass production of poplar trees under block plantations ranges from 7.4 tha^{-1} (1 year old) to 105.4 tha^{-1} (10 year old). Financial and economic assessments of yields of 8 years rotation of poplar-based agroforestry system on growing on bunds have shown that those were 8–12% less than the net study year value, and with a lower Benefit: Cost Ratio (BCR) of 2.18 to as low as 2.12. This was significantly greater than the BCR (1.34–1.42) of traditional cropping systems. In one study, an agroforestry system with 10 × 2 m spaced poplar trees and a sorghum-berseem crop rotation was used. The results showed 12% less than net returns, a greatest land equivalent ratio (LER) of 2.28, a BCR of 1:2.22, and an Internal Rate of Return (IRR) of 70%. This suggests that agroforestry based on poplar trees is more profitable than conventional cropping practices (Box 3.3; Grewal 1992; Sharma 1996; Dagar et al. 2001a; Tomar et al. 2004; Mishra et al. 2004; Lal 2004; Kareemulla et al. 2005; Gill et al. 2009; Chauhan and Johnson 2010; Gandhi and Dhiman 2010; Pathak et al. 2014; Chauhan et al. 2015; Panwar et al. 2017; Chavan and Dhillon 2019).

Box 3.3 Land Equivalent Ratio (LER), Benefit: Cost Ratio (BCR), Internal Rate of Return (IRR)[a]

- **Land Equivalent ratio: LER** = $L_A + L_B = \frac{Y_A}{S_A} + \frac{Y_B}{S_{B.}}$

where L_A and L_B are the LERs for the individual crops, Y_A and Y_B are the individual crop yields in intercropping, and S_A and S_B are their yields as sole crops.

So, LER can be calculated using the formula LER = $\sum$ (Ypi/Ymi), where Ypi is the yield of each crop or variety in the intercrop or polyculture, and Ymi is the yield of each crop or variety in the sole crop or monoculture.

- **Benefit: Cost Ratio: BCR = $PV_{Benefit}/PV_{Cost}$Benefit = Income—Cost**

The benefit-cost ratio (BCR) is calculated as the present value (PV) of benefits divided by the present value (PV) of costs.

If the BCR exceeds one, then the project might be a good candidate for acceptance.

- **Internal Rate of Return: IRR** (%) = $\frac{\text{Future Value}(\text{FV})^{\frac{1}{\text{Periods}}}}{\text{Present Value}} - 1$

NPV = Net present (today's) value of the expected cash flows - Today's value of invested cash

Return on investment (ROI) = (Total benefits - total costs) / total costs

The internal rate of return (IRR) is the maximum interest that could be paid for the project resources, leaving enough money to cover investment and operating costs, which would still allow the investor to break even. In other words, the IRR is the discount rate for which the present value of total benefits equals the present value of total costs:

PV(Benefits) – PV(Costs) = 0

In general, the IRR should be greater than the discount rate for a project to be accepted.

[a]Mead and Willey (1980), Shively (2012), https://www.wallstreetprep.com/knowledge/irr-internal-rate-of-return/; https://www.harvestreturns.com/blog/2018/6/15/3-ways-to-measure-agriculture-investment-returns.

3.6.2 Trans Himalayan Region

Trans Himalayan Region encompasses Ladakh, Lahaul-Spiti, and Kinnaur (Himachal Pradesh) areas in India. The area is endowed with (i) willow or salix (*Salix fragilis* and *S. alba*) and poplar (*Populus euphratica*, *P. alba*, *P. nigra*) based agroforestry models, and (ii) apple (*Pyrus malus*) based horti-medicinal AF models. Most cultivated fruit trees are apple and apricot (*Prunus armeniaca*). Most common agriculture crops are wheat, barley and vegetables and the common intercropping fodder crops were lucerne and clovers. Common leafy vegetables in this region are *sagani* or Jerusalem oak goosefoot (*Chenopodium botrys*), buckwheat (*Fagopyrum esculentum*), *khala* or long-leaved lettuce (*Lactuca dolichophylla*), dittander or perennial pepperweed (*Lepidium latifolium*), mountain sorrel (*Oxyria digyna*), *shoma* or patience dock or monk's rhubarb (*Rumex patientia*), *Zacchat* or northern nettle (*Urtica hyperborea* Jacquem. Ex. Wedd.) (Box 3.4; Singh and Gill 1992; Singh et al. 1993, 2008; Joshy 1997; Dagar et al. 2001b; Khan 2003; Saha et al. 2007; Dagar 2014; 2016; Raghuvanshi et al. 2019; Dagar and Tewari 2016; Fatima et al. 2022).

Box 3.4 Agroforestry Models in Western Himalayan Region and Indo-Gangetic Plains[a]

The AF models developed for this zone are followed in Jammu, Kashmir, Himachal Pradesh, Punjab, Haryana and the hilly region of Uttarakhand and in Indo-Gangetic Plains:

(i) Melia (*Melia composita*)—Aonla (*Emblica officinalis*) based agri-silvi medicinal model
(ii) Poplar (*Populus deltoides*) based agri-silviculture
(iii) Poplar (*Populus deltoides*) based silviculture-medicinal models
(iv) Melia (*Melia composita*) based agri-silviculture model
(v) Eucalyptus (*Eucalyptus tereticornis*)-Wheat (Triticum aestivum)/ Paddy (*Oryza sativa*)silvi-block model
(vi) Poplar (*Populus deltoides*)-wheat (*Triticum aestivum*) agroforestry model
(vii) Mulberry (*Morus alba*)based silvi-pastural AF System
(viii) Grewia (*Grewia optiva*)based agri-silvicultural AF System
(ix) Bakain (*Melia azedarach*) based agri-silvicultural AF System
(x) Ailanthus (*Ailanthus excelsa*) based agri-silvicultural AF system under rainfed condition
(xi) Shisham (*Dalbergia sissoo*) based agri-silvicultural and silvi-pastoral AF system
(xii) Aonla or Indian gooseberry (*Phyllanthus emblica*) based agri-horticultural AF system
(xiii) Bamboo based agri-silvicultural AF system
(xiv) Gmelina (*Gmelina arborea*) based argi-silvi-horticultural AF system
(xv) Lamboo (Dysoxylum binectariferum) based argi-silvi-horticultural AF system
(xvi) Kadamb (*Anthocephalus cadamba*) based agri-silvi-horticultural AF system
(xvii) Mango (*Mangifera indica*) based agri-silvi-horticultural AF system
(xviii) *Bambusa balcooa* based agri-silvi-horticultural AF system

[a]Jeet-Ram et al. (2011), Pathak et al. (2014), Panda et al. (2018), Handa et al. (2019), ICFRE (2020).

3.6.3 Commonly Followed Cropping Sequences in the Trans Himalayan Region in India

(i) Wheat–Potato/vegetables,
(ii) Barley—Potato/vegetables.

The Trans Himalayan Region in India is characterized by limited resources and an extreme climate due to the Himalayan mountain range. The region experiences mean annual precipitation of 80–300 mm, which is scanty and negligible in the thirsty laps of mountains. Agriculture in the region has evolved into a small-scale farming system, adapted to the unique and extreme environment. Families rely on subsistence agriculture based on principal crops like wheat, barley, and potato on their stone-built terraces. Weeding is performed by farm women in groups, with stones called *shangma* used for weed control. Productivity in this cold desert is centered on crops and animal husbandry, based on human labor, animal power, and hand-made tools.

Terraced fields on slopes of high mountains are irrigated with glacial meltwater and fertilized with undecomposed farmyard manure, human "night soil," and a starter dose of fertilizers like urea, DAP, and muriate of potash. Every house also has small vegetable gardens, small orchards of apple or apricot, and keep a few animals such as goats, sheep, cows, and *dzo* (a cross between a yak and a local breed of cow). Despite harsh climatic conditions, agriculture remains successful and ecologically sustainable. The festival called *Saka* denotes the onset of agriculture season in Ladakh around April, and *dzo* are honored with *chhang* (a barley based alcoholic beverage) and barley products. Prioritizing crop sowing is a competitional phenomenon, with girls being main stakeholders in the sowing process.

Box 3.5 Some Traditional Germplasms of Crops, Still Now Cultivated, Used in Folk Songs in Ladakh Revealing the Sustainability of Local Germplasms[a]

- Barley: *ne nak*, *yang-ma*, *yang-kar*, *sermo*, *tug-zur*
- Alfalfa: *buck-suk*
- Small and big wheat: *to-chen*
- Pea: *sner-mar*
- Mustard: *nyas-kara*
- Buck wheat: *ta-wo*

[a] Raghuvanshi et al. (2019).

Traditional crops (Box 3.5) replaced by cash crops lead to loss of agrobiodiversity, increased vulnerability to climatic and market changes, reduction of high altitude pastoral lands, and reduced capacity to produce food on small/marginal land holdings. Climate change affects all four dimensions of food security: food availability, food accessibility, food utilization, and food system stability. Vegetables are crucial in Ladakhi cuisine, with 310 ha of land producing 7544 tons annually. Weeding is minimal in remote villages, reducing economic yields by 6–82% in potato, 25–30% in peas, 70–80% in carrot, 67% in onion, 42–71% in tomato, and 61% in cauliflower. Vegetable cultivation is limited within April to October, and harvesting occurs from July to October. Winter vegetable production is limited to governmental institutes and small polyhouses, making agroforestry a viable solution. Stored vegetables provide income for growers, and womenfolk sell them in local markets, mainly on road side. The climate in the region is influenced by altitude variation, with temperatures dropping to −40° in winter and 14° in summer. The region receives minimal annual precipitation, over 320 sunny days, and high solar radiation of 6–7 kw m^{-2} (i.e., 7,32,000–8,54,000 lux). Trees could improve crop production by reducing incoming sunlight through canopy and increasing humidity under the shade. In those ways, the introduction of trees in farmland would be helpful for better crop production under agroforestry and also would be helpful toward more winter leafy vegetable productions by moderating the microclimate under canopies (Fatima et al. 2022; Raghuvanshi et al. 2019; Singh et al. 1993; Michael et al. 2020).

3.6.4 Eastern Himalayan Region

Eastern Himalayan Region is extended in Arunachal Pradesh, hills of Assam, Sikkim, Meghalaya, Nagaland, Manipur, Mizoram, Tripura, and Darjeeling district of West Bengal.

In this region, farmers use different tree species for various land uses, with priority species varying based on ethnic diversity and food habits of tribal communities. Forty promising species are cultivated in tropical and subtropical regions, 30 in temperate regions, and 28 bamboo species and 2 cane genera in agroforestry programs. Tree density is crucial for sloppy lands, with optimum 400 trees ha^{-1} for agri-horticulture and 200 plants ha^{-1} for agri-silviculture. The Alder-based AF system is a sustainable land use model, with *jhum* fields cultivated for 2 years within a 10 year cycle with 1:4 ratio of cropping to fallow, while Alder-based system allows crop harvests in two out of every 4 years. (l:1 ratio of cropping to fallow), producing up to 57 edible crops along with primary crops of rice, maize and potato. The cropping systems of Eastern Himalayan Region are small-scale mixed rainfed systems, influenced by traditional knowledge and experience. These systems are evolving due to socio-economic changes and rural society development but differ due to factors like agro-ecological conditions, demographic pressure, and market access. Social forestry dynamics can be promoted in mountain regions with thriving traditional tree-based systems, improving existing systems or developing new models (Box 3.6, Bhardwaj and Dev 1985; Borthakur 1992; Singh and Dhyani 1995; Dhyani et al. 1996; Bhatt 2003; Umashankar 2005; Sharma et al. 2007; Das et al. 2012; Sivakumar et al. 2014; Selvan and Kumar 2017).

Box 3.6 Agroforestry Models in Eastern Himalayan Region and Assam Bengal Plains[a]

(i) King chilli (*capsicum annuum*)—Arecanut (*Areca catechu*) based horti-spice AF model
(ii) Gmelina (*Gmelina arborea*) based agri-silvi agroforestry model
(iii) Agar (*Aquilaria malaccensis*) based agroforestry model
(iv) Agrisilviculture (multipurpose tree species + agricultural crops)
(v) Multistoried AFS (alder + tea + black pepper + agricultural crops between the tree rows)
(vi) Silvi-horti-pastoral agroforestry models:
  (a) Alder (*Alnus nepalensis*) + pineapple (*Ananas comosus*) + fodder grasses
  (b) Alder based Agri-silvi-horticultural agroforestry model
  (c) Alder + Orange (*Citrus reticulata*) fruit based agri-horti-silvi agroforestry model
(vii) Alder + Pear (*Pyrus communis*) fruit based agri-horti-silvi agroforestry model
(viii) Silvi-horti-pastoral (alder + pineapple + fodder grasses)
(ix) Coconut based horti-agri agroforestry models
(x) Arecanut—khasi mandarin based agroforestry models:
  (a) Khasi mandarin (*Citrus reticulata* Blanco.) + agricultural crops,
  (b) Assam lemon (*Citrus limon* L. Burm.) + agricultural crops
(xi) Lamboo (*Dysoxylum binectariferum*) based agri-silvi-horticultural AF system
(xii) Kadamb (*Anthocephalus cadamba*) based argi-silvi-horticultural AF system
(xiii) Mango (*Mangifera indica*) based agri-silvi-horticultural AF system
(xiv) Litsea or Indian laurel (*Litsea polyantha*) based agri-silvicultural AF system
(xv) Bambusa balcooa based agri-silvi-horticultural AF system

[a] Sharma et al. (2007), Selvan and Kumar (2017), Subba et al. (2017), Handa et al. (2019), ICFRE (2020).

3.6.5 Assam Bengal Plains

Assam Bengal plains constitute the N-E in Assam and N-E states and southern parts in West Bengal in India and Bangladesh of the great IGPs Region.

The inventory of the agroforestry systems in North bank plains of Assam revealed 11 types of traditional agroforestry systems. Homegardens being the most common agroforestry system in all the four districts followed by aquaforestry and commercial crops trailed on support trees. The least common agroforestry system is Litsea-based agroforestry system, and this system was not recorded from Darrang and

Sonitpur districts of Assam. *Bambusa tulda* and plant species like *Areca catechu*, *Musa* sp., *Camellia sinensis*, *Curcuma longa*, *Zingiber officinale*, *Ananas comosus*, and *Piper* sp. are spatially arranged in scattered or square planting pattern in various combinations. This pattern of cropping has been evolved over the years by the locals to fulfill their needs from the limited resources at their hand. In the current scenario of climate change, these practices will solve problems of depleting natural resources, malnutrition and livelihood. Documentation of these systems practiced in North bank plains of Assam has helped in creating a baseline data based on which further research and scientific interventions on plant species combination, spacing, above and below ground interactions can be carried out. These systems can then be replicated as model practices, thus improving the land management system and providing food security.

The AF systems in the Bengal plain primarily resemble those in the Assam plain due to their geographical proximity and climatic resemblance (Box 3.6, Tomar et al. 2012).

3.6.6 Plateaus Region

The Indian plateau region is a vast and diverse geographical area, covering nearly 30% of the country's landmass. It stretches from the Himalayan foothills in the north to the Deccan Plateau in the south, and from the Indo-Gangetic plain in the east to the Western Ghats and Arabian Sea in the west. The region's geography is characterized by a series of plateaus, hills, and valleys, with varying elevations and climates, including the Central Highlands, the Malwa Plateau, and the Deccan Plateau.

The plateaus Region is following various agroforestry models, including teak-turmeric, bamboo-based silvi-agri, Bach-paddy, Flemingia-lac, agri-lac culture, Babul-paddy, Sandalwood-Teak-Eucalyptus-Red sanders, *Melia dubia*, Ailanthus, Shisham, Aonla, Sapota-teak, Tamarind, Mango, Coconut, Gmelina, Eucalyptus, Jackfruit, and Acacia based AF models. These models aim to improve soil fertility, reduce pollution, and promote sustainable agriculture. Some of the most popular models include Ailanthus, Shisham, Aonla, Sapota, Tamarind, Mango, Coconut, Gmelina, Eucalyptus, Jackfruit, and Acacia (Box 3.7, Dagar 2016, Dagar and Tewari 2016).

Box 3.7 Agroforestry Models Followed in Plateaus Region[a]

(i) Teak (*Tectona grandis*) -Turmeric (*Curcuma domestica*) silvi-medicinal agroforestry model
(ii) Bamboo based silvi—agri agroforestry model
(iii) Bach (*Acorus calamus*)—Paddy (Oryza sativa) agri-medicinal agroforestry model
(iv) Flemingia (*Flemingia macrophylla* and *F. semialata*) based silvi-agri-lac agroforestry model

(continued)

Box 3.7 (continued)

(v) Agri-lac culture agroforestry model
(vi) Babul (*Acacia nilotica*)—Paddy (*Oryza sativa*) agroforestry model
(vii) Sandalwood-Teak-Eucalyptus-Red sanders based silvi-agri agroforestry model
(viii) Sandal wood (*Santalum album*) based agroforestry model
(ix) *Melia dubia* based agri-silvicultural agroforestry model
(x) Ailanthus (*Ailanthus excelsa*) based agri-silvicultural agroforestry model under rainfed condition
(xi) Shisham (Dalbergia sissoo) based agri-silvicultural and silvi-pastoral agroforestry model
(xii) Aonla or Indian gooseberry (*Phyllanthus emblica*) based agri-horticultural agroforestry model
(xiii) Sapota (*Manilkara zapota*)-teak (*Tectona grandis*) based horti-silvicultural agroforestry model for hill zone of Karnataka
(xiv) Sapota (*Manilkara zapota*)—teak (*Tectona grandis*) based horti-silvi-pastoral agroforestry model
(xv) Tamarind(*Tamarindus indica*) based silvi-horticultural agroforestry model
(xvi) Tamarind (*Tamarindus indica*) based agri-silvicultural agroforestry model
(xvii) Mango (*Mangifera indica*) based agri-horticultural agroforestry model
(xviii) Coconut (*Cocos nucifera*) based horti-pastoral agroforestry model
(xix) Gmelina (*Gmelina arborea*) based agri-silvi-horticultural agroforestry model
(xx) Eucalyptus (*Eucalyptus tereticornis*) based agri-silvi-horticultural agroforestry model
(xxi) Jackfruit (*Artocarpus heterophyllus*) based silvi-horticultural agroforestry model
(xxii) Acacia (*Acacia auriculiformis*) based silvi-horticultural agroforestry model

[a]Handa et al. (2019), ICFRE (2020).

3.6.7 Western Dry Region

Western Dry Region extends on hot desert, arid and semi-arid areas extending over Rajasthan, west of the Aravallis, Gujarat, and Dadar & Nagar Haveli.

The Western Dry Region has several agroforestry models, including MPTs like Anjan, Aonla, Mopane, Khejri, Bakain, Ailanthus, Shisham, Aonla, and *Melia dubia*. These models focus on various aspects of agriculture, such as rainfed conditions, silvi-pastoral and agri-horticultural systems, and agri-silvicultural AF systems. The models aim to improve soil health and biodiversity (Box 3.8; Dagar et al.

2001a, b; Tripathy and Barik 2003; Dagar et al. 2013, 2015; Dagar 2016; Dagar and Tewari 2016).

Box 3.8 Agroforestry Models Followed in Western Dry Region[a]

(i) Anjan (*Hardwickia binata*) based agroforestry model
(ii) Aonla or Indian gooseberry (*Phyllanthus emblica*) based agroforestry model
(iii) Mopane or turpentine tree (*Colophospermum mopane*) based agroforestry model
(iv) Khejri (*Prosopis cineraria*)- Ber or Indian Jujube (*Zizyphus mauritiana*) agroforestry model
(v) Bakain (*Melia azedarach*) based agri-silvicultural agroforestry model
(vi) Ailanthus (*Ailanthus excelsa*) based agri-silvicultural AF system under rainfed condition
(vii) Shisham (*Dalbergia sissoo*) based agri-silvicultural and silvi-pastoral AF system
(viii) Aonla or Indian gooseberry (*Phyllanthus emblica*) based agri-horticultural AF system
(ix) Melia dubia based agri-silvicultural agroforestry model

[a]Handa et al. (2019), ICFRE (2020).

Box 3.9 Agroforestry Models Followed in Coastal Plains and Ghats[a]

(i) Casuarina (*Casuarina equisetifolia*)—Maize (*Zea mays*) agrisilviculture AF model
(ii) Casuarina (*Casuarina equisetifolia*)—Moringa (*Moringa oleifera*)—Maize (*Zea mays*) agri-silvi-horticulture AF model
(iii) Akashmoni (*Acacia auriculiformis*)—Napier grass (*Pennisetum purpureum*) silvi-pasture AF model
(iv) Teak (*Tectona grandis*)—Green gram (*Phaseolus mungo*) agri-silviculture AF model
(v) *Acacia mangium*—Beans (*Vigna* species) agri-silviculture AF model
(vi) *Acacia mangium*—Pepper (*Piper nigrum*) silvi-horticulture AF model
(vii) *Casuarina* spp. based wind break AF model
(viii) Mango (*Mangifera indica*) based agri horticultural AF system
(ix) Arjun (*Terminalia arjuna*) based agri-silvicultural AF system
(x) Tamarind (*Tamarindus indica*) based agri-silvicultural AF system
(xi) Mango (*Mangifera indica*) based agri-horticultural AF system
(xii) Block plantation of sandalwood (*Santalum album*) AF system
(xiii) *Melia dubia* based AF system: Mini clonal technology for mass multiplication

(continued)

Box 3.9 (continued)

(xiv) Coconut (*Cocos nucifera*) based horti-pastural AF system
(xv) Gliricidia (*Gliricidia sepium*) and Subabul (*Leucaena leucocephala*) based silvi-pastural AF system
(xvi) Guava (*Psidium guajava*) with Foxtail buffalo grass (*Cenchrus ciliaris*) based horti-pastural AF system for degraded lands
(xvii) *Acacia mangium* based agri-silvicultural AF system
(xviii) Gmelina (*Gmelina arborea*) based agri-silvicultural AF system
(xix) Bamboo based agri-silvicultural AF system
(xx) Jackfruit (*Artocarpus heterophyllus*) and Akashmoni (*Acacia auriculiformis*) based silvi-horticultural AF system
(xxi) Mulberry (*Morus indica*) and Spiked Powder Puff (*Calliandra calothyrsus*) based Silvi-pastural AF System

[a]Handa et al. (2019), ICFRE (2020).

3.6.8 Coastal Plains and Ghats

Coastal Plains and Ghats cover areas of Coromandel and northern Circar coasts of Andhra Pradesh and Odisha, Malabar and Konkan coastal plains, and the Sahyadris.

Coastal Plains and Ghats have been the focus of various agroforestry models, including Casuarina-Maize, Casuarina-Moringa-Maize, Akashmoni-Napier grass, Teak-Green gram, *Acacia mangium*-Beans, *Acacia mangium*-Pepper, *Casuarina* spp.-Windbreak, Mango-Arjun, Tamarind, Block Plantation of Sandalwood, Melia dubia-based AF system, Coconut, Gliricidia, Subabul, Guava with Foxtail buffalo grass, *Acacia mangium*, Gmelina, Bamboo, Jackfruit, Mulberry, and Spiked Powder Puff. These models aim to improve soil fertility, reduce erosion, and promote sustainable agriculture. The use of mini clonal technology for mass multiplication, coconut cultivation, and agroforestry systems can help improve the quality of agricultural lands in these regions (Box 3.9; Yadav et al. 1983; Kumar and Nair 2004; Jena et al. 2011; Roy Chowdhury et al. 2011).

3.6.9 Islands of Andaman-Nicobar and Lakshadweep

Coconut (*Cocos nucifera*) based horti-pastural AF system is the sole popular AF system in those islands. There intercrops offer numerous advantages and are a useful agronomic solution for arable cropping. However, proposing generic crop technical strategies is challenging due to the multitude of possible production objectives and combinations of species, varieties, densities, structure, and organic manure strategies. This highlights the limitations of experiments and the value of modeling multi-species cropping systems. Modeling allows for evaluating intercrop performance under various conditions, determining suitable varietal characteristics, optimizing crop technical protocols, and devising a decision-aid model. However, this requires further mechanistic understanding of multi-species cropping systems and integrating this knowledge into current crop models or developing new ones to accurately represent inter and intraspecific competition. Dedicated intercropping systems should be linked with other strategies like perennial grasslands and agroforestry, by sharing knowledge and tools to create a generic model of the behavior of multi-species canopies (Dagar 1995; Launay et al. 2009; Barillot et al. 2012, 2014a, b; Dagar et al. 2014).

3.7 Phenological, Morphological and Physiological Adjustments Toward Yields of Cereals, Pulses Grains, Oilseeds; Fruits and Wood Trees Under Agroforestry Systems

A study has found that while some yield components of durum wheat were affected by agroforestry and variety, there was no significant interaction between these factors, limiting the selection of wheat cultivars adapted to agroforestry. However, yield was higher under AFS than under full sun (FS), and the only yield component negatively impacted by agroforestry was the number of grains per spike (Daniel et al. 2016).

Another field experiment has revealed representative results on grain and straw yields of green gram (*Vigna radiata*) and pearl millet (*Pennisetum typhoides*) with and without Khejri (*Prosopis cineraria*) based agroforestry in the Thar desert. The study evaluated tree-crop interactions, comparing different tree densities and intercrops like green gram and pearl millet. Results showed that Khejri did not compete with intercrops during the establishment phase. The optimal density for Khejri was found to be 417 stems per ha at 4 years of age, and pulse crops were better suited than pearl millet with or without *Prosopis cineraria*. Yields of green gram and pearl millet were measured under various tree density levels. Under 417 trees per ha producing 153 $kg.ha^{-1}$ grain and 1666 $kg.ha^{-1}$ straw of green gram, under 278 trees ha^{-1}, producing, 162 $kg.ha^{-1}$ grain and 1699 $kg.ha^{-1}$ straw of green gram, and under 208 trees per ha producing 193 $kg.ha^{-1}$ grain and 1803 $kg.ha^{-1}$ straw of green gram. On open sky condition, i.e., without agroforestry condition, green gram grain yield was 180 $kg.ha^{-1}$, while the straw yield was 1720 $kg.ha^{-1}$. Pearl millet yields were also measured under *Khejri* based agroforestry, under 417 trees with 133 $kg.ha^{-1}$ grain and 4330 $kg.ha^{-1}$ straw of pearl millet; under 278 trees pearl millet grain yield was 141 $kg.ha^{-1}$, and the straw yield was 4430 $kg.ha^{-1}$; under 208 trees pearl millet grain yield was 158 $kg.ha^{-1}$, and the straw yield was 5420 $kg.ha^{-1}$. On open sky condition, i.e. without agroforestry condition, pearl millet grain yield was 165 $kg.ha^{-1}$, while the straw yield was 5470 $kg.ha^{-1}$. From the results of that experiment, it is evident that *Khejri* can significantly improve the yields of green gram and pearl millet crops. Under 208 trees both grain and straw yields of green gram were maximum, while maximum grain and straw yields of pearl millet were obtained without agroforestry (Gupta et al. 1998).

3.8 Concluding Remarks

Agroforestry, the integration of trees into agricultural landscapes, is gaining attention as a sustainable approach to cereal-based cropping systems. This practice promotes ecological balance, improves crop yields, and enhances farm resilience. Traditional cereal-based cropping systems often rely on synthetic fertilizers, pesticides, and irrigation, which can have negative environmental impacts. Agroforestry offers a holistic approach by combining trees with other crops or livestock, creating more diverse and resilient ecosystem. Benefits include soil conservation, improved crop yields, biodiversity conservation, climate change mitigation, farm resilience, and economic benefits. To design agroforestry systems for sustainable cereal-based cropping, farmers should consider tree species selection, crop selection, system design, soil management, and irrigation management. By incorporating trees into these systems, farmers can create a more diverse and resilient ecosystem, reducing dependence on external inputs and promoting ecological balance.

Light, an electromagnetic radiation, is essential for crop production, particularly in the 400–700 nm wavelength range. This light is needed to boost plant metabolism and photosynthesis. Farms aim to ensure enough optimum light by minimizing light stress, wasted energy, and electricity costs. Photometric units like foot-candles and lux are commonly used to measure light intensity, but a single measurement is not enough due to variations in light levels throughout the day. The Daily Light Integral (DLI) is the amount of photosynthetically Active Radiation (PAR) a plant receives daily, expressed in moles.m^{-2}.d^{-1} (moles per m^2per day). Plants are classified based on DLI, with low light plants accommodated with 217–435 lux, moderate light plant 435–870 lux, and high light plants 870–1739 lux to maximum 2217 lux. Thus, Comparing range of required DLI against full sunlight, it can be agreed that planting crops under tree canopy can provide beneficial effects on crop growth, as it is relatively low compared to full sunlight.

In western parts of the Indo-Gangetic Plains (IGPs) a popular agroforestry tree species is poplar due to its short rotation, compatibility with agricultural crops, strong demand in wood-based industries, deciduous nature, and high profitability. Poplar trees are cultivated in block plantations, in alley cropping, and border plantings, with a rotation age of 6–9 years. Wheat is the most preferred crop under poplar, as it is a winter crop. As trees age, competition for resources shifts, but the system yield increases. After 8 years of rotation, poplar can yield 150–200 m^3ha^{-1} of wood under block plantations and 12–20 m^3ha^{-1} under border plantations. Spacing poplar trees five by four meters apart gives optimal performance in agri-silviculture systems. Financial and economic assessments show that eight-year rotation-based poplar-agri-silviculture yields are 8%, 10%, and 12% less than net study year value, with a lower Benefit: cost Ratio (BCR) of 2.8. This is significantly greater than traditional cropping systems.

The Trans Himalayan Region in India faces limited resources and extreme climate due to the Himalayan mountain range. Agriculture in the region is adapted to the unique environment, with families relying on subsistence farming based on wheat, barley, and potato on stone-built terraces. Weeding is performed by farm women in groups, and productivity is centered on crops and animal husbandry. Terraced fields are irrigated with glacial meltwater and fertilized with undecomposed farmyard manure, human "night soil," and fertilizers. Each house has small vegetable gardens, orchards, and a few animals. Despite harsh conditions, agriculture remains successful and ecologically sustainable. Climate change affects food security in four dimensions: food availability, food accessibility, food utilization, and food system stability. Vegetables are crucial in Ladakhi cuisine, but weeding is minimal in remote villages, reducing economic yields. Trees could improve crop production by reducing sunlight and increasing humidity under the shade.

India, being known for its agroforestry practices, has evaluated successful models across various ecological regions. These models can be seen in snow clad mountainous areas, hot deserts, plain lands, rugged hilly areas, coastal lands, and high altitudes. However, citing all aspects of these systems is challenging. This chapter has discussed yearlong experimental results from those agroforestry models in India, focusing on their effects on grain yields in various agro-ecological regions.

Intercrops are a beneficial agronomic solution for arable cropping, but proposing generic crop technical strategies is challenging due to various production objectives and combinations of species, varieties, densities, structure, and organic manure strategies. Modeling multi-species cropping systems allows evaluating performance, optimizing technical protocols, and devising decision-aid models. Linking intercropping systems with perennial grasslands and agroforestry can create a generic model.

References

Barillot R, Combes D, Chevalier V, Fournier C, Escobar-Gutierrez A (2012) How does pea architecture influence light sharing in virtual wheat–pea mixtures? A simulation study based on pea genotypes with contrasting architectures. AoB Plants 2012. https://doi.org/10.1093/aobpla/pls038

Barillot R, Combes D, Pineau S, Huynh P, Escobar-Gutiérrez A (2014a) Comparison of the morphogenesis of three genotypes of pea (Pisum sativum) grown in pure stands and wheat-based intercrops. AoB Plants. https://doi.org/10.1093/aobpla/plu006

Barillot R, Escobar-Gutiérrez AJ, Fournier C, Huynh P, Combes D (2014b) Assessing the effects of architectural variations on light partitioning within virtual wheat-pea mixtures. Ann Bot. https://doi.org/10.1093/aob/mcu099

Bhardwaj KKR, Dev SP (1985) Production and decomposition of Sesbania cannabina (Retz.) Pers. in relation to its effect on the yield of wetland rice. Trop Agric 62:233–336

Bhatt BP (2003) Agroforestry for sustainable mountain development in N.E.H. Region. In: Rawat MSS (ed) Central Himalaya environment and development: potentials, actions and challenges. Transmedia Publisher, Uttaranchal, pp 206–223

Borthakur DN (1992) Agriculture of the northeastern region with special reference to hill agriculture. Beece Prakashan, Guwahati

Chauhan BS, Johnson DE (2010) Relative importance of shoot and root competition in dry-seeded rice growing with jungle rice (*Echinochloa colona*) and ludwigia (*Ludwigia hyssopifolia*). Weed Sci 58:295–299

Chauhan SK, Saini KS, Saralch HS et al. (2015) Wheat and barley crop performance under different sowing methods under poplar based agroforestry system. Indian J Ecol 42:528–530

Chavan SB, Dhillon RS (2019) Doubling farmers' income through Populus deltoides-based agroforestry systems in northwestern India: an economic analysis. Curr Sci 117:219–226. https://doi.org/10.18520/cs/v117/i2/219-226

Dagar JC (1995) Agroforestry systems for the Andaman & Nicobar Islands. Int Tree Crops J 8:107–128

Dagar JC (2014) Greening salty and waterlogged lands through agroforestry systems for livelihood security and better environment. In: Dagar JC, Singh AK, Arunachalam A (eds) Agroforestry systems in India: livelihood security & ecosystem services, Advances in agroforestry, vol 10. Springer, New York, pp 273–332

Dagar JC (2016) Agroforestry: four decades of research development. Indian J Agrofor 18(2):1–32. https://www.researchgate.net/publication/316155721

Dagar JC, Tewari JC (2016) Agroforestry research developments: anecdotal to modern science. In: Dagar JC, Tewari JC (eds) Agroforestry research developments. Nova Publishers, New York, pp 1–45

Dagar JC, Sharma HB, Shukla YK (2001a) Raised and sunken bed technique for agroforestry on alkali soils of Northwest India. Land Degrad Dev 12:107–118

Dagar JC, Singh G, Singh NT (2001b) Evaluation of forest and fruit trees used for rehabilitation of semiarid alkali soils in India. Arid Land Res Manag 15:115–133

Dagar JC, Tomar OS, Minhas PS, Kumar M (2013) Lemongrass (Cymbopogon flexuosus) productivity as affected by salinity of irrigation water, planting method and fertilizer doses on degraded calcareous soil in a semi-arid region of Northwest India. Indian J Agric Sci 83(7):734–738

Dagar JC, Pandey CB, Chaturvedi CS (2014) Agroforestry: a way forward for sustaining fragile coastal and Island agro-ecosystems. In: Dagar JC, Singh AK, Arunachalam A (eds) Agroforestry systems in India: livelihood security & ecosystem services, Advances in agroforestry, vol 10, pp 185–223

Dagar JC, Yadav RK, Dar SR, Ahamad S (2015) Liquorice (Glycyrrhiza glabra): a potential salt-tolerant, highly remunerative medicinal crop for remediation of alkali soils. Curr Sci 108:1683–1687

Daniel I-A, Dufour L, Dupraz C, Lauri P-E, Gosme M (2016) Effect of agroforestry on phenology and components of yield of different varieties of durum wheat. 3rd European Agroforestry Conference—Montpellier, 23–25 May 2016. Tree-crop competition and facilitation (oral) Abstract number 109

Das A, Ram Krushna GI, Choudhury BU, Munda GC, Patel DP, Ngachan SV, Ghosh PK, Tripathi AK, Das S, Kumar M (2012) Natural resource conservation through indigenous farming systems: wisdom alive in North East India. Indian J Tradit Knowl 11(3):505–513

Dhyani SK, Chauhan DS, Kumar D, Kushwaha RV, Lepcha ST (1996) Sericulture based agroforestry systems for hilly areas of north-East India. Agrofor Syst 34:1–12

Fatima S, Bhat GM, Islam MA, Rather TA, Megna Rashid M, Dar MUD (2022) Traditional agroforestry systems practiced in Leh district of Ladakh union territory-India. Pharma Innov J SP-11(9):2946–2952. https://www.thepharmajournal.com/

G2V Optics Inc. (2020) Horticulture lighting for commercial growers. g2voptics.com, info@g2voptics.com. http://www.g2voptics.com/horticulture-lighting/

Gandhi JN, Dhiman RC (2010) Performance of wheat intercropped under different tree spacings of poplar (*Populus deltoides* marsh) plantations. Indian J Agrofor 12:28–31

Gill R, Singh B, Kaur N (2009) Productivity and nutrient uptake of newly released wheat varieties at different sowing times under poplar plantation in North-Western India. Agrofor Syst 76:579–590

Grewal SS (1992) Watersheds management project, Relmajra (Punjab), annual report, 1991–92. CSWCRTI, Dehra Dun, pp 142–143

Gupta GN, Singh G, Kawacha GR (1998) Performance of *Prosopis Cineraria* and associated crops under varying spacing regimes in the arid zone of India. Agrofor Syst 40:149–157

Handa AK, Dev I, Rizvi RH, Kumar N, Ram A, Kumar D, Kumar A, Bhaskar S, Dhyani SK, Rizvi J (eds) (2019) Successful agroforestry models for different agro-ecological regions in

India. Jointly published by the Central Agroforestry Research Institute (CAFRI), Jhansi, and the South Asia Regional Programme (based in New Delhi) of World Agroforestry (ICRAF)
ICFRE (2020) Agroforestry models developed by Indian Council of Forestry Research and Education (a compilation). ICFRE, Dehradun. Website: www.icfre.org
Jeet-Ram DJC, Khajanchi L, Singh G, Toky OP, Tanwar VS, Dar SR, Chauhan MK (2011) Bio-drainage to combat waterlogging, increase farm productivity and sequester carbon in canal command areas of Northwest India. Curr Sci 100(11):1673–1680
Jena SK, Sahoo N, Roy Chowdhury S, Mohanty RK, Kundu DK, Behera MS, Patil DU, Kumar A (2011) Reclamation of coastal waterlogged wasteland through biodrainage. J Indian Soc Coastal Agric Res 29(2):57–62
Joshy D (1997) Indigenous technical knowledge in Nepal. In: Pongsapich A, Lesile RN (eds) Indigenous technical knowledge for land management in Asia. International Board for Soil Research and Management, Bangkok, pp 45–61
Kareemulla K, Rizvi RH, Kumar K et al. (2005) Poplar agroforestry systems of western Uttar Pradesh in northern India: a socioeconomic analysis. For Trees Livelihoods 15:375–381. https://doi.org/10.1080/14728028.2005.9752537
Khan SA (2003) Studies on sodic soils through agroforestry system for central plain zone of Uttar Pradesh, India. Paper in XII World Forestry Congress, Quebec City. http://www.fao.org/docrep/article/wfc/xii/1009-b5.htm
Kumar BM, Nair PKR (2004) The enigma of tropical homegardens. Agrofor Syst 61:135–152
Lal P (2004) Integrated development of farm forestry plantation and wood based industries. Indian For 130:71–78
Launay M, Brisson N, Satger S, Hauggaard-Nielsen H, Corre-Hellou G, Kasynova E, Ruske R, Jensen ES, Gooding MJ (2009) Exploring options for managing strategies for pea-barley intercropping using a modeling approach. Europ J Agron 31:85–98. https://doi.org/10.1016/j.eja.2009.04.002
Mead R, Willey RW (1980) The concept of a 'land equivalent ratio' and advantages in yields from intercropping. Exp Agric 16(3):217–228. https://doi.org/10.1017/s0014479700010978
Michael PR, Johnston DE, Moreno W (2020) A conversion guide: solar irradiance and lux illuminance. J Meas Eng 8(4):153–166. https://doi.org/10.21595/jme.2020.21667
Mishra A, Sharma SD, Pandey R, Mishra L (2004) Amelioration of highly alkali soil by trees in northern India. Soil Use Manag 20:325–332
Mõttus M, Sulev M, Baret F, Lopez-Lozano R, Reinart A (2012) Photosynthetically active radiation: measurement and modeling. In: Meyers RA (ed) Encyclopedia of sustainability science and technology. Springer, pp 7902–7932. https://doi.org/10.1007/978-1-4419-0851-3_451
Nair PKR (1993) An introduction to agroforestry. Kluwer Academic Publishers, Dordrecht
Nguyen TH, Catacutan D (2012) History of agroforestry research and development in Viet Nam: analysis of research opportunities and gaps. Working paper 153. World Agroforestry Centre (ICRAF) Southeast Asia Regional Program, Hanoi, Viet Nam. https://doi.org/10.5716/WP12052.PDF. 32p
Panda S, Dhara PK, Sarkar S, Das NC (2018) Chapter 35: Agroforestry—a multipurpose multi-storeyed renewable plant treasure with medicinal values giving multireturn for livelihood and benefits over climate change. In: Pandey CB, Gaur MK, Goyal RK (eds) Climate change and agroforestry adaptation, mitigation and livelihood security, 1st edn. nipa New India Publishing Agency, New Delhi, pp 627–637. https://www.researchgate.net/publication/324947113_Agroforestry_-_a_multipurpose_multistoreyed_renewable_plant_treasure_with_medicinal_values_giving_multireturn_for_livelihood_and_benefits_over_climate_change
Panwar P, Chauhan S, Kaushal R et al. (2017) Carbon sequestration potential of poplar-based agroforestry using the CO2FIX model in the indo-Gangetic region of India. Trop Ecol 58:439–447
Pathak PS, Dagar JC, Kaushal R, Chaturvedi OP (2014) Agroforestry inroads from the traditional two-crop systems in heartlands of the Indo-Gangetic plains. In: Agroforestry systems in India: livelihood security & ecosystem services. Springer, New Delhi, pp 87–116

Raghuvanshi MS, Dorjay Singh RK, Manjunatha BL, Moharana PC, Spalbar E, Landol S, Saxena A (2019) Ladakh traditional farming: an approach to resource utilization under changing climate. Int J Curr Microbiol App Sci 8(9):654–666. https://doi.org/10.20546/ijcmas.2019.809.079

Roy Chowdhury S, Kumar A, Brahmanand PS, Ghosh S, Mohanty RK, Jena SK, Sahoo N, Panda GS (2011) Application of biodrainage for reclamation of waterlogged situations in deltaic Orissa, Research bulletin 53. Directorate of Water Management, Bhubaneswar, p 32

Saha R, Tomar JMS, Ghosh PK (2007) Evaluation and selection of multipurpose trees for comparing soil hydro-physical behavior under hilly ecosystem of north East India. Agrofor Syst 69:239–247

Selvan T, Kumar S (2017) Chapter 2: Agroforestry in the north-eastern Himalayas. In: Gupta SK, Pankaj Pawar P, Rajesh Kaushal R (eds) Agroforestry for increased production and livelihood security. New India Publishing Agency, New Delhi, pp 23–40. https://www.researchgate.net/publication/318768828_Agroforestry_in_the_North-Eastern_Himalayas

Sharma KK (1996) Agroforestry in farming systems development. Indian For 122:547–559

Sharma R, Xu J, Sharma G (2007) Traditional agroforestry in the eastern Himalayan region: land management system supporting ecosystem services. Trop Ecol 48(2):1–12. ISSN 0564-3295. http://www.tropecol.com/

Shively G (2012) An overview of benefit-cost analysis. January 2012. https://www.researchgate.net/publication/255661807_An_Overview_of_Benefit-Cost_Analysis

Singh BP, Dhyani SK (1995) Significance of jackfruit in restoration of soil fertility. Annual Report. ICAR Research Complex, Umiam

Singh G, Gill HS (1992) Ameliorative effect of tree species on characteristics of sodic soils at Karnal. Indian J Agric Sci 62:142–146

Singh K, Panda MC, Jhakaral KK (1993) Proceeding of international symposium. Indian Soc Weed Sci, Hisar 1:365–368

Singh YP, Sharma DK, Singh G, Nayak AK, Mishra VK, Singh R (2008) Alternate land use management for sodic soils, CSSRI technical bull no. 2/2008. CSSRI, Karnal, p 16

Sivakumar PS, Thirugnanvel A, Chakruno M (2014) Understanding the mechanisms of achieving food and nutritional security through traditional tuber crops foods: a case study among Konyak Tribes of Nagaland, India. J Root Crop 40(2):49–57. Indian Society for Root Crops

Soemarwoto O (1987) Homegardens: a traditional agroforestry system with a promising future. Section 3, Chap. 10. In: Steppler HA, PKR N (eds) Agroforestry a decade of development. International Council for Research in Agroforestry, ICRAF House, Nairobi. ISBN 92 9059 036 X

Subba B, Dhara PK, Panda S (2017) Agri-horti-silvi based models for sustainable land use and conservation in hill zone of West Bengal. Journal of Hill Agriculture 8(1):81–86. https://doi.org/10.5958/2230-7338.2017.00021.0

Tomar OS, Dagar JC, Singh YP (2004) Forest and fruit trees for alkali soils. Indian Farm 53(11):44–47

Tomar JMS, Das A, Puni L, Chaturvedi OP, Munda GC (2012) Shifting cultivation in north-eastern region of India—status and strategies for sustainable development. Indian Forester 138(1):52–62

Tripathy RS, Barik SK (2003) Shifting cultivation in north East India. In: Bhat BP, Bujarbaruah KM, Sharma YP, Patiram (eds) Proc. approaches for increasing agricultural productivity in hill and mountain eco system. ICAR Research Complex for NEH Region, Umiam, pp 317–322

Umashankar (2005) Indigenous agroforestry tree species for conservation and rural livelihood. In: Bhatt BP, Bujarbaruah KM (eds) Agroforestry in north East India: opportunities and challenges. ICAR Research Complex for NEH Region, Umiam, pp 149–174

Viswanath S, Lubina PA, Subbanna S, Sandhya MC (2018) Traditional agroforestry systems and practices: a review. Adv Agric Res Technol J II(1):18–29

Yadav JSP, Bandyopadhyay AK, Bandyopadhyay BK (1983) Extent of coastal saline soils in India. J Indian Soc Coast Agric Res 1:1–6

Websites

https://www.nexsel.tech/how-to-measure-light-intensity-for-plants.php

https://www.harvestreturns.com/blog/2018/6/15/3-ways-to-measure-agriculture-investment-returns

https://www.wallstreetprep.com/knowledge/irr-internal-rate-of-return/

Agroforestry: Helping Soil Organic Carbon Built-up

4

Abstract

The integration of agroforestry systems into agricultural landscapes offers a promising strategy for enhancing sustainability, particularly through the accumulation of soil organic carbon (SOC). By promoting microbial activity and nutrient cycling, agroforestry reduces dependency on chemical inputs such as fertilizers and pesticides, thus, yielding both economic and environmental benefits. This approach improves soil health by increasing organic matter through tree litter, root biomass, and decaying wood, which supports microbial communities crucial for nutrient cycling and soil structure. Agroforestry also mitigates soil erosion by stabilizing the soil with tree roots and canopy cover, which enhances soil conservation and agricultural productivity. Furthermore, agroforestry systems can adapt to various climatic and ecological contexts, making those an effective tool for carbon sequestration and climate adaptation. Despite its advantages, the adoption of agroforestry faces barriers such as economic viability, market access, and insufficient policy support. To overcome these challenges, targeted interventions including improved infrastructure, supportive policies, and farmer education are necessary. Future research should focus on quantifying the benefits of agroforestry across diverse systems and climates, while community engagement and farmer empowerment are essential for successful implementation. Ultimately, agroforestry offers a multifaceted solution for sustainable agricultural practices, contributing to soil health, biodiversity, and climate resilience, thereby supporting both human and ecological well-being.

Keywords

Agroforestry systems · Silvopastoral · Agrosilvopastoral · Agrosilvicultural · Sequential agroforestry · Ecological · Agroecological · Agricultural · Drought· Flood· Food security · Biodiversity conservation · Fertilizer trees · Ecosystem services · Soil erosion · Soil organic matter · Soil health · Root exudates· Amino

S. Panda, *Agroforestry*, https://doi.org/10.1007/978-981-96-6855-7_4

acid · Sugar · Carbon credit · Soil organic carbon (SOC) · Soil microbiome · Mycorrhizal fungi · Drought resistance · Organic certification · Climatic regions · Tropical · Temperate · Mediterranean climate · Sequential agroforestry · Soil quality indicators · Water quality · Water retention · SMAF · Kendall's rank correlation · Soil and water conservation · Resilience · Regenerative agriculture · Cycle of growth and decay

Abbreviations

AF: Agroforestry
B/C: Benefit/Cost
CEC: Cation exchange capacity
EC: Electrical conductivity
FMNR: Farmer managed natural regeneration
LER: Land equivalent ratio
N: Nitrogen
P: Phosphorus
RR: Risk ratio (in meta-analysis)
SMAF: Soil management assessment framework
SOC: Soil organic carbon
SOM: Soil organic matter

4.1 Soil Organic Carbon (SOC) Built-up Through Agroforestry

Various studies aimed at understanding the existing patterns and linkages between agroforestry and soil properties across different climatic zones have yielded highly valuable insights. These investigations have not only highlighted the regulatory factors that pose threats to the success of agroforestry practices but have also identified key elements that promote positive outcomes. As a result, agroforestry systems are increasingly recognized as promising agroecological approaches for sustainable agricultural intensification.

Agroforestry encompasses diverse systems that integrate crops, animals, and trees—often referred to as agrosilvopastoral systems—or solely crops and trees, known as agrosilvicultural systems, all on the same plot of land. This multifaceted approach operates at the nexus of agriculture and forestry, allowing it to provide a multitude of benefits. One of the primary advantages of agroforestry is its capacity to bolster food security and enhance revenue generation for farmers. By diversifying agricultural production, these systems not only increase the resilience of farming operations but also create multiple streams of income, making livelihoods more stable. In addition to economic benefits, agroforestry plays a critical role in safeguarding biodiversity. By incorporating a variety of plant species and creating complex habitats, agroforestry systems support a wide range of flora and fauna, thereby promoting ecological balance. This biodiversity is vital for ecosystem health, as it enhances pollination, pest control, and resilience against diseases.

Agroforestry systems also significantly enhance ecosystem services, particularly in terms of soil management. The integration of trees into agricultural landscapes has profound effects on soil quality indicators, including physical, chemical, and biological properties of soil. For example, tree roots contribute to improved soil structure and fertility. They stabilize the soil, preventing erosion and promoting the retention of nutrients. Moreover, the leaf litter and organic matter produced by trees play a crucial role in nutrient cycling, enriching the soil and benefiting not just the trees but also the crops planted alongside them (Elias 2002).

A key advantage of agroforestry is its ability to augment soil organic carbon (SOC) content. Increased SOC is vital for maintaining soil health, enhancing fertility, and supporting agricultural productivity over the long-term. The presence of deep-rooted trees enables access to nutrients that may be unavailable to shallow-rooted crops, thereby enriching the soil for all plants involved in the system.

Water management is another critical aspect where agroforestry systems excel. These systems can enhance water retention in the soil, helping to mitigate the impacts of drought, which is increasingly relevant in the face of climate change. The presence of trees reduces surface runoff, allowing for greater water infiltration and percolation into the ground. This, in turn, supports the recharge of aquifers, making water more available for both agricultural and domestic uses, especially in arid and semi-arid regions where water scarcity poses significant challenges.

Studies conducted across various climates, *viz.* tropical, temperate, and Mediterranean regions have contributed to an improved understanding of the effects of agroforestry on soil quality and the factors influencing these outcomes. Factors such as climatic conditions, plant species selection, and soil and water management practices are crural in determining the effectiveness of agroforestry systems. For instance, in tropical climates, where biodiversity is inherently high, the selection of specific tree and crop combinations can lead to synergistic effects that enhance soil fertility and agricultural productivity. In contrast, temperate regions may benefit from different species that are more suited to their cooler climates and soil types.

As research continues to evolve, it becomes increasingly clear that agroforestry is not merely a method of land use; it is a holistic approach that interlinks agriculture, forestry, and environmental stewardship. This integration allows for the sustainable intensification of agriculture, ensuring that food production can meet the demands of a growing global population while preserving the health of our ecosystems.

Thus, the multifaceted benefits of agroforestry underscore its potential as a sustainable agricultural practice that can adapt to various climatic challenges. By enhancing soil quality, improving water management, and supporting biodiversity, agroforestry systems contribute to the resilience and sustainability of agricultural landscapes worldwide. The insights gained from ongoing studies will be essential in refining these systems and maximizing their benefits for both farmers and the environment, paving the way for a more sustainable future in agriculture (Smith et al. 2012; Souza et al. 2012; Lagerlof et al. 2014; Zuazo et al. 2014; Ravi et al. 2015;

Torralba et al. 2016; Fahmi et al. 2018; Cherubin et al. 2019; Liu et al. 2019; Zhu et al. 2019; Guo et al. 2020; Guillot et al. 2021; Rosati et al. 2021; Amare et al. 2022; Beule et al. 2022; Fahad et al. 2022; Mesfin and Haileselassie 2022; Bracken et al. 2023; Ngaba et al. 2024).

4.2 Understanding Agroforestry and Soil Properties across Climatic Zones: Insights and Regulatory Factors Influencing SOC Built-up

Agroforestry, a land-use management system that integrates trees and shrubs into agricultural landscapes, has garnered significant attention for its potential to enhance agricultural productivity, to improve soil health, and to contribute to environmental sustainability. Various studies examining the patterns and linkages between agroforestry practices and soil properties across different climatic zones have yielded essential insights that inform both agricultural practices and policy decisions. These studies have not only highlighted the regulatory factors that threaten the success of agroforestry systems but also have identified elements that promote positive outcomes.

Research across diverse climatic zones—ranging from tropical to temperate regions—has illustrated how agroforestry can influence soil properties. Key findings indicate that agroforestry systems can enhance soil structure, increase organic matter content, and improve nutrient cycling. For example, studies have shown that the presence of tree roots can improve soil aeration and water infiltration leading to better crop yields. In tropical regions, the integration of leguminous trees into farming systems has been particularly effective in fixing nitrogen, thereby reducing the need for synthetic fertilizers. Despite the promising benefits, several regulatory factors pose threats to the successful implementation of agroforestry systems. One significant challenge is the lack of supportive policies that recognize the long-term benefits of agroforestry. In many regions, agricultural policies prioritize conventional farming practices, often neglecting the ecological benefits of integrating trees into farming systems. This policy disconnect and can discourage farmers, from adopting agroforestry practices, as they may not receive incentives or recognition for their efforts in enhancing soil health and biodiversity.

Moreover, these studies underscore the importance of specific tree species and their interactions with soil microorganisms. Certain species can enhance microbial diversity and activity, which in turn contributes to improved soil fertility and resilience against erosion. These findings suggest that selecting appropriate tree species based on local conditions can optimize the benefits of agroforestry, tailoring practices to enhance soil properties specific to each climatic zone (Seiter et al. 1999).

Additionally, land tenure issues can hinder the adoption of agroforestry. Farmers may be reluctant to invest in long-term practices if they lack secure land rights. In regions where land is communally held or subject to conflicting claims, the absence of clear ownership can deter investment in agroforestry systems that require years to establish before yielding tangible benefits. Environmental factors, such as climate variability and extreme weather events, also play a crucial role in the success of agroforestry. Studies have shown that agroforestry systems can mitigate some impacts of climate change such as soil erosion and nutrient depletion. However, extreme weather can challenge the resilience of these systems, particularly if farmers are unprepared for shifts in climate patterns. Therefore, a deeper understanding of local climatic conditions and their variability is essential for designing agroforestry systems that can withstand such challenges. Furthermore, education and extension services play a vital role in promoting agroforestry. Training programs that demonstrate the benefits of agroforestry—such as improved soil health, increased crop resilience, and additional sources of income from tree products—can empower farmers to transition to these practices. Additionally, showcasing successful case studies within similar climatic zones can motivate farmers to adopt agroforestry systems. Collaboration between governments, non-governmental organizations, and research institutions is also critical. Such partnerships can facilitate the development of policies that support agroforestry, provide financial incentives, and ensure the availability of resources and training for farmers. By fostering an environment that values agroforestry, stakeholders can collectively work toward sustainable agricultural practices that enhance soil properties and overall ecosystem health. On a more optimistic note, several factors have been identified that can promote the success of agroforestry initiatives. Community engagement and local knowledge are paramount. Studies have shown that involving local communities in the planning and implementation of agroforestry systems leads to higher acceptance and more successful outcomes. Farmers who understand the ecological and economic benefits of integrating trees into their farming systems are more likely to adopt these practices.

So, the intricate linkages between agroforestry and soil properties across various climatic zones have been illustrated through extensive research. While regulatory factors pose significant challenges to the successful implementation of agroforestry systems, there are numerous opportunities to promote positive outcomes. By addressing policy gaps, ensuring secure land tenure, and fostering community engagement and education, stakeholders can harness the benefits of agroforestry. Ultimately, the continued study and understanding of these dynamics will be essential in promoting sustainable agricultural practices that not only improve soil health but also contribute to broader environmental and economic resilience (Uthappa et al. 2017; Handa et al. 2019; Zinngrebe et al. 2020; Sh et al. 2022; Sekhar et al. 2024; Xiao et al. 2024; Table 4.1).

4.2.1 The Impact of Agroforestry on Soil Quality Indicators Including SOC Built-up: A Comprehensive Overview

Agroforestry, the integration of trees and shrubs into agricultural landscapes, has garnered significant attention for its potential to enhance soil quality. The multifaceted benefits of agroforestry extend beyond mere tree-crop interactions; they encompass conspicuous physical, chemical, and biological soil properties. Recent

Table 4.1 Factors influencing soil improvement induced under agroforestry land use practices[a]

Factors	Findings
General soil fertility	Agroforestry (AF) improves physical, chemical, and biological soil conditions.
Erosion control	AF reduces soil erosion through tree canopy and litter (up to 90–150 tons of carbon sequestration per hectare.)
Soil fertility decline	Declining soil fertility due to the loss of organic matter, nutrient deterioration, and acidification.
Types of AF practices	Various practices such as fallows, hedgerows, and alley cropping enhance soil fertility.
Leguminous trees	Leguminous trees improve soil fertility through nitrogen fixation and biomass transfer.
Soil aggregation improvement	AF systems increase soil aggregation, enhancing water infiltration and retention. (higher aggregation compared to non-AF systems)
Leaf litter contribution	Leaf litter from AF systems enriches soil fertility by providing organic matter.
Soil nutrient maintenance	AF systems maintain nutrient status and promote biodiversity, increasing soil fauna and flora.
Water infiltration improvement	AF systems lead to better soil porosity and increased water infiltration, compared to bare soil
Impact on non-point source pollution	AF practices help in reducing agricultural runoff pollution through nutrient uptake by deeper roots.
Biodiversity support	AF systems enhance biodiversity by providing habitats and corridors for wildlife.
Ecosystem services	AF contributes to ecosystem services like carbon sequestration and water purification.

[a]Acharya and Kafle (2009), Ajayi et al. (2009), Eyasu (2002), Juo (2004), Kandji et al. (2006), Kessler (1992), Mooney et al. (1987), Nair (1993, 2008), Okigbo and Lal (1985), Rao et al. (1998), Udawatta (2002), Young (1989), Xiao et al. (2024)

studies conducted across diverse climatic regions—tropical, temperate, and Mediterranean—have illuminated the complex interplay between agroforestry practices and soil quality indicators, emphasizing the role of climatic conditions, plant species selection, and soil and water management strategies (Tables 4.1 and 4.2).

4.2.1.1 Impact of Agroforestry on Soil Chemical Properties

The chemical properties of soil, including nutrient availability and pH levels, are significantly influenced by agroforestry practices. Trees in agroforestry systems can contribute organic matter through leaf litter and root biomass, enriching the soil with essential nutrients. In tropical climates, species such as leguminous trees are particularly beneficial as they can fix atmospheric nitrogen, enhancing soil fertility. This nitrogen enrichment is crucial for sustainable agricultural productivity, as it reduces the need for synthetic fertilizers, which can have detrimental effects on environment.

In temperate and Mediterranean regions, the interaction between trees and crops can also lead to improved nutrient cycling. The seasonal shedding of leaves and the

Table 4.2 Global SOC content range and potential under agroforestry in various climatic regions

Study reference	Climatic regions	Agroforestry system type	Soil organic carbon content (g.kg^{-1})	SOC sequestration potential (Mg. ha^{-1}.year^{-1})	Key findings
Lorenz and Lal (2014)	Various	Mixed species systems	10–30	2–10	Highlights the role of agroforestry in enhancing SOC through root biomass and litter inputs.
Mayer et al. (2022)	Temperate	Silviculture, alley cropping	20–50	1.5–5	Meta-analysis shows significant SOC increases in temperate agroforestry systems, emphasizing management practices.
Shi et al. (2018)	Various	Diverse agroforestry	15–40	2–8	Comprehensive meta-analysis demonstrating the importance of agroforestry for SOC stocks and its potential for climate mitigation.
Cardinael et al. (2017)	France	Diverse (six sites)	25–60	3–7	Survey results indicate increased SOC stocks in agroforestry compared to conventional systems across multiple sites.
Uthappa et al. (2017)	Global	Diverse agroforestry including homegardens	20–250	2–10	Discussed agroforestry as a sustainable solution for climate challenges.
Gupta et al. (2023)	Various	Various (general review)	15–45	2–6	Discussed the overall role of agroforestry in improving SOC and resilience against climate change.

decomposition of organic matter enrich the soil with nutrients, while the tree canopy can influence soil pH and nutrient availability by affecting moisture retention and microbial activity. Consequently, the integration of agroforestry can create a more balanced nutrient profile, ultimately enhancing crop yield and soil quality (Table 4.3).

Table 4.3 Impact of agroforestry on mitigation of soil pollution[a]

Geographical area	Agroforestry system	Reduction in nitrate (%)	Reduction in phosphate (%)	Reduction in potassium (%)	Reduction in ammonium (%)	Reduction in nitrite (%)	Reduction in herbicides (%)	Reduction in agrochemicals (%)
Greece	Potato - poplar	90%	100%	86%	97%	85%	>80%	100%
Greece	Maize - poplar	77%	100%	73%	97%	77%	–	–
Greece	Maize-olive	–	–	–	–	–	>80%	–
Italy	Various systems	60–100%	100%	–	–	–	–	–
USA	Pecan-cotton	30–70% higher N uptake at 0.9 m	–	–	–	–	–	–
India	Sequential agroforestry (within fields of *Brassica napus*, *Cajanus cajan*, *Crotalaria juncea*, *Gossypium arboreum*)	–	–	–	–	–	–	Significant selenium reduction
General (simulated)	Enhanced tree integration	31% (nutrients)	–	–	–	–	–	47% (pesticides)
Riparian buffers	Various perennial species	Higher phytoremediation	–	–	–	–	–	Greater than in other systems

[a]Allen et al. (2020), Licht and Isebrands (2005), Borin et al. (2010), Greipsson (2011), Wang et al. (2011), Atangana et al. (2014), Dawson et al. (2014), Tsonkova et al. (2014), De'Longe et al. (2016), Gikas et al. (2016), Pavlidis and Tsihrintzis (2017), Pavlidis et al. (2018, 2020, 2021), Hussain et al. (2020)

4.2.1.2 Impact of Agroforestry on Soil Biological Properties

The biological properties of soil, which include microbial activity, earthworm populations, and overall biodiversity, are critical for maintaining soil health. Agroforestry systems have been shown to foster a diverse microbial community and increase the abundance of beneficial organisms. In tropical regions, the presence of trees can create a more favorable microclimate for soil organisms, enhancing microbial biomass and activity. These organisms play vital roles in organic matter decomposition and nutrient cycling, further contributing to soil fertility.

In temperate and Mediterranean climates, the interactions between tree roots and soil organisms can improve soil structure and nutrient dynamics. For example, mycorrhizal fungi associated with tree roots can enhance nutrient uptake for both trees and crops, creating a synergistic relationship that boosts overall soil health. Moreover, the promotion of biodiversity within agroforestry systems can lead to greater resilience against pests and diseases, reducing the reliance on chemical inputs.

4.2.2 Environmental Factors Created Through Agroforestry Helping Toward SOC Built-up

The benefits of agroforestry on soil quality indicators are influenced by several factors, including climatic conditions, plant species selection, and effective soil and water management practices.

4.2.2.1 Climatic Conditions

The regional climate significantly impacts the effectiveness of agroforestry systems. In tropical regions, for example, high rainfall and temperatures can accelerate organic matter decomposition, necessitating careful management to maintain soil fertility. Conversely, in Mediterranean climates, where water scarcity is prevalent, agroforestry can enhance water retention and reduce evaporation, making it an effective strategy for soil conservation (Navarro-Pedreño et al. 2021).

4.2.2.2 Plant Species Selection

The choice of tree and shrub species is crucial for optimizing soil quality. Species that are compatible (or acclimatized after introduction in the new area) with local conditions and that have beneficial traits, such as nitrogen fixation or deep rooting systems, can enhance the overall productivity and sustainability of agroforestry systems. The selection process must consider the specific ecological and agricultural context to maximize benefits.

4.2.2.3 Soil and Water Management

Effective management practices are essential for maximizing the potential of agroforestry. Strategies such as cover cropping, mulching, and controlled grazing can enhance the positive impacts of trees on soil quality. Moreover, proper water management practices, including the integration of rainwater harvesting and irrigation, can optimize soil moisture levels, supporting both trees and crops during critical growth periods.

4.3 Influence of Agroforestry on Soil Organic Carbon Built-up

Agroforestry significantly enhances soil quality by promoting the build-up of soil organic carbon (SOC). The integration of trees with crops or livestock improves organic matter input through leaf litter, root biomass, and other plant residues. These additions enhance soil structure, water retention, and nutrient cycling, leading to increased SOC content. The diverse plant species in agroforestry systems create a more stable and resilient soil environment, fostering microbial activity that accelerates the decomposition of organic matter and enhances carbon sequestration. Consequently, agroforestry serves as a sustainable land management practice that improves soil health while mitigating climate change through increased soil carbon storage.

4.3.1 Impact of Agroforestry on Soil Quality with Improved Soil Organic Carbon Content

Agroforestry, the integration of trees and shrubs into agricultural landscapes, has gathered attention for its potential to enhance environmental sustainability, improve agricultural productivity, and contribute to climate change mitigation. One of the critical benefits of agroforestry systems is their influence on soil organic carbon (SOC) levels. SOC plays a vital role in soil health, fertility, and the global carbon cycle, making its management crucial for sustainable land use practices.

The degradation of soil quality due to intensive agricultural practices is a global concern, with significant implications for food security, biodiversity, and climate stability. Traditional monoculture systems often lead to soil erosion, nutrient depletion, and a decline in soil organic matter, adversely affecting agricultural productivity. In contrast, agroforestry offers a holistic approach to land management, combining the benefits of forestry with agricultural practices. This deliberation aims to elucidate the impact of agroforestry on soil quality, particularly in terms of improving soil carbon content, which is crucial for enhancing ecosystem resilience and combating climate change (Tables 4.2, 4.3, 4.4, 4.9 and 4.11).

4.3.1.1 Definitions and Concepts

1. Agroforestry: A land-use management system that combines agricultural crops with trees and shrubs. It encompasses various practices, including alley cropping, silvopasture, and forest farming.
2. Soil Quality: Refers to the ability of soil to perform its functions, including supporting plant growth, regulating water flow, and storing nutrients.

3. Soil Carbon Content: The amount of organic and inorganic carbon stored in the soil, a key indicator of soil health and fertility.

4.3.1.2 Impact of Agroforestry on SOC Content in Six Agriculturally Prominent World Geographical Regions

Agroforestry, the integration of trees and shrubs, fishery and livestock into agricultural landscapes, varies significantly across six geographical regions (Table 4.2).

4.3.1.2.1 Tropical Region

In places like Southeast Asia and Africa, agroforestry often involves intercropping food crops with shade trees, enhancing biodiversity and improving soil health. Numerous studies in tropical agroforestry systems have shown significant increases in soil carbon content. For example, the use of shade trees in coffee plantations has been linked to higher soil organic matter and improved nutrient availability (De Souza et al. 2011, Ayala-Montejo et al. 2022, Manson et al. 2022, Bracken et al. 2023).

4.3.1.2.2 Sub-tropical Region

Agroforestry in subtropical regions integrates trees and shrubs with crops and livestock, enhancing biodiversity and improving soil health. This practice promotes sustainable land use by optimizing land productivity, conserving water, and reducing erosion. It also provides additional income through timber, fruits, and nuts, while improving resilience against climate variability. By fostering symbiotic relationships between different species, agroforestry can significantly contribute to food security and ecosystem health in subtropical areas.

4.3.1.2.3 Temperate Region

In Europe and North America, practices include alley cropping and silvopasture, where trees are planted alongside crops or livestock, optimizing land use and providing shelter. Research in temperate agroforestry systems, such as silvopasture, indicates improved soil quality and increased carbon stocks due to the integration of trees and livestock.

4.3.1.2.4 Arid Region

In areas like the Middle East and parts of Africa, agroforestry focuses on drought-resistant tree species to combat desertification and improve water retention.

4.3.1.2.5 Cold Region

In northern climates, agroforestry can include windbreaks and shelterbelts, protecting crops from harsh winds and improving microclimates.

4.3.1.2.6 Mediterranean Region

Agroforestry in the Mediterranean region integrates trees and shrubs with crops and livestock, promoting biodiversity, soil health, and sustainable land use. This practice enhances resilience to climate change by improving water retention and reducing erosion. It also supports local economies through diversified income sources and helps preserve traditional landscapes and ecosystems. Key examples include olive groves intercropped with herbs and fruit trees, which optimize land use while maintaining cultural heritage of the region.

From the overall observations, it is obvious to note that agroforestry promotes sustainable land use management, and thus, enhances resilience to climate change. In this way, AF is potential to support local economies across diverse environments (Table 4.4).

4.3.1.3 Agroforestry in Historical Perspectives

In historical context, practices of agroforestry can be classified mainly into two phases (Steppler and Nair 1987; Panda et al. 2018):

4.3.1.3.1 Traditional Practices

Agroforestry has been practiced for centuries, particularly in indigenous and traditional farming systems. These practices often involve the cultivation of crops alongside trees, which provide shade, improve soil structure, and enhance biodiversity. Historical evidence indicates that such systems have supported sustainable land use and agricultural productivity for generations.

4.3.1.3.2 Modern Developments

In recent decades, there has been a resurgence of interest in agroforestry, driven by the need for sustainable agricultural practices and the recognition of the benefits of tree-crop interactions. Research has increasingly focused on quantifying the benefits of agroforestry systems in terms of soil quality and carbon sequestration.

4.3.2 Mechanisms of SOC Accumulation in Agroforestry

Mechanisms of soil quality improvement in agroforestry are dependent on five main factors: (i) Increased biomass input, (ii) Improved soil structure, (iii) Enhanced microbial activity, (iv) Reduced soil erosion, and (v) Nutrient cycling.

4.3.2.1 Increased Biomass Input

(i) Trees and shrubs contribute significant organic matter through leaf litter, dead branches, and root biomass. This additional organic matter decomposes over time, enriching the soil with SOC. (ii) The diverse plant species in agroforestry systems provide continuous biomass input throughout the year, unlike monoculture systems that may have seasonal biomass inputs.

4.3.2.2 Improved Soil Structure

The root systems of trees can enhance soil aggregation and porosity, promoting better infiltration and retention of water. Improved soil structure facilitates microbial activity, leading to more efficient decomposition of organic matter and higher SOC levels.

Table 4.4 Impact of agroforestry on soil function improvement and modulation of microclimate[a]

Study area	Agroforestry system	Soil function improvement	Data/ percentage change/ statistic value
Southwest Cameroon	Agro-silvopastoral system	Increased soil fertility	53% of farmers perceive success; 57% report highest soil fertility
Bangladesh	Jack fruit-based agroforestry	Reduced soil temperature; increased soil moisture; increased nitrogen	3.37–9.25% reduction in soil temperature; 10–20% increase in moisture; 9–19% increase in nitrogen
Himachal Pradesh, India	Agri-horticulture, agri-silviculture	Enhanced carbon density and physical soil properties	Not specified
Various regions	Mixed tree systems (coffee, Cordia)	Higher N, P, K, SOC, and OM concentrations	N (0.17–0.26%), SOC up to 1.82%
Sudan	*Acacia senegal* + sorghum	Improved crop yields and soil properties	Not specified
USA	*Castanea mollisima - Asima trilobala*	Higher total nitrogen stock and SOC levels	SOC higher at 24-year-old system; specific increases not stated
Haryana, India	Poplar windbreak	Increased nutrient levels (N, P, K)	N (365.2 kg/ha), P (19.7 kg/ha), K (357.3 kg/ha) at 2 m distance
Ethiopia	*Faidherbia albida*	Reduced allelopathy; improved nutrient status	Not specified
Rajasthan, India	*Prosopis cineraria* systems	Higher N, P, and soil microbial biomass	N (32.1—42.4 μg/g), P (11.6–15.6 μg/g), microbial biomass (262–320 μg/g)
USA (corn-soy rotation)	Agroforestry buffers	Increased SOC, total N, available P, exchangeable K	SOC (11.9–91.5% increase), N (22.2–125% increase), P (31–71% increase)
Mexico	Avocado with shaded coffee	Decrease in soil respiration; increase in SOC	Not specified

[a]Kho et al. (2001), Schroth and Burkhard (2003), Mohamed (2005), Augustine et al. (2007), Narendra (2008), Lacombe et al. (2009), Yadav et al. (2011), Shoga'a Aldeen et al. (2013), Lorenz and Lal (2014), Gebru (2015), Nath et al. (2015), Sonmez et al. (2016), Bisht et al. (2017), Dhaliwal et al. (2017), Sirohi and Bangarwa (2017), Chatterjee et al. (2018), De Stefano and Jacobson (2018), Riyadh et al. (2018), Yengwe et al. (2018), Diallo et al. (2019), Sarmiento-Soler et al. (2019), Akoto et al. (2020), Akpalu et al. (2020), Douglas et al. (2020), Gusli et al. (2020), Rizwan et al. (2020), Salve and Bhardwaj (2020), Surki et al. (2020), Wartenberg et al. (2020), Rodríguez et al. (2021), Salceda et al. (2021), Tsufac et al. (2021a, b), Ayala-Montejo et al. (2022), Mesfin and Haileselassie (2022), Rivest et al. (2022), Sirohi et al. (2022), Xiao et al. (2024)

4.3.2.3 Enhanced Microbial Activity

The diverse root exudates from various species stimulate microbial populations in the soil. A healthier microbial community accelerates the breakdown of organic matter, enhancing SOC formation.

4.3.2.4 Reduced Soil Erosion

Agroforestry systems can significantly reduce soil erosion by stabilizing the soil with tree roots. Reduced soil erosion minimizes the loss of topsoil, which is rich in organic carbon, thereby maintaining higher SOC levels.

4.3.2.5 Nutrient Cycling

Trees can access nutrients from deeper soil layers and, through leaf litter and root turnover, return these nutrients to the surface soil. This nutrient cycling contributes to higher biomass productivity, further increasing SOC content in soil.

4.3.3 Benefits of Increased Soil Organic Carbon through Agroforestry

4.3.3.1 Soil Fertility

Higher SOC levels improve soil fertility by enhancing nutrient availability and retention. This leads to better crop yields and increases agricultural productivity.

4.3.3.2 Water Retention

SOC improves the water-holding capacity of soil, making agroforestry systems more resilient to drought and fluctuations in rainfall patterns, which is crucial under changing climatic conditions.

4.3.3.3 Biodiversity Promotion

Agroforestry systems that enhance SOC can support greater biodiversity, including soil organisms, which play essential roles in nutrient cycling and soil health.

4.3.3.4 Climate Change Mitigation

Increasing SOC levels in agroforestry systems contributes to carbon sequestration, helping mitigate climate change. SOC acts as a long-term carbon sink, reducing the overall concentration of greenhouse gases in the atmosphere (ICAI 2009).

4.3.4 Factors Influencing SOC in Agroforestry

4.3.4.1 Species Composition

The types of trees and shrubs integrated into the AF system affect the quantity and quality of organic matter returned to the soil. Fast-growing species with high litter production may enhance SOC more effectively than slower-growing species.

4.3.4.2 Management Practices

Practices such as cover cropping, reduced tillage, and organic amendments can significantly influence SOC dynamics. Agroforestry systems managed with sustainable practices tend to accumulate more SOC compared to those subjected to conventional farming methods.

4.3.4.3 Soil Type and Climate

The initial SOC levels, soil texture, and local climatic conditions (temperature and precipitation patterns) can influence how effectively agroforestry systems enhance SOC. Some soils may have a higher capacity for SOC accumulation than others.

4.3.4.4 Land Use History

Previous land use and management practices affect current SOC levels. Areas previously subjected to intensive agriculture may have lower SOC, but transitioning to agroforestry can help restore soil carbon levels over time.

4.3.5 Challenges and Limitations in Agroforestry Toward Enhancing SOC Content

4.3.5.1 Time Lag

The benefits of agroforestry on SOC are often not immediate. It can take years to decades for significant SOC accumulation to be realized, which may discourage some farmers from adopting these systems.

4.3.5.2 Initial Investment

Establishing agroforestry systems can require higher initial investments in terms of labor and resources. Farmers may be hesitant to invest without immediate financial returns.

4.3.5.3 Complexity of Management

The integration of trees into agricultural systems adds complexity to management practices. Farmers may need additional training and support to effectively manage agroforestry systems.

4.3.5.4 Market Access

For agroforestry to be economically viable, farmers need access to markets for the products they produce, including timber, fruits, or nuts. Limited market access can hinder the adoption of agroforestry practices.

4.4 Impact of Agroforestry on Increased Biomass Input into Soil Helping Toward SOC Built-up

Agroforestry, the integration of trees and shrubs into agricultural landscapes, plays a crucial role in enhancing soil health and productivity. One of its significant benefits is the increased biomass input into the soil, which has far-reaching implications for soil fertility, carbon sequestration, and overall ecosystem sustainability. Thus, it is important to explore the mechanisms through which agroforestry enhances biomass input, its impacts on soil quality, and the broader environmental implications (Sh et al. 2022; Sekhar et al. 2024).

4.4.1 Impact of Biomass Input in Agroforestry

The mechanisms of biomass input in agroforestry involving diverse planting systems are multi-faceted. By combining a variety of trees, shrubs, and crops, agroforestry not only enhances the quantity and quality of biomass but also contributes to improved soil health and agricultural sustainability. This diversity leads to more resilient ecosystems capable of adapting to environmental changes while providing essential resources for both the plants and the surrounding community.

4.4.1.1 Impact of Diverse Planting Systems in Agroforestry on Increased Biomass Input into Soil

This involves two mechanisms: (i) Agroforestry systems typically involve a mixture of trees, shrubs, and crops. This biodiversity leads to a variety of root structures, leaf litter, and organic matter that contribute to soil biomass. (ii) Different plant species have varying rates of growth and decomposition, which means that biomass input can occur over extended periods, maintaining a more consistent supply of organic matter.

The diverse planting approach in Agroforestry systems, characterized by the integration of trees, shrubs, and crops within the same agricultural landscape, significantly enhances biomass input through several key mechanisms:

4.4.1.1.1 Biodiversity and Root Structures on Increasing Soil Biomass

The presence of a variety of plant species in agroforestry systems leads to a rich array of root structures. Different plants establish root systems that can penetrate various soil layers, improving soil aeration and water retention. These diverse roots contribute to:

(i) **Soil biomass:** As roots die and decompose, they add organic matter to the soil, enhancing its fertility.
(ii) **Nutrient cycling:** Various root systems access different soil nutrients, leading to a more efficient use of resources and minimizing nutrient depletion.

4.4.1.1.2 Leaf Litter Enhancing Soil Organic Matter Content

Diverse plant species produce varying types and amounts of leaf litter, which plays a crucial role in biomass input:

(i) **Decomposition rates:** Different species have unique leaf characteristics, affecting their decomposition rates. Some leaves break down quickly, while others take longer, providing a sustained source of organic matter over time.
(ii) **Soil cover:** Continuous leaf litter accumulation helps protect the soil surface from erosion, promotes moisture retention, and enhances microbial activity, all contributing to a healthy soil ecosystem.

4.4.1.1.3 Plant Growth Rates and Biomass Production Influencing SOC Built-up

The presence of multiple plant species in agroforestry systems means that biomass input can occur over extended periods:

(i) **Complementary growth patterns:** Fast-growing species can provide rapid biomass input, while slower-growing species contribute to long-term biomass accumulation. This diversity ensures that organic matter is consistently added to the soil of AF system throughout the year.
(ii) **Seasonal variation:** Different species may peak in growth at different times, allowing for a steady supply of biomass and organic matter, which is critical for maintaining soil health.

4.4.1.2 Influence of Leaf Litter and Pruning in Agroforestry on Increased Biomass Input into Soil

Biomass input from leaf litter and pruning is a crucial aspect of agroforestry system, the integration of trees and shrubs into agricultural landscapes, enhances both biodiversity and productivity, and thus, sustainably contributes to soil health and fertility. This involves two mechanisms: (i) The leaves, branches, and fruits that fall from trees contribute significantly to soil organic matter. In many agroforestry systems, farmers also engage in pruning, where branches are trimmed and left on the ground, further enhancing biomass input. (ii) The decomposition of this organic material enriches the soil with nutrients, improving its structure and fertility.

Thus, the mechanisms of biomass input through leaf litter and pruning in agroforestry systems are integral to sustaining soil health. By contributing organic matter and nutrients, these practices enhance soil structure and fertility, ultimately supporting more resilient and productive agricultural landscapes.

4.4.1.2.1 Contribution of Leaf Litter and Pruning

Trees in agroforestry systems naturally shed leaves, branches, and fruits. This organic matter, known as leaf litter, plays a vital role in maintaining soil quality. When leaves and other plant materials fall to the ground, they create a layer of organic debris that not only provides habitat for soil organisms but also protects the soil from erosion and moisture loss.

In addition to natural litterfall, farmers often engage in pruning practices. Pruning involves the selective cutting of branches and foliage to enhance tree growth and productivity. The pruned materials are frequently left on the ground, contributing additional biomass. This practice serves dual purposes, as it improves tree health by removing excess growth and increases the organic matter available to the soil.

4.4.1.2.2 Decomposition and Soil Enrichment

The organic materials from leaf litter and pruned branches undergo decomposition through the action of microorganisms, fungi, and detritivores. This process breaks down complex organic compounds into simpler forms, releasing essential nutrients like nitrogen, phosphorus, and potassium back into the soil.

As decomposition progresses, the organic matter transforms into humus, which improves soil structure by enhancing aggregation and porosity. This results in better water retention, aeration, and root penetration, creating a more favorable environment for plant growth. The increased nutrient availability from decomposed materials further enriches the soil, leading to improved fertility and productivity of crops grown in agroforestry systems.

4.4.1.3 Mechanisms of Increased Biomass Input into Soil Through Diverse Root Systems in Agroforestry

The integration of trees and shrubs in agroforestry systems through their root mechanisms significantly enhances biomass input. Their ability to access nutrients from deep in the soil, contribute organic matter, and stabilize soil structure not only supports the health and productivity of the entire ecosystem but also promotes sustainable agricultural practices. By understanding these mechanisms, farmers and land managers can better design and manage agroforestry systems for optimal productivity and environmental benefits. This involves two mechanisms: (i) Trees and shrubs have deep and extensive root systems that can access nutrients and moisture unavailable to shallower-rooted crops. When these plants die or shed roots, they add substantial organic matter to the soil. (ii) The presence of roots also helps to stabilize soil structure and reduce erosion, further contributing to soil health.

4.4.1.3.1 Access to Nutrients and Moisture Under Agroforestry

This involves two key mechanisms:

(i) **Deep and extensive root systems:** Trees and shrubs typically possess deep and extensive root systems that enable them to access nutrients and moisture from deeper soil layers in agroforestry plots. This is particularly beneficial for alley crops under agroforestry in areas where shallower-rooted crops may struggle to obtain adequate resources, especially during dry periods.

(ii) **Organic matter contribution:** When trees and shrubs die, or when they shed roots and leaves, they contribute substantial organic matter to the soil. This process, known as litterfall, enriches the soil with carbon and other nutrients, enhancing its fertility. The decomposition of this organic matter not only provides a slow release of nutrients but also improves soil structure and increases its water retention capacity.

4.4.1.3.2 Soil Stabilization and Erosion Reduction through Practices of Agroforestry

This involves two mechanisms:

(i) **Soil structure stabilization:** The presence of extensive root systems plays a critical role in stabilizing soil structure. Roots bind soil particles together, creating a more cohesive soil matrix. This stabilization helps prevent soil erosion from agroforestry plots, particularly on slopes or in areas prone to heavy rainfall.

(ii) **Improved soil health:** By reducing erosion and promoting soil structure, root systems contribute to overall soil health. Healthy soil supports microbial activity and increases the availability of nutrients, further benefiting both tree and crop growth. Additionally, the improved infiltration of water reduces surface runoff, allowing for better moisture retention in the root zone of alley crops under agroforestry.

4.4.1.4 Mechanisms of Increased Biomass Input in Soil Under Agroforestry Involving Soil Microbial Activity

Agroforestry systems, which integrate trees and shrubs with agricultural crops and livestock, play a significant role in enhancing soil health and productivity. A critical aspect of this process is the input of biomass, both from above-ground (e.g., leaf litter, branches) and below-ground (e.g., root exudates, decomposing roots) sources. This biomass input enhances soil microbial activity and contributes to nutrient cycling and soil fertility.

The mechanisms of biomass input in agroforestry systems, on the other hand, significantly enhance soil microbial diversity and activity. This increased microbial action is vital for nutrient cycling, leading to improved soil fertility over time. By fostering a rich microbial ecosystem, agroforestry practices not only enhance agricultural productivity but also contribute to sustainable land management and ecosystem health. This involves two mechanisms: (i) Increased biomass input from both above and below ground enhances microbial diversity and activity. Soil microorganisms break down organic matter, releasing nutrients and promoting a healthier soil ecosystem. (ii) This enhanced microbial activity is crucial for nutrient cycling and can lead to improved soil fertility over time. These two mechanisms are depicted in some detail as follows.

4.4.1.4.1 Increased Biomass Input and Microbial Diversity in Soils Under Agroforestry Systems—Four Causes

(i) **Sources of biomass:** In agroforestry systems, diverse plant species contribute organic matter through leaf litter, dead wood, and root systems. This diversity provides a range of substrates for soil microorganisms in agroforestry systems.

(ii) **Soil microbial diversity and activity:** Increased biomass inputs lead to higher microbial diversity, as different microorganisms specialize in breaking down various types of organic matter. This diversity enhances overall microbial activity, as a greater variety of microbes can thrive in the presence of abundant organic resources in soils under agroforestry.

(iii) **Nutrient release:** As soil microorganisms decompose organic matter, they release essential nutrients, such as nitrogen, phosphorus, and potassium. This process not only provides nutrients for plants but also fosters a more dynamic and resilient soil ecosystem involving agroforestry.

(iv) **Soil structure improvement:** Microbial activity contributes to the formation of soil aggregates, improving soil structure and aeration. This enhanced structure supports root growth and water infiltration, further benefiting plant health through agroforestry practices.

4.4.1.4.2 Influence of Soil Microbial Activity on Enhancing Nutrient Cycling

Influence of soil microbial activity on enhancing nutrient cycling—four types of interrelationships:

(i) **Nutrient cycling:** The breakdown of organic matter by microorganisms is crucial for nutrient cycling. Microbes convert complex organic compounds into simpler forms that plants can readily absorb. This continuous cycling process helps maintain soil fertility over time in agroforestry systems.

(ii) **Long-term soil fertility:** As microbial populations thrive; they contribute to the long-term stability of soil nutrients. Enhanced microbial activity leads to increased organic matter decomposition, which in turn enriches the soil with nutrients and improves its capacity to retain moisture in soils under agroforestry systems.

(iii) **Symbiotic relationships:** Many soil microorganisms form symbiotic relationships with plant roots (e.g., mycorrhizal fungi), further facilitating nutrient uptake and enhancing plant growth. These interactions can significantly improve nutrient availability, particularly in nutrient-poor soils through agroforestry systems.

(iv) **Resilience to stress:** A diverse and active microbial community can improve soil resilience to environmental stressors, such as drought or nutrient depletion. Healthy soil ecosystems are better equipped to withstand changes and maintain productivity through agroforestry practices.

4.4.2 Impacts of Agroforestry Through Increased Biomass Input on Soil Quality

4.4.2.1 Soil Fertility

(i) The addition of organic matter through biomass inputs directly improves soil fertility by increasing nutrient availability, particularly nitrogen, phosphorus, and potassium. (ii) Organic matter also promotes the formation of soil aggregates, improving water retention and aeration, which are vital for healthy plant growth.

4.4.2.2 Carbon Sequestration

(i) Agroforestry systems can sequester significant amounts of carbon in both plant biomass and soil organic matter. This is particularly important in the context of climate change mitigation. (ii) The continuous input of biomass contributes to long-term carbon storage in the soil, which can help offset greenhouse gas emissions.

4.4.2.3 Soil Structure and Erosion Control

(i) The presence of diverse root systems helps bind soil particles together, improving soil structure and reducing erosion risk. This is especially beneficial in sloped or vulnerable landscapes. (ii) Improved soil structure enhances water infiltration and retention, reducing runoff and promoting sustainable water management practices.

4.4.3 Broader Environmental Implications of Agroforestry Through Increased Biomass Input into Soil

4.4.3.1 Biodiversity Enhancement

(i) Agroforestry systems promote biodiversity at multiple levels, including genetic, species, and ecosystem diversity. This can lead to a more resilient agricultural landscape that can better withstand pests and diseases. (ii) Increased biodiversity in agroecosystems can also support pollinator populations, which are crucial for many crop species.

4.4.3.2 Water Management

(i) The integration of trees into agricultural systems can improve water retention and quality. Trees can enhance the hydrological cycle, facilitating better water management and reducing the need for irrigation. (ii) Healthy soils with increased organic matter have improved water-holding capacity, which can be crucial in arid and semi-arid regions.

4.4.3.3 Sustainable Agricultural Practices

(i) Agroforestry supports sustainable agricultural practices by reducing reliance on chemical fertilizers and pesticides, as the natural nutrient cycling and pest management offered by diverse ecosystems can mitigate the need for external inputs. (ii) This approach aligns with the principles of regenerative agriculture, promoting long-term sustainability and resilience in food production systems.

4.5 Impact of Agroforestry on Improved Soil Structure Through SOC Built-up

Agroforestry, the integration of trees and shrubs into agricultural landscapes, offers numerous environmental benefits, particularly in enhancing soil structure. Soil structure refers to the arrangement of soil particles and the pores between them, which influences water retention, aeration, root penetration, and microbial activity. Here is a detailed discussion on how agroforestry impacts soil structure positively (Table 4.5).

4.5.1 Impact of Root Systems in Agroforestry on Soil Aggregation

4.5.1.1 Root Architecture

Trees and shrubs in agroforestry systems develop extensive root networks that penetrate deep into the soil. These roots physically disrupt compacted layers, creating channels that enhance soil porosity. Additionally, the diverse root structures of various plant species contribute to the formation of soil aggregates, which are clusters of soil particles bound together by organic matter, minerals, and microbial activity.

Table 4.5 Impact of agroforestry on soil physical conditions towards managing soil erosion[a]

Geographical area	Agroforestry system	Infiltration rate improvement	Runoff rate reduction	Soil erosion rate	Nutrient loss	Other findings
USA	Agroforestry *vs.* control	75% higher than monoculture	2655 m^3 ha^{-1} (AF) *vs.* 5598 m^3 ha^{-1} (control)	1.85 kg ha^{-1} *vs.* 7.47 kg ha^{-1}	Lowered by 75%	Ground cover reduces erodibility
Brazil	2-year-old agroforestry	Not specified	Not specified	Similar to mature forest	Not specified	Erosion lower than continuous cropping
Tanzania	Traditional agroforestry	Not specified	Not specified	Similar to terracing	Not specified	Effective in controlling erosion
Kenya	*Senna siamea* hedgerows	Not specified	Reduced from 100 to 20 mm	Minimized soil loss to 2 Mg ha^{-1}	Not specified	Significant reduction in soil loss
Throughout the globe	Various systems	Not specified	57% lower than monoculture	50% lower than monoculture	Not specified	Perennial plants increase litter cover, reducing erosion

[a]Kinama et al. (2007), Udawatta et al. (2010), Van Stan et al. (2016), Wilson and Lovell (2016), Beliveau et al. (2017), Kamau et al. (2017), Montagnini (2019), Vignozzi et al. (2019), Cao et al. (2020), Liang (2020), Molina Cuasapaz (2020), Muchane et al. (2020), Purwaningsih et al. (2020), Suprayogo et al. (2020), Surki et al. (2020), van Noordwijk et al. (2020), Du et al. (2022), Parwada et al. (2022)

4.5.1.2 Biological Activity of Roots

Roots exude organic compounds (e.g., sugars, amino acids) that feed soil microbes and fungi, fostering a diverse soil microbiome. This biological activity of roots is crucial for soil aggregation, as microbes and roots produce sticky substances (i.e., root exudates) that bind soil particles together, enhancing the overall soil structure.

4.5.2 Impact of Organic Matter Input in Agroforestry on Soil Aggregation

4.5.2.1 Effect of Leaf Litter and Organic Residues in Agroforestry on Soil Aggregation

The presence of trees leads to increased litterfall, which adds organic matter to the soil. Decomposing leaves, twigs, and other plant materials enrich the soil with nutrients, improving its structure. Organic matter enhances soil aggregation, increases water retention, and promotes a healthy soil environment (Table 4.6).

4.5.2.2 Effect of Crop Residue Management in Agroforestry on Soil Aggregation

In agroforestry systems, residues from both trees and crops can be managed to maintain ground cover. This practice reduces soil erosion and enhances soil structure by promoting organic matter accumulation.

4.5.3 Microclimatic Effects of Agroforestry on Improved Soil Structure

4.5.3.1 Temperature and Moisture Regulation in Agroforestry on Improved Soil Structure

Trees provide shade, which helps regulate soil temperature and moisture levels. Stable temperatures and adequate moisture promote microbial activity, which is essential for organic matter decomposition and nutrient cycling. Improved moisture retention contributes to better soil structure, reducing compaction and enhancing root growth.

4.5.3.2 Wind and Water Erosion Control Through Agroforestry on Improved Soil Structure

Tree canopies can reduce wind speed and protect the soil surface from erosive forces. This protection minimizes soil erosion, which can degrade soil structure and lead to loss of topsoil. By stabilizing the soil and reducing erosion, agroforestry systems maintain soil integrity.

Table 4.6 Impact of agroforestry on improving global degraded soil and soils in arid areas[a]

Geographical area	Agroforestry system	Soil function improvement	Numerical Data / percentage /statistic
Colombian Amazon	Trees in pastures	Formation of biogenic macroaggregates, carbon storage	Enhanced carbon rates, higher soil aggregates
Bangladesh	Agroforestry *vs.* reforestation	Soil organic matter (SOM), nutrient availability	SOM: 4.75%, available P: 12.17 $\mu g\ g^{-1}$, exchangeable K: 0.39 $mg\ kg^{-1}$
USA (Fayette)	Agroforestry *vs.* no-till cropping	SOC and N stocks, microbial biomass	15% higher SOC and N stocks compared to no-till
Zambia	Fertilizer trees (*Faidherbia albida*)	Nutrient cycling, crop yield sustenance	Leaf litter: 1.6–3.8 $Mg\ ha^{-1}$, N: 34–83 $kg\ ha^{-1}$, P: 1.8–4.3 $kg\ ha^{-1}$
Zambia	Agroforestry systems	Soil erosion reduction, improved nutrient storage	50% lower soil erosion, 21% more SOC, 13% more N storage
Ghana	*Faidherbia albida* parkland	SOC and soil N rates	SOC: 0.55% (mid-canopy) to 0.34% (5 m from edge)
Indonesia	Mixed agroforestry (*Hevea*, *Gmelina*)	Chemical fertility, improved soil structure	Higher macroporosity, root length and weight in topsoil
Ghana	Bamboo agroforestry	Improved soil properties	Enhanced CEC, soil moisture, pH, N, available P. and K
India (northwest)	Poplar-based agroforestry	SOC, EC, pH, and nutrient availability	Significant increases in SOC, available N, P, and K
Niger	*Faidherbia albida* canopy	Nutrient availability and crop yield	P: 30% higher, N: 200% higher, millet yield: 2.5 times higher
Tanzania	Gliricidia-based system	Adoption rates and investment cost	Peak adoption: 67.6% in 12 years
Philippines	Various land uses	Soil quality index (SMAF scores)	Agroforestry: 76.21%, agriculture: 49.43%. Forest: 76.39%
Mediterranean (Iberian)	Agrosilvopastoral system	Soil C stock growth	Annual growth rate: 11% in C stock

[a]Shanmughavel et al. (2000), Kho et al. (2001), Al Mamun et al. (2014), Sirohi and Bangarwa (2017), Yengwe et al. (2018), Diallo et al. (2019), Diel et al. (2019), Llorente and Moreno (2019), Akoto et al. (2020), Akpalu et al. (2020), Rizwan et al. (2020), Swamila et al. (2020), Muchane et al. (2020), Saputra et al. (2020), Rodriguez et al. (2021), Chowdhury et al. (2022), Eddy and Yang (2022), Matos et al. (2022a, b), Ramirez et al. (2022)

4.5.4 Effect of Agroforestry on Soil pH Modulation and Nutrient Cycling on Improved Soil Structure

Agroforestry practices play a significant role in soil pH modulation and nutrient cycling, contributing to the improvement of soil structure. Trees in agroforestry systems enhance soil fertility through processes like nitrogen fixation, organic matter decomposition, and the recycling of nutrients from deeper soil layers. Nitrogen-fixing species, in particular, enrich the soil with bioavailable nitrogen, promoting plant growth and soil health. The deep root systems of trees also help prevent soil erosion, improve water infiltration, and facilitate the redistribution of nutrients. These combined effects result in a more balanced soil pH and improved nutrient availability, which collectively enhance soil structure and sustainability in agricultural landscapes (Table 4.7).

4.5.4.1 Effect of Nutrient Availability Through Agroforestry on Improved Soil Structure

Agroforestry systems often feature trees that can fix nitrogen, such as legumes. Increased nitrogen levels can enhance soil microbial activity, leading to improved soil structure. Additionally, the organic matter from decomposed tree litter contributes to nutrient cycling, which benefits the overall health of the soil.

4.5.4.2 Effect of pH Modulation Through Agroforestry on Improved Soil Structure

The organic matter and root activity from trees can influence soil pH, making it more conducive to microbial activity. A balanced pH is crucial for optimal nutrient availability and soil structure, as it supports the diverse community of organisms that contribute to soil health.

4.5.5 Sustainability and Long-Term Benefits of Agroforestry on Improved Soil Structure

Agroforestry enhances soil structure by promoting biodiversity, increasing organic matter, and improving water retention, leading to long-term agricultural sustainability and resilience, through supporting two ecosystem services of soil conservation and carbon sequestration.

4.5.5.1 Soil Conservation Through Agroforestry on Improved Soil Structure

Agroforestry promotes sustainable land management practices that protect soil from degradation. By enhancing soil structure, these systems improve soil resilience to extreme weather events, such as heavy rainfall and droughts, thus ensuring long-term agricultural productivity.

Table 4.7 Impact of species selection, density, and manure application on soil function under agroforestry[a]

Study area	Agroforestry system	Impact on soil function	Numerical data/ percentages/statistic
Various	Trees *vs*. shrubs	Trees create more nutrient-dense soil than shrubs.	Favor trees over shrubs for soil improvement.
Various	*Gliricidia*	Increase in soil organic matter (SOM) improves nutrients but not sufficient alone.	Requires management to reduce erosion; benefits observed in 3 years.
Various	Tree diversity	Boosts soil organic carbon (SOC) in topsoil; limited impact on deeper Jayers.	Increased topsoil SOC but minimal deeper layer improvement.
Various	Crop diversification	Limited additional benefits for soil health compared to monoculture.	Crop diversification does not significantly enhance soil health.
Various	Niche complementarity	Different root depths enhance resource-use efficiency and reduce water loss.	Improved water intake and reduced competition.
Pastures	Poplar *vs*. open pastures	Greater total soil carbon mass in open pastures compared to poplar systems.	11–18% greater in open pastures vs. poplar.
Pastures	*Alder*-based agroforestry	*Alder* systems show better soil health than open pastures.	2–6% greater soil health in alder systems *vs*. open pastures
FMNR parklands	*Faidherbia albida* and *Piliostigma reticulatum*	Higher soil nutrient levels compared to other species and open areas.	Significantly higher Na, Ca, P, Mg, K, NH4-N, OC, pH.
Sudan Savannah Zones	*F. Albida* density	Recommended increase in tree density for soil fertility improvements.	Target densities: 59 trees/ha in Sudan, 37 trees/ha in Guinea.
Northwest India	Poplar-based intercropping	Higher soil N, P, and K levels in specific geometries.	Best results at 5 m × 4 m spacing.
China	Rubber agroforestry	Avoid dense planting of herb species to prevent negative impacts on soil moisture.	Increased leaching due to root structure.
Haryana, India	Wheat-poplar agroforestry	Improved soil health with specific wheat varieties.	Highest pH, EC, N, K with UP-2572; highest SOC and P with DBW-711.

(continued)

Table 4.7 (continued)

Study area	Agroforestry system	Impact on soil function	Numerical data/ percentages/statistic
Central Uganda	Coffee-banana system	Higher SOM and soil C stocks than banana sole cropping; limited P availability.	Suggested application: 20 mg ha-1 well-composted manure yearly.
Ghana	*F. Albida* management	Combine inorganic P with organic materials for better soil management.	Higher N/P ratio in tree biomass.

[a]Augustine et al. (2007), Zake et al. (2015), Bai et al. (2016), Temitope and Thonda (2016), Bisht et al. (2017), Sirohi and Bangarwa (2017), Diallo et al. (2019), Sarmiento-Soler et al. (2019), Akpalu et al. (2020), Douglas et al. (2020), Kannur et al. (2020), Kumar et al. (2020), Saputra et al. (2020), Wartenberg et al. (2020), Eddy and Yang (2022), Manson et al. (2022), Sh et al. (2022), Sirohi et al. (2022), Zhu et al. (2022), Sekhar et al. (2024)

4.5.5.2 Carbon Sequestration Through Agroforestry on Improved Soil Structure

Trees in agroforestry systems sequester carbon, contributing to soil organic carbon levels. Increased organic carbon enhances soil structure, leading to improved water retention, nutrient availability, and a healthier ecosystem.

4.5.6 Challenges and Considerations Regarding Effect of Agroforestry on Improved Soil Structure

While the benefits of agroforestry on soil structure are significant, there are challenges that need to be addressed.

4.5.6.1 Competition for Resources in Agroforestry on Improved Soil Structure

Trees may compete with crops for water and nutrients, which could negatively impact yields if not managed properly. Selecting appropriate species and maintaining a balance between trees and crops is crucial.

4.5.6.2 Management Practices in Agroforestry on Improved Soil Structure

Effective agroforestry requires knowledge and skill in managing inter-species interactions between diverse plant species and understanding their interactions with soil. Without proper management, the positive impacts on soil structure may not be realized.

4.6 Impact of Agroforestry on Enhanced Soil Microbial Activity Helping Toward SOC Built-up

Agroforestry, the integration of trees and shrubs into agricultural landscapes, has emerged as a sustainable land-use practice that not only enhances biodiversity but also significantly influences soil health. One of the critical aspects of soil health is microbial activity, which plays a pivotal role in nutrient cycling, organic matter decomposition, and overall soil fertility. Thus, it is highly pertinent to explore the mechanisms through which agroforestry systems enhance soil microbial activity, the associated benefits, and the implications for agricultural sustainability.

4.6.1 Mechanisms of Enhanced Soil Microbial Activity in Agroforestry—Five Factors

Agroforestry systems enhance soil microbial activity through several interconnected mechanisms. (i) Increased organic matter input from diverse plant species enriches the soil, promoting microbial growth and activity. (ii) Improved soil structure enhances aeration and water retention, creating favorable conditions for microbial life. (iii) Effective nutrient cycling is facilitated by the symbiotic relationships between plants and microorganisms, and thus, optimizing nutrient availability. (iv) The diversity of microbial habitats in agroforestry systems supports a wide range of microbial communities, enhancing resilience and functionality. (v) Additionally, mycorrhizal associations play a crucial role in nutrient uptake and plant health, further bolstering microbial interactions (Table 4.8).

4.6.1.1 Increased Organic Matter Input Through Agroforestry—Two Factors

Agroforestry systems significantly boost soil microbial activity by increasing organic matter through diverse sources such as leaf litter, root biomass, and decaying tree residues. The variety of plant species present creates a rich substrate, promoting microbial diversity and enhancing activity. Additionally, the presence of perennial plants ensures a continuous input of organic matter year-round, contrasting with conventional annual cropping systems that often leave soil bare and limit organic matter availability. Thus, combination of two factors of (i) plant and soil microbial biomass and diversity and (ii) management foster a vibrant microbial community, essential for soil health and ecosystem functioning.

(i) **Plant and soil microbial biomass and diversity factors:** Agroforestry systems, which integrate trees and shrubs with crops or livestock, enhance organic matter levels in several ways: (a) Leaf litter: The leaves that fall from trees contribute organic matter to the soil. As they decompose, they enrich the soil with nutrients and improve its structure. (b) Root biomass: Trees and shrubs have extensive root systems that not only stabilize the soil but also add organic matter through root die-off and decomposition. This process enhances soil fer-

Table 4.8 Impact of agroforestry systems on soil biota[a]

Geographical area	Agroforestry system	Key findings	Data/ percentages/statistic
Rajasthan, India	Various agroforestry systems	Higher microbial diversity compared to monoculture, favorable conditions for soil microflora.	Increased microbial diversity rates reported.
Germany	Alley cropping	Microbial biomass confined to rows of trees; composition of bacterial communities differs from arable land.	Higher bacterial biomass near trees.
USA	*Castanea mollisima—Asimina triloba*	Intermediate level of microbial biomass C compared to secondary forest and corn-soybean rotation.	1.99 Mg ha^{-1} (agroforestry), 1.7 Mg ha^{-1} (Forest), 1.36 Mg ha^{-1} (rotation)
China	Moso bamboo systems	Changes in soil microbial communities influenced by intercropping; certain bacteria increased while others decreased.	Increase in Gemmatimonadetes and Acidobacteria
Brazil	Various agroforestry systems	Increased sporulation and species diversity for arbuscular mycorrhizae and glomalin after 8 years.	Enhanced glomalin production, increased litter quality.
Belgium	Walnut-based alley cropping	Greater soil organic carbon (SOC) and nutrient content near trees, increased macrodetritivore abundance.	Higher SOC near trees.
China	Poplar shelterbelts	Increased diversity of ectomycorrhizal fungi, lower harmful fungi compared to farmlands.	Basidiomycota increased from 14.72–19.18%
Temperate regions	Temperate alley cropping	Increase in number of soil microorganisms, potential spread of benefits into crop rows.	Increased soil microbial numbers noted.

[a]Dossa et al. (2009), Banerjee et al. (2016), Beuschel et al. (2019), Zhang et al. (2019), Li et al. (2020), Marsden et al. (2020), Pardon et al. (2020), Beule and Karlovsky (2021), Ayala-Montejo et al. (2022), Beule et al. (2022), Choudhary and Rijhwani (2022), Eddy and Yang (2022), Matos et al. (2022a, b), Zhu et al. (2022, 2023a)

tility and water retention. (c) Decaying tree residues: The remnants of branches and other tree parts also decompose over time, further contributing to the organic matter content of the soil. (d) Plant diversity: The variety of species in agroforestry systems creates a complex environment that supports diverse microbial communities. Different plants release various organic compounds into the soil, which serve as food for microbes. This diversity can lead to higher microbial activity, which is crucial for nutrient cycling and overall soil health. (e) Microbial diversity and activity: A diverse microbial community can enhance soil processes such as decomposition, nutrient mineralization, and disease suppression. This activity supports plant growth and can lead to improved crop yields. Overall, agroforestry promotes a synergistic relationship between plants and soil organisms, leading to enhanced soil health and productivity.

(ii) **Management factors:** The contribution of perennial plants to continuous organic matter input throughout the year stems from their growth habits and life cycles. This can be represented with breakdown in five factors: (a) Continuous growth: Perennial plants live for multiple years, meaning they grow, flower, and produce biomass year after year. This results in a steady input of organic matter (like leaves, stems, and roots) into the soil, enhancing soil health. (b) Soil coverage: Unlike annual cropping systems, which often leave soil bare during fallow periods, perennial plants provide ongoing ground cover. This minimizes erosion, reduces weed competition, and helps maintain soil structure. (c) Root systems: Perennial plants often develop deep and extensive root systems. These roots contribute organic matter through root turnover and exudates, which feed soil microbes and enhance nutrient cycling. (d) Enhanced soil biology: The continuous organic matter input supports a diverse soil microbiome, which plays a crucial role in nutrient availability and soil health. Healthy soil biology can improve the resilience of ecosystems to stressors. (e) Carbon sequestration: The persistent biomass of perennials can sequester carbon, mitigating climate change impacts while enriching soil organic matter. In contrast, conventional annual cropping systems typically involve planting and harvesting of crops once a year, often leading to periods when the soil is bare. This can result in reduced organic matter input, increased erosion, and decreased soil fertility over time. Thus, perennial systems are often viewed as more sustainable in promoting soil health and agricultural resilience.

4.6.1.2 Soil Microbial Mediated Improvement in Soil Structure Under Agroforestry—Two Mechanisms

Agroforestry systems, by integrating trees and crops, significantly enhance soil microbial activity through improved soil structure. (i) Tree roots bind soil particles, forming aggregates that enhance aeration and water infiltration, creating a conducive environment for microbial growth. (ii) Additionally, increased soil porosity promotes air and water movement, facilitating microbial respiration and efficient nutrient transport. Those, two mechanisms together contribute to healthier soils and improved ecosystem functions in agroforestry landscapes.

(i) **Tree roots effects—four factors**: The critical role of tree roots in maintaining soil health and structure can be viewed as per the effects of four factors: (a) Binding soil particles: Tree roots extend through the soil, physically intertwining with soil particles. This binding action creates aggregates—clumps of soil particles that stick together. (b) Improved soil structure: These aggregates enhance the overall structure of the soil, making it less prone to erosion and more stable. Good soil structure allows for better arrangement of soil components, which is crucial for various biological processes. (c) Aeration and water infiltration: Well-structured soil has more spaces (or pores) between particles, allowing air and water to flow more freely. This aeration is vital for root respiration and the health of soil organisms, while improved water infiltration helps prevent runoff and allows water to reach deeper soil layers. (d) Favorable microbial activity: Healthy soil structure and adequate aeration create an ideal environment for microorganisms. These microbes are essential for nutrient cycling, decomposition of organic matter, and overall soil fertility. Thus, the interplay between tree roots and soil particles not only stabilizes the soil but also fosters an environment that supports plant and microbial life, contributing to ecosystem health (Allen et al. 2004).

(ii) **Enhanced soil porosity effects—four factors**: Enhanced soil porosity promotes the movement of air and water, facilitating microbial respiration and nutrient transport. The importance of soil structure in maintaining healthy ecosystems can be viewed as per following four factors: (a) Soil porosity: This refers to the presence of small spaces or pores within the soil. Enhanced porosity means there are more and larger pores, which allows for better air and water movement. (b) Air movement: Improved soil porosity enables better air circulation within the soil. This is crucial for microbial respiration, as microorganisms need oxygen to break down organic matter and carry out their metabolic processes. Without adequate air flow, anaerobic conditions can develop, which can hinder microbial activity and lead to the accumulation of harmful substances. (c) Water movement: Enhanced porosity allows water to infiltrate the soil more easily and reduces surface runoff. This is beneficial for plants and soil organisms, as it ensures that water reaches the root zone, promoting healthy growth. Additionally, proper water movement helps prevent waterlogging, which can be congenial for plant health and soil structure. (d) Nutrient transport: With better air and water movement, nutrients dissolved in the soil water can be transported more effectively to plant roots and microorganisms. This means that plants can access essential nutrients more readily, supporting their growth and productivity. On an overall consideration, it can be viewed that enhanced soil porosity plays a vital role in maintaining a balanced soil ecosystem, promoting both microbial health and plant growth through improved air and water dynamics.

4.6.1.3 Nutrient Cycling Mediated by Soil Microbial Activity Under Agroforestry

Agroforestry systems leverage the benefits of integrating trees and crops, significantly influencing soil microbial activity and nutrient cycling. The inclusion of nitrogen-fixing species, such as legumes, enhances soil nitrogen levels, fostering microbial growth and stimulating processes like nitrification and denitrification. Additionally, the dynamic nutrient exchange between trees and crops creates a rich soil environment that supports diverse microbial communities. This interaction not only boosts nutrient availability but also promotes various metabolic processes essential for sustainable agricultural productivity.

4.6.1.3.1 Actions of Nitrogen-Fixing Plants Influencing Soil Microbial Activity Under Agroforestry—Seven Types

Nitrogen-fixing plants, particularly legumes, play a crucial role in agroforestry systems by enhancing soil fertility through their ability to convert atmospheric nitrogen into a form that plants can utilize. This process is facilitated by symbiotic bacteria, primarily from the genus Rhizobium, which reside in root nodules of leguminous plants. This activity of nitrogen-fixing plants can be classified in seven types. These actions are (a) Nitrogen fixation: Nitrogen-fixing plants convert atmospheric nitrogen (N_2) into ammonia (NH_3) through a process known as biological nitrogen fixation. This ammonia can be assimilated by plants and microorganisms, enriching the soil with bioavailable nitrogen. (b) Soil enrichment: As these plants grow, they contribute organic matter to the soil through leaf litter, root biomass, and decaying plant material. This organic matter enhances soil structure, increases microbial activity, and promotes nutrient cycling. (c) Nutrient cycling: The decomposition of fallen leaves and dead roots by soil microorganisms releases nitrogen and other nutrients back into the soil, making them available for other crops. This process supports the overall nutrient cycling within the agroecosystem. (d) Reduction of fertilizer dependency: By naturally replenishing soil nitrogen, nitrogen-fixing plants reduce the need for synthetic fertilizers, lowering costs for farmers and minimizing environmental impacts associated with chemical fertilizers. (e) Improvement of soil health: Nitrogen-fixing species can enhance soil microbial diversity and activity, which is essential for nutrient cycling and soil health. Healthy soil contributes to better crop yields and resilience against pests and diseases. (f) Alley cropping and integration: In agroforestry systems such as alley cropping, nitrogen-fixing trees or shrubs are planted alongside cash crops. This arrangement allows for simultaneous nutrient contributions and crop production, optimizing land use and improving overall farm productivity. (g) Carbon sequestration: Through photosynthesis, nitrogen-fixing plants capture carbon dioxide, contributing to carbon sequestration. This process helps mitigate climate change while simultaneously improving soil fertility. Thus, incorporating nitrogen-fixing plants into agroforestry systems significantly enhances nutrient cycling, soil health, and agricultural sustainability. By fostering a symbiotic relationship with soil microbes and contributing organic matter, these plants create a resilient agroecosystem that benefits both crops and the environment. As agricultural practices evolve toward more sustainable methods, the role of nitrogen-fixing species will continue to be essential in promoting soil fertility and overall ecosystem health.

4.6.1.3.2 Interactions Between Trees and Crops Influencing Soil Microbial Activity under Agroforestry

Agroforestry systems, which integrate trees with agricultural crops, are increasingly recognized for their potential to enhance sustainability and productivity. One of the critical aspects of these systems is the interaction between trees and alley crop species, which plays a significant role in nutrient cycling.

Impact of Tree and Alley Crop Interactions in Nutrient Dynamics in Agroforestry—Four Key Points

(i) **Nutrient cycling mechanisms - three types**: (a) Tree root systems: Trees often have deeper and more extensive root systems compared to annual crops. These roots can access nutrients and water from deeper soil layers, bringing them closer to the surface through litter fall and root turnover. (b) Nitrogen fixation: Certain tree species, particularly legumes, can fix atmospheric nitrogen through symbiotic relationships with nitrogen-fixing bacteria. This process enriches the soil with available nitrogen, benefiting adjacent crops that rely on this essential nutrient. (c) Organic matter addition: Fallen leaves, fruits, and other organic materials from trees contribute to soil organic matter. This organic matter improves soil structure, enhances water retention, and provides a reservoir of nutrients as it decomposes.

(ii) **Nutrient competition and complementarity: (a) Competition for resources**: While trees provide benefits, they can also compete with alley crops for light, water, and nutrients. Understanding this balance is crucial for optimizing the arrangement of species within the agroforestry system. **(b) Complementary growth**: Trees can create a microclimate that benefits understorey crops by providing shade and reducing evaporation. Certain crops may thrive under tree canopies, benefiting from reduced competition and improved moisture retention.

(iii) **Soil health and biodiversity—two effects**: **(a) Enhanced soil fertility**: The dynamic interactions between trees and crops can lead to improved soil health. Increased microbial activity, driven by organic inputs from trees and root interactions, fosters a more fertile environment for crops. **(b) Biodiversity support**: Agroforestry systems promote biodiversity, which is essential for nutrient cycling. Diverse plant communities can enhance resilience against pests and diseases, ultimately supporting more sustainable nutrient management.

(iv) **Management practices for optimizing interactions between trees and crops-three types**: **(a) Selection of species**: Choosing compatible tree and crop species is crucial. Fast-growing nitrogen-fixing trees can support nutrient cycling while providing shade and protection for alley crops. **(b) Spatial arrangement**: Designing the layout of trees and crops can mitigate competition. For example, planting trees in rows with sufficient spacing allows crops to access light and moisture while benefiting from the nutrient input from trees. **(c) Regular monitoring**: Assessing soil health and nutrient levels over time can help in adapting management practices to ensure optimal interactions between trees and crops. Thus, it is necessary inference that interactions between trees and alley crop species are fundamental to nutrient cycling in agroforestry systems. By enhancing soil fertility, supporting biodiversity, and fostering beneficial nutrient dynamics, these interactions can lead to improved agricultural productivity and sustainability. Careful selection and management of species and spatial arrangements are essential to maximize these benefits, ultimately contributing to more resilient agroecosystems.

4.6.1.4 Vegetation Structures and Diverse Array of Niches Under Agroforestry Influencing Soil Microbial Habitat Diversity

Agroforestry systems, characterized by their complex vegetation structures, create a diverse array of niches that support a rich variety of microorganisms. This enhanced microbial habitat diversity fosters unique microbial communities associated with different plant species, leading to increased overall microbial diversity. Such microbial habitat diversity not only enhances resilience against disturbances and diseases but also ensures a wide array of microbial functions, contributing to the stability and productivity of the ecosystem. In this context, understanding the mechanisms behind enhanced soil microbial activity in agroforestry can provide insights into sustainable land management practices.

4.6.1.4.1 Complex Vegetation Structure in Agroforestry Influencing Soil Microbial Habitat Diversity—Six Key Points

Agroforestry systems, which integrate trees and shrubs with crops and livestock, create a complex vegetation structure that significantly influences soil health and microbial diversity. This complexity provides various niches for microorganisms, which can thrive due to six key causes in different microenvironments: **(a) Diverse plant species**: Different plant species interact with the soil in unique ways, such as varying root structures and exudates (substances secreted by roots). These factors contribute to the formation of distinct microbial communities, as certain microbes are adapted to specific plant types. **(b) Microhabitats**: The layering of vegetation—canopy, understory, and ground cover—creates microhabitats with varying light, moisture, and nutrient conditions. For instance, shaded areas may support different microbial communities compared to open, sunlit areas. **(c) Root interactions**: The root systems of different plants can influence soil composition and structure. Mycorrhizal fungi, for example, may associate with certain tree species, enhancing nutrient uptake for both the plant and surrounding microbes. **(d) Organic matter contribution**: The presence of trees contributes organic matter through leaf litter, fallen branches, and root decay. This organic matter enriches the soil, providing food for a diverse range of microorganisms. **(e) Increased resilience**: Higher microbial diversity contributes to ecosystem resilience. Diverse microbial communities can enhance nutrient cycling, disease resistance, and overall soil health, benefiting the entire agroecosystem. **(f) Intercropping benefits**: Different plants can suppress soil-borne diseases or pests, indirectly supporting a more robust microbial community that may help protect crops. Thus, the complex vegetation structure in agroforestry systems fosters a rich array of niches for microorganisms. This leads to higher overall microbial diversity, which is crucial for maintaining healthy soils and promoting sustainable agricultural practices. As such, agroforestry not only supports plant growth but also enhances the microbial ecosystem that underpins soil health and productivity.

4.6.1.4.2 Plant Diversity in Agroforestry Influencing Soil Microbial Habitat Diversity—Three Types of Benefits

Plant diversity refers to the variety of plant species in a given ecosystem. This diversity is crucial for maintaining ecological balance and resilience against disturbances - such as natural disasters, climate change, and disease outbreaks. A diverse plant community can enhance the stability and health of ecosystems, offering a multitude of benefits, commonly classified under three groups:

(i) **Resilience of plant diversity against disturbances—three types**: (a) Resource availability: Diverse plant species utilize different resources (e.g., light, nutrients, and water) in varying ways. This complementary resource use helps ensure that ecosystems can maintain productivity even when some species are stressed or removed. (b) Adaptive responses: Different plant species have evolved various strategies to cope with environmental stresses. When disturbances occur, a wider range of species can respond adaptively, allowing the ecosystem to recover more quickly. (c) Soil health: Diverse root structures and growth forms contribute to soil structure and nutrient cycling. This can improve soil health, enhance water retention, and reduce erosion, making ecosystems more resilient to physical disturbances.

(ii) **Protection against diseases—two ways** under agroforestry: The effect of plant diversity in agroforestry for protection against diseases can be of two types: (a) Microbial interactions: A diverse plant community supports a variety of microbial populations in the soil and surrounding environment. These microbes play essential roles in nutrient cycling, disease suppression, and overall soil health. (b) Pathogen resistance: A variety of plant species can help prevent the spread of diseases. If one species is susceptible to a particular pathogen, others may be resistant, thus preventing the disease from decimating the entire ecosystem.

(iii) **Modulation of microbial functions under agroforestry**: Different plant species can influence the microbial community composition. A diverse array of plants can enhance microbial functions, such as decomposition and nutrient cycling, which contribute to ecosystem stability and health. The overall observations show that plant diversity plays a critical role in enhancing ecosystem resilience against disturbances and diseases. By supporting a wide range of microbial functions, diverse plant communities help maintain ecological balance, promote soil health, and ensure stability in the face of environmental challenges. Thus, protecting and promoting plant diversity, as usually followed through agroforestry practices, is essential for sustainable ecosystems and the services they provide.

4.6.1.5 Mycorrhizal Associations Under Agroforestry Helping Toward SOC Built-up

In agroforestry systems, trees and mycorrhizal fungi engage in a mutually beneficial relationship. Mycorrhizal fungi form symbiotic associations with the roots of trees, which helps enhance nutrient uptake. These fungi extend their hyphae (thread-like structures) into the soil, increasing the surface area for absorption. This allows trees to access more nutrients, particularly phosphorus and nitrogen, which are essential for their growth.

4.6.1.5.1 Mutually Beneficial Ecosystem Between Agroforestry Trees, Mycorrhiza, and Alley Crops Helping Toward SOC Built-up

In agroforestry systems, the symbiotic relationship between trees, mycorrhizal fungi, and crops forms a dynamic and mutually beneficial ecosystem. Trees engage in photosynthesis, converting sunlight into energy and producing carbohydrates that serve as vital nourishment for mycorrhizal fungi. These fungi, in turn, establish intricate networks of mycelium that expand the root systems of trees and neighboring crops, effectively increasing their access to essential nutrients and water in the soil. This symbiotic relationship enhances nutrient cycling within the soil, leading to improved availability of vital elements like nitrogen, phosphorus, and potassium. As the trees provide carbohydrates, the fungi assist in breaking down organic matter, facilitating the release of these nutrients. This creates a fertile environment for crops grown alongside the trees. When crops are interplanted with trees, they benefit from the mycorrhizal networks. These networks not only improve nutrient uptake but also enhance the soil structure, promoting better aeration and water retention. As a result, crops can achieve greater resilience against environmental stressors such as drought and pests, which appear to be increasingly critical in the face of climate change. Furthermore, the presence of trees contributes to overall ecosystem health. They provide habitat for beneficial insects and wildlife, improve biodiversity, and enhance soil organic matter through leaf litter and root biomass. This biodiversity plays a crucial role in pest management and pollination, further supporting agricultural productivity. Overall, the collaboration between trees, fungi, and crops in agroforestry systems creates a sustainable agricultural model. It leads to enhanced soil health, improved crop yields, and increased resilience, ultimately contributing to food security and environmental sustainability. This intricate web of relationships highlights the importance of integrating ecological principles into agricultural practices for a more sustainable future.

4.6.1.5.2 Role of Mycorrhiza in Tree and Crop Growth Helping Toward SOC Built-up

Mycorrhizal fungi attach to the roots of the trees and extend their hyphae (fungal filaments) into the soil. This significantly increases the surface area available for nutrient absorption, allowing for more efficient uptake of essential elements such as phosphorus and nitrogen. As a result, both the trees and the crops benefit from improved access to these nutrients, which can lead to better growth and higher yields. Mycorrhizal fungi form symbiotic relationships with the roots of trees and many other plants, significantly enhancing their ability to absorb nutrients from the soil.

4.6.1.5.3 Structure and Function of Mycorrhiza

(i) **Hyphal network**: Mycorrhizal fungi produce hyphae, which are long, thread-like structures. These hyphae extend far beyond the root zone of the host plant, creating an extensive network in the soil. This increases the effective surface area available for nutrient and water absorption.

(ii) **Nutrient uptake**: The mycorrhizal association primarily improves the plant's access to essential nutrients like phosphorus and nitrogen.

(a) **Phosphorus**: Mycorrhizal fungi are particularly adept at extracting phosphorus from the soil, which is often present in forms that are unavailable to plants. The fungi solubilize this phosphorus, making it more accessible to the tree roots.

(b) **Nitrogen:** In addition to phosphorus, certain mycorrhizal fungi can also assist in nitrogen uptake either by enhancing the absorption of nitrogen compounds or through associations with nitrogen-fixing bacteria.

(iii) **Water absorption:** The hyphal network also increases the plant's ability to absorb water, especially in dry conditions, improving overall drought resistance.

4.6.1.5.4 Mutual Benefits of Symbiotic Association of Mycorrhiza, Tree and Crop Under Agroforestry Through SOC Built-up

(i) **For trees:** The relationship allows trees to access a broader range of nutrients and water, which can enhance their growth, resilience, and overall health. Trees with mycorrhizal associations often exhibit increased root biomass, leading to better stability and nutrient storage.

(ii) **For fungi:** In return, the fungi benefit from the carbohydrates produced by the trees through photosynthesis. This organic carbon serves as a vital energy source for the fungi, facilitating their growth and reproduction.

4.6.1.5.5 Impact of Symbiotic Association of Mycorrhiza, Tree and Crop Under Agroforestry on Crop Yields Through SOC Built-up

(i) **Improved growth:** The enhanced nutrient uptake leads to better growth rates and overall plant vigor. This is particularly beneficial in agricultural settings, where crops can yield significantly more produce due to the improved nutrient availability.

(ii) **Soil health:** The presence of mycorrhizal fungi contributes to soil structure and health. They help bind soil particles together, improving aeration and water retention. Healthy soil ecosystems also support diverse microbial communities that further enhance nutrient cycling.

(iii) **Sustainability:** By fostering mycorrhizal relationships, farmers can reduce the need for synthetic fertilizers, leading to more sustainable agricultural practices. This not only lowers costs but also minimizes environmental impacts associated with fertilizer runoff.

Above all, it is generally accepted that mycorrhizal fungi play a crucial role in the nutrient dynamics of forest and agricultural ecosystems. Their ability to extend the root systems reach through hyphal networks enhances nutrient and water absorption, benefiting both trees and crops. This symbiotic relationship is essential for promoting healthy plant growth, improving yields, and supporting sustainable land management practices.

Additionally, these fungi contribute to soil health by improving microbial biomass and activity. The presence of mycorrhizal associations can stimulate the growth of a diverse microbial community in the soil, which is vital for maintaining soil structure and fertility. Because the enhanced microbial activity also aids in the accumulation of soil organic carbon (SOC), a key indicator of soil health that contributes to long-term sustainability in agricultural practices. In this way, the mycorrhizal associations formed in agroforestry systems are essential for improving nutrient uptake, fostering soil health, and promoting overall ecosystem productivity.

4.6.2 Implications for Agricultural Sustainability Through Agroforestry Mediated SOC Built-up—Four Benefits

The positive impact of agroforestry on soil microbial activity has far-reaching implications for sustainable agriculture. Agroforestry represents a transformative approach to agriculture that harmonizes the cultivation of crops and trees, yielding numerous benefits for sustainable agricultural practices. One of the most significant advantages of agroforestry is its positive impact on soil organic carbon (SOC) accumulation and microbial activity, which are critical components of soil health. By enhancing these aspects, agroforestry can lead to four key benefits: First, increased crop yields result from healthier soils that support robust plant growth and resilience. Second, as microbial activity improves nutrient cycling, the reliance on chemical fertilizers and pesticides diminishes, thereby reducing production costs and minimizing environmental damage. Third, agroforestry plays a vital role in biodiversity conservation, fostering diverse ecosystems essential for maintaining ecological balance. Finally, by promoting better soil health and enhancing carbon sequestration, agroforestry contributes significantly to climate change mitigation efforts. Collectively, these benefits underscore the potential of agroforestry as a cornerstone for achieving agricultural sustainability in an increasingly challenging environmental landscape.

4.6.2.1 Increased Crop Yields Under Agroforestry Through SOC Built-up

Agroforestry, the integration of trees and shrubs into agricultural landscapes, offers a holistic approach to farming that promotes environmental health, biodiversity, and economic viability. One of the most significant implications of this practice is the potential for increased crop yields, primarily driven by the enhancement of soil organic carbon (SOC). SOC plays a crucial role in soil fertility, water retention, and overall ecosystem resilience. Soil organic carbon refers to the carbon component of organic compounds in the soil, derived from decomposed plant and animal materials. Its build-up is vital for maintaining soil health and productivity. Increased SOC levels improve soil structure, enhance nutrient availability, and boost microbial activity, all of which contribute to higher agricultural productivity. This SOC built-up under agroforestry catalyses agricultural crop production system in the following five ways (Table 4.9).

(a) **Diverse root systems:** The presence of trees and deep-rooted shrubs in agroforestry systems helps improve soil structure and aeration. These plants contribute organic matter through leaf litter, fallen branches, and root biomass, and thus enhancing SOC levels.
(b) **Nutrient cycling:** Trees play a pivotal role in nutrient cycling by capturing nutrients from deeper soil layers and making them available to crops through

Table 4.9 Ecosystem services of agroforestry based on field trials in sub-Saharan Africa[a]

Key findings	Details
Effect of agroforestry on crop yield	Agroforestry significantly increased crop yield, with the average crop yield being almost twice as high as in non-agroforestry systems.
Effect on soil fertility	Agroforestry improved soil fertility by a factor of 1.2 compared to non-agroforestry systems.
Effect on erosion control	Agroforestry systems were five times better at controlling runoff and nine times better at controlling soil loss compared to non-agroforestry systems.
Effect on water regulation	Agroforestry systems enhanced water regulation, with water infiltration being three times higher in agroforestry systems compared to controls.
General benefits across ecological and Management conditions	Positive effects of agroforestry were found across a wide range of ecological conditions, management practices, and soil types, suggesting its broad applicability in sub-Saharan Africa.
Exceptions and limitations	Some agroforestry practices, such as hedgerows, and certain soil types showed negative effects. However, these were exceptions rather than the norm.
Agroforestry's role for smallholder farms	Smallholder farmers in sub-Saharan Africa, who make up 80% of farms, are likely to benefit from agroforestry systems that provide both ecosystem services (e.g., soil fertility, erosion control) and economic products (e.g., fruits, fodder, fuel).
Positive effects across contexts	The research demonstrated that agroforestry practices, when properly managed, offer significant ecosystem service benefits across a variety of ecological, management, and biological contexts in sub-Saharan Africa.
Advances in agroforestry knowledge	The widespread positive results are partly due to advances in agroforestry knowledge, such as better matching of tree species to ecological conditions and more effective management practices.
Publication bias	Kendall's rank correlation showed no bias for crop yield, total nitrogen (N), soil organic carbon (SOC), and available phosphorus (P). However, bias was found for water regulation and erosion control. Studies showing nonsignificant effects for these services were less likely to be published.

[a]Kuyah et al. (2019)

leaf drop and root decay. This process not only enriches the soil but also reduces the need for chemical fertilizers.

(c) **Microbial activity**: An increase in organic matter fosters a diverse community of soil microorganisms. These microbes decompose organic material, further enriching the soil with nutrients and improving its structure, leading to better crop growth.

(d) **Enhanced water retention**: Improved SOC levels contribute to better soil moisture retention. Agroforestry systems, with their tree cover, mitigate evaporation and allow crops to access moisture more efficiently, especially during dry spells (Padovan et al. 2018).

(e) **Erosion control**: The root systems of trees and shrubs help anchor the soil, reducing erosion and nutrient loss. This stability is vital for maintaining productivity over the long term, particularly in vulnerable agricultural areas.

4.6.2.1.1 Impacts of SOC Built-up on Crop Yields Under Agroforestry—Three Key Effects

Higher productivity sustainably potential through times is the direct result of increased SOC and improved soil health is a significant boost in crop yields. Studies have shown that farms practicing agroforestry can experience 20–50% yield increase as compared to conventional monoculture systems due to three key effects of agroforestry: (a) Resilience to climate variability: Agroforestry systems are often more resilient to climate extremes. The increased organic matter and improved soil structure allow for better water management and nutrient availability, helping crops withstand droughts and floods. (b) Biodiversity enhancement: The integration of trees encourages a diverse ecosystem that supports beneficial insects and pollinators, which can further enhance crop yields through improved pollination and pest control. (c) Sustainability and economic viability: Higher crop yields contribute to the economic sustainability of farming operations. Farmers benefit not only from increased production but also from reduced input costs associated with fertilizers and water, leading to a more profitable and sustainable agricultural model. Based on overall views, it can be conjectured that increased crop yields as a key implication of agricultural sustainability under agroforestry through SOC build-up exemplifies the synergistic benefits of integrating trees into farming systems. By enhancing soil health and fostering a sustainable agricultural environment, agroforestry not only supports food security but also contributes to environmental resilience, making it a crucial practice for future agricultural development (Table 4.10).

4.6.2.2 Impacts of SOC Built-up Under Agroforestry Helping Toward Reduced Chemical Dependency in Agricultural Crop Production

As microbial activity enhances nutrient cycling, reliance on chemical fertilizers and pesticides can decrease, resulting in lower production costs and environmental impact. The integration of agroforestry practices has the potential to significantly transform agricultural landscapes, particularly through the enhancement of soil organic carbon (SOC) levels. This increase in SOC not only boosts microbial

Table 4.10 Impact of agroforestry on soil productivity and its management[a]

Geographical area	Agroforestry system	Impact on soil health	Yield increase (%/ Numerical)
Rwanda	*Faidherbia albida* (tree canopy)	Improved microclimate	Maize yield increased by 150%
Iran	Almond intercropping with barley/ wheat	Higher soil fertility	Barley: 39%, wheat: 35% increase in grain yields
Bangladesh	Malta-based agroforestry	Increased soil fertility	B/C ratio: 2.92 (Malta-broccoli-turmeric)
Ghana	Alley cropping with *Bambusa balcooa*	Enhanced soil nutrient availability	LER: Cowpea 1.54, maize 1.38, cassava 1.19
Sudan	Parkland agroforestry	Improved soil productivity	Increased yields for sesame, pearl millet, sorghum
China	Wheat+ Ziziphus system	Enhanced light usage	Income increased from $414 to $980 per ha
India (Bundelkhand)	Bamboo + sesame-chickpea system	Higher resource capture and efficiency	LER: 1.95–2.14
Guatemala	*Gliricidia sepium* -alleys	Higher total N and soil C contents	Not specified, but improved soil health noted
Niger	*Ziziphus mauritiana* with *Pennisetum*/ *Vigna*	Improved water use efficiency	Millet: 47% (without fertilizer), 17% (with FYM)

[a]Vandermeer (1989), Neufeldt et al. (2009), Haglund et al. (2011), Gebru (2015), Ibrahim et al. (2015), Sileshi (2016), Ndoli et al. (2017), Bado and Bationo (2018), Fahrni et al. (2018), Yengwe et al. (2018), Zhang et al. (2018), Akoto et al. (2020), Das et al. (2020), Dev et al. (2020), Do et al. (2020), Mantino et al. (2020), Marsden et al. (2020), Surki et al. (2020), Bado et al. (2021), Tsufac et al. (2021a, b), Beule et al. (2022), Eddy and Yang (2022)

activity but also promotes efficient nutrient cycling within the ecosystem. As these biological processes flourish, farmers can reduce their reliance on chemical fertilizers and pesticides, leading to a myriad of benefits. Lower production costs emerge as chemical inputs diminish, while environmental impacts are mitigated through more sustainable practices. This transition fosters a healthier agricultural system, promoting both productivity and ecological balance. So, it is obvious that SOC build-up under agroforestry contributes to these outcomes, highlighting its critical role in the future of sustainable agriculture, because of the following reasons: (a) Agroforestry systems promote biodiversity and enhance soil organic matter through the incorporation of various tree species alongside crops. The leaves, roots, and decaying matter from trees contribute significantly to SOC levels. (b) Agroforestry systems foster diverse microbial communities due to the varied plant species and root structures and facilitating nutrient cycling. (c) AF system reduces chemical dependency because of (i) lower fertilizer use due to enhancement in SOC content improves nutrient-holding capacity of soil, and thus, allowing it to retain and supply essential nutrients naturally. Also, this decreases farmers' reliance on chemical fertilizers, leading to reduced input costs. (ii) Likewise helping to minimize the need for chemical pesticides toward pest management, because enhanced biodiversity in agroforestry systems contributes to natural pest control, helping beneficial insects and predators thrive in these environments.

4.6.2.3 Economic Benefits of SOC Built-up Under Agroforestry

Economic benefits of SOC built-up under agroforestry can be enumerated based on cost reduction and market opportunities. Agroforestry, the integration of trees and shrubs into agricultural landscapes, presents a multitude of economic benefits that extend beyond mere environmental advantages. One of the most significant aspects of agroforestry is its ability to enhance soil organic carbon (SOC) levels, which can lead to substantial cost reductions for farmers and new market opportunities. By improving soil health, agroforestry systems not only increase crop yields but also reduce the need for synthetic fertilizers and pesticides, ultimately lowering production costs. Furthermore, as sustainability becomes a priority for consumers and businesses alike, farmers implementing agroforestry practices can tap into emerging markets for carbon credits and ecofriendly products. This introduction explores how the economic advantages of SOC built-up under agroforestry can transform agricultural practices and contribute to a more sustainable and profitable future for farmers.

Cost reduction This can be achieved by decreasing reliance on chemical inputs which can translate to lower production costs for farmers' and thereby, savings on fertilizers and pesticides can significantly improve profit margins. Agroforestry, the integration of trees and shrubs into agricultural landscapes, presents a sustainable approach to farming that not only enhances biodiversity but also significantly contributes to soil health.

Market opportunities Market opportunities due to shift of consumer demand toward sustainably produced goods, by which agroforestry practices can enhance marketability, offering farmers a competitive edge. In recent years, the agricultural landscape has been increasingly shaped by a growing consumer demand for sustainably produced goods. As a result, agroforestry, an integrated approach that combines agriculture and forestry, has emerged as a compelling practice that not only enhances biodiversity but also offers substantial economic benefits. The systematic enumeration of these benefits, particularly the Strengthened Operating Capital (SOC) built up through agroforestry, reveals a multifaceted opportunity for farmers. By leveraging market trends favouring sustainability, agroforestry practices can significantly enhance the marketability of agricultural products, providing farmers with a competitive edge in a rapidly evolving marketplace. This explores how the implementation of agroforestry cannot only contribute to ecological resilience but also unlock new avenues for economic growth, positioning farmers to meet the demands of conscientious consumers while fostering sustainable agricultural practices.

4.6.2.3.1 Enumeration of Economic Benefits of Soil Organic Carbon (SOC) Built-up Under Agroforestry

The enumeration of economic benefits of soil organic carbon (SOC) built-up under agroforestry necessarily requires considering five key points of (i) cost reduction through decreased reliance on chemical inputs, (ii) improved soil health and productivity, (iii) long-term sustainability and economic viability, (iv) access to premium markets, and (v) community and societal benefits. This enumeration also requires consideration of environmental impact of agroforestry practices (Fagerholm et al. 2016, Wilson and Lovell 2016) giving due importance on (a) soil health improvement, as the higher SOC levels lead to improved soil structure, water retention, and erosion control, contributing to long-term soil health; and (b) reduced pollution as lower chemical inputs result in reduced runoff and pollution of local waterways, protecting ecosystems and promoting biodiversity.

Cost Reduction Through Decreased Reliance on Chemical Inputs—Two Ways

As the agricultural sector faces increasing pressure to enhance sustainability and reduce environmental impacts, agroforestry has emerged as a promising solution. This integrated land-use system combines agriculture and forestry practices, fostering biodiversity while providing economic benefits. One of the most significant advantages of adopting agroforestry is its potentiality for cost reduction through decreased reliance on chemical inputs. Specifically, two critical ways, viz., (i) fertilizer savings and (ii) pesticide reduction, highlight how agroforestry can lead to more sustainable and economically viable farming practices.

(i) **Fertilizer savings**: SOC enhances soil fertility by improving nutrient availability and retention. This reduces the need for synthetic fertilizers. This can be achieved by adopting fertilizer trees in AF system (Sileshi et al. 2014). Examples of some of those fertilizer trees are Gliricidia G(*liricidia sepium*), Leucaena (*Leucaena leucocephala*), Faidherbia (*Faidherbia albida*), Sesbania (*Sesbania sesban*), Albizia (*Albizia lebbeck*), Moringa (*Moringa oleifera*), Calliandra (*Calliandra calothyrsus*), various Acacia species, Neem (*Azadirachta indica*), Gmelina (*Gmelina arborea*), Alder (*Alnus spp.*); and certain species of Corymbia and Eucalyptus (with associated actinobacteria), and some other trees suitable for agroforestry. Along with these, nitrogen-fixing agricultural crops can also decrease nitrogen requirements in agriculture through agroforestry practices. Examples of some those nitrogen-fixing agricultural crops are leguminous plants like common beans P(*haseolus vulgaris*), peas (*Pisum sativum*), chickpeas (*Cicer arietinum*), lentils (*Lens culinaris*), soybeans (*Glycine max*), cowpeas (*Vigna unguiculata*), alfalfa (*Medicago sativa*), etc. (Mahmud et al. 2020). Adopting nitrogen fixing strategies in cereal can also serve the purpose of decreasing nitrogen use in crop production (Guo et al. 2023). Studies have shown that farms practicing agroforestry can cut fertilizer costs by up to 30–71%. This direct cost saving translates into lower production costs for farmers (Fadl 2013, Muchane et al. 2020, Fahad et al. 2022, Tsufac et al. 2022).

(ii) **Pesticide reduction**: Healthy soils with high SOC levels support diverse ecosystems that can naturally control pests. Agroforestry practices encourage beneficial insects and microorganisms, reducing the need for chemical pesticides. This not only lowers input costs but also minimizes health risks and environmental impacts associated with pesticide use (Tsufac et al. 2022, Fahad et al. 2022).

4.6.2.3.2 Improved Soil Health and Productivity Due to SOC Built-up Under Agroforestry

Agroforestry, the integration of trees and shrubs into agricultural landscapes, has emerged as a pivotal practice in promoting sustainable land management. One of its most significant contributions is the (i) enhancement of soil organic carbon (SOC), which plays a crucial role in improving soil health and productivity. Increased SOC levels lead to enhanced soil structure, characterized by improved water retention and aeration. This not only fosters robust root development but also bolsters plant growth, resulting in higher crop yields. As farmers can produce more without escalating input costs, their profit margins expand, contributing to the economic viability of agricultural operations. Moreover, agroforestry systems rich in SOC demonstrate remarkable (ii) resilience to climate variability. In an era marked by increasing climate extremes, such as prolonged droughts and intense flooding, these systems provide a buffer against the unpredictable impacts of climate change. By safeguarding farmers' investments and ensuring stable production, high SOC levels mitigate the risk of economic loss during adverse weather conditions, including droughts and floods. Thus, the build-up of SOC under agroforestry practices not only enhances immediate agricultural productivity but also fortifies the long-term sustainability of farming systems in the face of climate challenges. This resilience protects farmers' investments and ensures stable production, reducing the risk of economic loss during adverse weather conditions. In this way, benefits of SOC built-up in agroforestry perform its critical role in promoting both soil health and agricultural resilience (Gupta et al. 2023; Quandt et al. 2023; Rolo et al. 2023).

4.6.2.3.3 Long-Term Sustainability and Economic Viability Due to SOC Built-up Under Agroforestry - Two Ways

Agroforestry presents a compelling pathway to enhance long-term sustainability and economic viability in agricultural systems. By integrating trees and shrubs into traditional farming practices, agroforestry not only fosters the accumulation of soil organic carbon (SOC) but also significantly boosts profit margins through the marketing of diverse agroforestry products. This practice creates a multifaceted income stream for farmers, reducing reliance on single crops and mitigating economic risks. As a result, agroforestry emerges as a powerful strategy for promoting environmental health while ensuring the financial stability of farming communities, paving the way for resilient and sustainable agricultural landscapes through the following two ways.

(i) **Increased profit margins from marketing of agroforestry produces**: The cumulative effects of lower input costs, improved yields, and resilience result in significantly higher profit margins for farmers. By investing in SOC build-up through agroforestry, farmers can achieve sustainable profit growth over the long term (Fatima et al. 2024, Sarveswaran et al. 2023, Thiesmeier and Zander 2023).

(ii) **Diversification of Income Sources through agroforestry**: Agroforestry can provide multiple income streams through the cultivation of timber, fruit, nuts, and other tree products alongside traditional crops. This diversification reduces economic risks and enhances financial stability for farmers (Fatima et al. 2024, Thiesmeier and Zander 2023, Mukhlis et al. 2022).

4.6.2.3.4 Access to Premium Markets Catalyses Economic Benefits of SOC Built-up Under Agroforestry—Two Ways

Access to premium markets plays a pivotal role in enhancing the economic benefits of soil organic carbon (SOC) accumulation in agroforestry systems. This dual advantage arises primarily through two key mechanisms: (i) carbon credit and (ii) carbon credit opportunities. Organic certification not only opens doors to higher-value markets but also promotes sustainable farming practices that improve soil health and productivity. Meanwhile, carbon credit opportunities provide financial incentives for farmers to adopt agroforestry practices that sequester carbon, thereby contributing to climate change mitigation. Together, these pathways create a synergistic effect, driving economic growth while fostering environmental sustainability in agricultural landscapes. These two ways are as follows: (i) Organic Certification: Agroforestry practices that promote SOC accumulation often align with organic farming principles. Farmers can access premium markets for organic produce, which typically offer higher' prices than conventional crops. This can significantly enhance profit margins. (ii) Carbon Credit Opportunities: Farmers practicing agroforestry can participate in carbon trading markets by selling carbon credits generated through increased SOC levels. This presents an additional revenue stream that can further improve economic returns (Brathikan 2024, Lal 2023, Minoli et al. 2023, Jones et al. 2022, ICAI 2009).

4.6.2.3.5 Community and Societal Benefits Generated Through Economic Benefits of SOC Built-up Under Agroforestry—Two Strategies

Agroforestry systems play a pivotal role in fostering community resilience and driving societal progress by leveraging economic benefits that arise from sustainable land-use practices. This multifaceted approach not only contributes to job creation and the revitalization of local economies but also enhances food security for vulnerable populations. By integrating trees and shrubs into agricultural landscapes, agroforestry promotes biodiversity, improves soil health, and optimizes resource use, all of which are essential for sustaining livelihoods. Through two key strategies, *viz*., **(i) Job creation and local economies** and **(ii) Enhanced food security**, agroforestry serves as a catalyst for economic development while simultaneously addressing pressing social challenges as follows.

(i) **Job creation and local economies**: Agroforestry systems can create jobs in rural areas through agroforestry management, harvesting, and processing.

Increased farm profitability can stimulate local economies and support community development.

(ii) **Enhanced food security**: By increasing agricultural productivity and profitability, agroforestry contributes to greater food security for communities. Stable food supply chains can further enhance economic stability in rural areas. By exploring these strategies, we can better understand how the economic advantages of agroforestry translate into broader community and societal benefits, fostering a more sustainable and equitable future (Muthuri et al. 2023, Sudomo et al. 2023, Sahoo et al. 2020).

Thus, enumeration of economic benefits from SOC built-up under agroforestry reveals a comprehensive framework for enhancing agricultural sustainability and profitability. By reducing reliance on chemical inputs, farmers not only lower production costs but also pave the way for improved soil health, increased resilience, and greater financial security. As the global agricultural landscape continues to face challenges from climate change and resource depletion, the integration of agroforestry practices offers a viable solution that aligns economic and environmental goals, ultimately leading to a more sustainable agricultural future.

4.7 Impact of Soil Type and Climate on Agroforestry Influencing Soil Organic Carbon (SOC) Built-up

Agroforestry, the practice of integrating trees and shrubs into agricultural landscapes, presents a multifaceted approach to enhancing environmental sustainability. This system not only promotes biodiversity and improves soil health, but also fortifies resilience against climate change. A key component of agroforestry's environmental benefits lies in its influence on soil organic carbon (SOC) stocks, which are crucial for maintaining soil fertility and overall ecosystem health. The interaction between soil type and climate significantly affects SOC dynamics, particularly in terms of how different soil textures, such as clay and sandy soils, impact carbon accumulation and decomposition. For instance, clay-rich soils, with their superior moisture retention and nutrient-holding capacity, often support higher levels of organic matter decomposition, leading to greater SOC stocks. In contrast, sandy soils, characterized by their lower water retention, tend to experience faster decomposition rates, hindering SOC accumulation. Furthermore, soil pH is an essential factor influencing nutrient availability and microbial activity, complicating the SOC dynamics within agroforestry systems. The depth and structure of the soil also play critical roles in carbon sequestration, as deeper soils typically harbor more organic matter and demonstrate better aggregation, promoting long-term carbon storage. Understanding these intricate relationships among soil type, climate, and SOC is vital for developing effective agroforestry practices. This exploration will illuminate how to optimize agroforestry systems to enhance agricultural productivity while contributing to climate change mitigation and improving soil health across diverse environmental contexts.

4.7.1 Soil Type or Texture and SOC Dynamics or Soil Organic Matter (SOM) Decomposition Influencing SOC Built-up Under Agroforestry

Understanding the dynamics of soil organic carbon (SOC) is crucial for sustainable agricultural practices, particularly in agroforestry systems. Soil type and texture play pivotal roles in influencing SOC dynamics and the decomposition of soil organic matter (SOM). Different soil textures, ranging from sandy to clayey soils, affect moisture retention, aeration, and microbial activity, all of which are integral to the processes of SOC build-up and degradation. In agroforestry, where trees and crops coexist, the interactions between diverse root systems and varying litter inputs can significantly enhance SOC accumulation. This introduction explores how the interplay between soil characteristics and agroforestry practices shapes SOC dynamics, highlighting the importance of managing these factors to optimize soil health and enhance carbon sequestration in agricultural landscapes.

4.7.1.1 Soil Texture and SOM Decomposition Influencing SOC Built-up Under Agroforestry

Soil texture plays a crucial role in determining the physical and chemical properties of soils, particularly when comparing clay and sandy soils. Clay soils, with their fine particles and high surface area, exhibit greater nutrient and moisture retention, while sandy soils, composed of larger particles, tend to drain quickly but may struggle to retain essential nutrients. Understanding these differences is vital, especially in the context of soil organic matter (SOM) decomposition, which significantly influences soil organic carbon (SOC) build-up. Under agroforestry systems, where trees and crops are cultivated in proximity, the interactions between soil texture and SOM dynamics become even more pronounced. This introduction will explore how the varying characteristics of clay and sandy soils affect SOM decomposition processes and, consequently, the accumulation of SOC in agroforestry settings, highlighting the potential for sustainable land management practices that enhance soil health and carbon sequestration. These two key factors are as follows: **(i) Clay sandy soils**: Clay soils generally have a higher capacity for SOC retention due to their larger surface area and ability to form stable aggregates. Sandy soils, with larger particles and lower cohesion, tend to have lower SOC levels. Agroforestry practices in clay-rich areas can lead to significant SOC increase by improving soil structure and moisture retention. **(ii) Organic matter decomposition**: Different soil types influence the decomposition rates of organic matter. For instance, finer-textured soils may slow down decomposition due to lower aeration, allowing for more SOC accumulation over time.

4.7.1.2 Soil pH and Nutrient Availability Influencing SOC Built-up Under Agroforestry

Soil organic carbon (SOC) is a critical component of soil health, playing a vital role in nutrient cycling, water retention, and overall ecosystem functioning. In agroforestry systems, where agricultural practices are integrated with tree cultivation, the dynamics of SOC can be significantly influenced by soil pH and nutrient availability. Understanding the relationship between these factors is essential for optimizing land management practices and enhancing carbon sequestration. Soil pH affects nutrient solubility and microbial activity, thereby influencing the availability of essential elements for plant growth. As agroforestry systems vary in their composition and management, the interplay between soil pH, nutrient availability, and SOC build-up presents a complex yet important area of study. This introduction aims to explore how these elements interact to promote soil health and sustainability in agroforestry landscapes, highlighting the implications for agricultural productivity and climate change mitigation. These two key factors are as follows: **(i) Soil pH**: Soil pH affects microbial activity and the breakdown of organic matter. Neutral to slightly acidic soils typically favor higher microbial activity, enhancing SOC turnover and accumulation. **(ii) Soil nutrient availability**: Nutrient-rich soils can promote robust plant growth, increasing biomass input into the soil through leaf litter and root exudates, which are critical for SOC enhancement.

4.7.2 Soil Depth and Structure Influencing SOC Built-up Under Agroforestry

Soil organic carbon (SOC) plays a crucial role in maintaining soil health, fertility, and overall ecosystem functioning. In recent years, the significance of SOC has gained increasing attention in the context of sustainable agricultural practices, particularly agroforestry systems. Agroforestry, the integration of trees and shrubs with crops and/or livestock, offers numerous environmental benefits, including enhanced biodiversity, improved soil structure, and increased resilience to climate change. One of the key factors influencing the accumulation and stability of SOC in these systems is the depth and structure of the soil. This can be classified in two factors as follows: **(i) Soil depth**: Deeper soils generally have a greater potential for carbon sequestration as they can store more organic matter. Agroforestry systems can contribute to deeper rooting systems, leading to greater SOC accumulation. **(ii) Soil structure**: Soil structure influences water infiltration and retention, affecting plant growth and biomass production. Improved structure in agroforestry systems can enhance these processes, further influencing SOC dynamics.

Soil depth refers to the vertical extent of the soil profile, while soil structure pertains to the arrangement of soil particles and the spaces between them. Both aspects can significantly affect the physical, chemical, and biological processes that govern SOC dynamics. Deeper soils often provide a greater volume for organic matter accumulation, allowing for more substantial SOC storage. Conversely, the soil structure, characterized by the size, shape, and distribution of soil aggregates, can influence water infiltration, aeration, and root penetration, all of which are essential for supporting plant growth and enhancing organic matter inputs from both crops and trees.

Agroforestry systems often introduce a complex interplay of root structures and organic inputs that can alter soil properties in profound ways. For instance, tree roots can enhance soil porosity and facilitate the movement of nutrients, while leaf litter and other organic debris contribute to surface organic matter. These interactions can lead to variations in SOC levels across different soil depths, with implications for carbon sequestration potential and soil health. Understanding how soil depth and structure affect SOC accumulation in agroforestry systems is vital for optimizing land management practices aimed at enhancing carbon storage and promoting sustainable agricultural productivity.

Moreover, the variability in soil characteristics within agroforestry landscapes can lead to differential SOC responses. Factors such as climate, topography, and land management practices all play a role in shaping the soil environment and, consequently, the capacity for SOC build-up. By investigating these relationships, researchers and practitioners can identify best practices for enhancing SOC storage in agroforestry systems, thereby contributing to climate change mitigation and the development of resilient agricultural landscapes.

4.7.3 Climatic Effects Coupled with SOC Dynamics Influencing SOC Built-up Under Agroforestry

The intricate relationship between climate dynamics and soil organic carbon (SOC) built-up in agroforestry systems is a pivotal area of study, particularly in the face of global climate change. Agroforestry, a land-use management system that combines agricultural crops and tree species, not only contributes to biodiversity and ecosystem services but also plays a significant role in carbon sequestration. Understanding the climatic effects, specifically how air temperature and soil moisture regimes interact under varying climate zones and precipitation patterns, offers crucial insights into optimizing SOC storage in these multifunctional landscapes (Table 4.11).

4.7.3.1 Air Temperature Coupled with Soil Moisture Regimes Influencing SOC Built-up Under Agroforestry

This can summarily be presented into following two classes of influences:

(i) **Climate zones:** In tropical regions, higher temperatures can accelerate decomposition rates, which might limit SOC accumulation unless countered by increased biomass production. Conversely, in temperate zones, cooler temperatures may lead to slower decomposition and potentially higher SOC accumulation if biomass input remains consistent.

(ii) **Precipitation patterns:** Moisture availability is critical in dry regions where limited water can restrict plant growth, leading to lower biomass input and SOC. Agroforestry systems that improve moisture retention (e.g., through shade or improved soil structure) can enhance SOC even in arid conditions.

Air temperature serves as a critical driver in biological and chemical processes within the soil ecosystem. Warmer temperatures generally accelerate organic matter

Table 4.11 Ecosystem services of agroforestry under various soil types and climate on SOC built-up

Ecosystem services	Impact of soil types	Impact of climate	Geographical areas	Key findings
Soil fertility enhancement	Clayey soils improve nutrient retention, while sandy soils may lead to nutrient leaching.	Tropical climates with high rainfall promote nutrient cycling but may also cause leaching.	Tropical regions (e.g., sub-Saharan Africa, Southeast Asia)	Don et al. (2024) found that agroforestry on fertile clay soils in tropical regions supported better soil nutrient cycling compared to monoculturefarming.
Carbon sequestration	Organic-rich soils in temperate and tropical regions enhance soil carbon storage.	Warm, humid climates promote faster decomposition and carbon cycling, but sequestration rates may vary with forest type.	Tropical and temperate regions (e.g., Kenya, Brazil, China)	Kuyah et al. (2019) reported higher carbon sequestration in agroforestry systems with deep, fertile soils, compared to systems on more degraded soils in semi-arid regions.
Biodiversity support	Diverse soil textures (loamy, clayey) support a variety of plants, which, in turn, support diverse wildlife.	Tropical climates generally support higher species diversity, but semi-arid climates show lower biodiversity in agroforestry systems.	Southeast Asia, sub-Saharan Africa, Central America	An et al. (2023) found that agroforestry systems in diverse soils with adequate water availability maintained higher species diversity compared to monocultures.
Water regulation	Soils with high organic matter enhance water retention and reduce surface runoff	Wet, humid climates increase the likelihood of agroforestry systems providing watershed services.	Africa, Latin America, Southeast Asia	Don et al. (2024) highlighted that agroforestry systems in regions with clay or loamy soils provided better water retention, reducing flood risks in regions with high rainfall.
Erosion control	Deep, loamy soils are effective in reducing erosion, whereas sandy soils are more prone to erosion.	Regions with high rainfall (e.g., tropical zones) may experience more soil erosion, which agroforestry systems help mitigate.	Tropical and temperate zones (e.g., Amazon, East Africa)	Agroforestry systems, particularly on clayey and loamy soils, significantly reduced soil erosion compared to conventional agriculture in high-rainfall areas (Kuyah et al. 2019, Zhu et al. 2023b)

(continued)

Table 4.11 (continued)

Ecosystem services	Impact of soil types	Impact of climate	Geographical areas	Key findings
Food security	Fertile soils support higher crop yields and a more stable food supply.	Agroforestry in temperate and tropical climates boosts food security by providing diverse, year-round yields.	Southeast Asia, sub-Saharan Africa, South America	An et al. (2023) found that agroforestry systems were particularly effective in enhancing food security in regions with degraded soils by improving water retention and reducing crop failure risks.
Climate regulation	Soil moisture levels and carbon content influence local climate stability.	Temperate and tropical climates with agroforestry provide better microclimate stabilization, reducing temperature extremes.	Global (tropical, temperate regions)	Don et al. (2024) emphasized that agroforestry in tropical regions not only sequestered more carbon but also moderated temperature fluctuations and improved local climate regulation, especially in regions with loamy soils.
Pollination services	Soils that support diverse plant growth (e.g., nutrient rich soils enhance pollinator habitat)	Moderate climates with consistent flowering seasons support better pollinator diversity.	Global (mainly temperate and subtropical areas)	Kuyah et al. (2019) observed that agroforestry systems in both temperate and tropical regions attracted more pollinators, with soils high in organic matter and stable moisture conditions enhancing pollinator populations.
Nutrient cycling	Soil texture (clay, silt, sand) and organic matter content play crucial roles in nutrient availability.	Warm, humid climates with high plant diversity improve nutrient cycling through higher decomposition rates.	Tropical, subtropical, temperate regions	An et al. (2023) found that in tropical regions, agroforestry systems with fertile soils supported higher rates of nutrient cycling due to the presence of diverse plant species and a high rate of organic matter decomposition.

(continued)

Table 4.11 (continued)

Ecosystem services	Impact of soil types	Impact of climate	Geographical areas	Key findings
Pest and disease control	Soils with higher microbial diversity can help in controlling pests and diseases.	Agroforestry in regions with stable climates reduces the need for pesticides by fostering natural pest control systems.	Southeast Asia, Africa, Central America	Kuyah et al. (2019) noted that agroforestry systems in tropical areas reduced the prevalence of pest outbreaks by enhancing soil health, which supported natural predators and reduced pesticide dependency

decomposition, which can either enhance or diminish SOC levels depending on the balance between organic inputs from plant biomass and the rates of decomposition. In temperate regions, where seasonal temperature fluctuations are more pronounced, the impact of warming can lead to a shorter duration of soil saturation during spring, potentially affecting microbial activity and nutrient cycling. Conversely, in tropical climates, elevated temperatures may lead to increased microbial respiration, which could significantly influence SOC dynamics by accelerating the loss of stored carbon.

Simultaneously, soil moisture is a key determinant of plant growth and microbial activity, both essential for SOC accumulation. Variations in soil moisture regimes, driven by precipitation patterns and temperature changes, create a complex interplay that affects the availability of water for plants and microorganisms. For instance, in regions experiencing increased drought frequency due to climate change, reduced soil moisture can hinder plant growth, thereby limiting organic matter inputs to the soil. In contrast, excessive moisture can lead to anaerobic conditions, further affecting microbial processes and organic matter decomposition rates. The balance between these competing influences varies widely across different climate zones, necessitating a nuanced understanding of how local conditions interact with broader climatic trends.

Moreover, the effects of precipitation patterns, such as changes in intensity and distribution, compound the influence of air temperature and soil moisture on SOC dynamics. In regions where rainfall becomes erratic, periods of drought can be interspersed with heavy downpours, leading to soil erosion and nutrient leaching, which can severely impact SOC build-up. Agroforestry practices, by incorporating deep-rooted tree species, can enhance soil structure and moisture retention, mitigating some of these adverse effects. However, the effectiveness of these practices is contingent upon the specific climatic context, underscoring the importance of local adaptation strategies.

Thus, from this discourse, it is apparent that the interplay between climatic factors, *viz.*, air temperature, soil moisture regimes, and precipitation patterns, significantly influences SOC dynamics in agroforestry systems. By delving into this complex relationship, we can enhance our understanding of SOC built-up under diverse environmental conditions, thereby informing sustainable land management practices that support carbon sequestration and resilience against climate change.

4.7.3.2 Seasonality Coupled with Tree Phenology Influencing SOC Built-up Under Agroforestry

Agroforestry systems represent a dynamic intersection of agriculture and forestry, where the deliberate integration of trees and shrubs into farming landscapes can yield multiple ecological and economic benefits. One of the key advantages of these systems is their potential to enhance soil organic carbon (SOC) stocks, a crucial factor in mitigating climate change, improving soil health, and boosting agricultural productivity. A deeper understanding of how seasonality and tree phenology, e.g. the timing of biological events in trees, interact to influence SOC accumulation can provide valuable insights into optimizing agroforestry practices.

Seasonality plays a critical role in plant growth cycles, directly impacting the amount of organic matter returned to the soil. The timing of leaf emergence, flowering, and fruiting varies significantly among tree species, leading to distinct patterns of biomass production and decomposition. In agroforestry systems, this variability can create a mosaic of organic matter input throughout the year. For instance, if different tree species within the system exhibit staggered growth patterns, organic matter can be continuously returned to the soil, maintaining microbial activity and promoting decomposition processes. This ongoing cycle of growth and decay not only enhances soil fertility but also contributes to the long-term accumulation of SOC.

Moreover, the strategic selection of tree species with complementary growth cycles can further optimize biomass input across seasons. By integrating species that flourish at different times of the year, agroforestry systems can maintain a steady supply of organic matter. This diversity not only maximizes biomass production but also supports a wider range of soil organisms, enhancing soil structure and nutrient cycling. For example, while one species may be actively growing in the spring, another may contribute biomass in the fall, ensuring that organic material is consistently available for decomposition. This continuous input is vital for SOC accumulation, as the presence of organic matter is fundamental for the formation of stable soil aggregates, which play a key role in carbon sequestration.

Understanding the interplay between seasonality and tree phenology in agroforestry is essential for developing sustainable practices that not only increase SOC but also improve resilience to climate variability. By tailoring agroforestry designs to leverage the unique growth patterns of various species, farmers can optimize their systems for enhanced productivity and environmental health. The potential for increased SOC through thoughtful integration of seasonally complementary tree species offers a promising avenue for research and application, with far-reaching implications for sustainable agriculture and climate change mitigation.

Thus, the synergy between seasonal dynamics and tree phenology within agroforestry systems can significantly influence SOC build-up. By focusing on the timing of plant growth cycles and strategically incorporating diverse species, agroforestry presents a powerful approach to enhance soil health, support biodiversity, and contribute to climate resilience, making it a vital component of sustainable land management strategies.

4.7.4 Synergistic Effects of Soil Type, Climate, and Agroforestry Practices Influencing SOC Built-up Under Agroforestry

Soil type is the foundational factor in determining the physical and chemical properties of a landscape, influencing water retention, nutrient availability, and microbial activity. Different soil textures, such as sandy, clayey, and loamy soils, exhibit distinct capacities for SOC storage. For instance, clay soils, with their higher cation exchange capacity, often support greater organic matter retention compared to sandy soils. Moreover, the inherent mineral composition of the soil can affect decomposition rates and the stabilization of organic matter. Therefore, understanding the specific interactions between various soil types and their capacity to accumulate SOC under agroforestry systems is essential for optimizing land management practices.

Climate is another critical factor affecting SOC dynamics. Temperature and precipitation patterns govern plant growth, decomposition rates, and soil microbial activity. In regions with abundant rainfall, for example, the decomposition of organic materials may be accelerated, potentially leading to lower SOC levels unless counterbalanced by the biomass inputs from agroforestry practices. Conversely, in arid climates, water scarcity can limit plant growth and organic matter input, posing challenges for SOC accumulation. Additionally, climate change introduces further complexities, altering precipitation patterns and temperature regimes, which can impact both the growth of tree species and the overall resilience of agroforestry systems.

Thus, the synergistic effects of soil type, climate, and agroforestry practices underscore the complexity of SOC dynamics in agroforestry systems. By carefully considering the interplay of these factors, specifically through strategic agroforestry design and thoughtful tree species selection, land managers and farmers can enhance SOC accumulation. This not only contributes to improved soil health and agricultural productivity but also plays a vital role in mitigating climate change through carbon sequestration. As we continue to explore and understand these relationships, the potential for agroforestry to provide sustainable solutions to contemporary agricultural challenges becomes increasingly evident.

4.7.4.1 Synergistic Effect of Agroforestry Design and Tree Species Selection Influencing SOC Built-up Under Agroforestry

Agroforestry design and tree species selection are fundamental components that can significantly influence SOC accumulation. The design of agroforestry systems, such as the spatial arrangement of trees and crops, can enhance light interception, promote nutrient cycling, and optimize water use, thereby increasing overall biomass production. Different tree species also exhibit varying growth rates, rooting depths, and litter quality, all of which play roles in SOC dynamics. For instance, deep-rooted species can access nutrients from lower soil layers and contribute to deeper SOC accumulation, while species with high litter quality may enhance microbial activity and decomposition rates. The selection of appropriate tree species, therefore, becomes a crucial consideration in designing agroforestry systems that maximize SOC build-up. This selection is, thus, summarily of two types as follows.

(i) **Choosing appropriate tree and understory species**: This is crucial that matches the local soil type and climate. Deep-rooted species may be favored in deeper soils to maximize carbon inputs and water use efficiency.
(ii) **Mixed-species** involved in agroforestry can improve resource use and resilience, leading to increased biomass and SOC across diverse environments (De Souza et al. 2011).

Box 4.1 Synergistic Effect of Management Practices Influencing SOC Built-up Under Agroforestry[a]

Examples of agroforestry design:

Alley cropping: This involves planting rows of trees (such as *Leucaena leucocephala* or *Gliricidia sepium*) alongside agricultural crops. This design allows for complementary growth and improved soil health.

Silvopasture: Integrating trees like *Quercus* spp. with pasture for livestock. This method enhances biodiversity and nutrient cycling while providing shade and forage.

Forest farming: Cultivating high-value crops (e.g., medicinal herbs) under a managed forest canopy. Species like *Acer rubrum* or *Carya* spp. can be utilized for their market value and ecological benefits.

Windbreaks and shelterbelts: Rows of trees planted to protect crops from wind and erosion, with species such as *Populus* spp. or *Salix* spp. enhancing microclimate conditions.

Examples of tree species selection:

Nitrogen-fixing trees: Species like *Albizia* and *Acacia* improve soil fertility through nitrogen fixation, enhancing SOC levels.

Deep-rooted trees:. Species such as *Eucalyptus* or *Pinus* can improve soil structure and prevent erosion, contributing to SOC.

Fruit trees: Incorporating trees like Mango (*Mangifera indica*) or *Citrus* spp. can provide economic returns while improving soil organic matter.

Shade trees: *Ficus* spp. or *Gliricidia* can provide shade for understory crops, improving moisture retention and soil quality.

These designs and species selections can synergistically improve soil organic carbon levels, promoting sustainability in agroecosystems.

[a]Murthy et al. (2016).

4.7.4.2 Synergistic Effect of Management Practices Influencing SOC Built-up Under Agroforestry

Soil organic carbon (SOC) is a crucial component of soil health and fertility, playing a pivotal role in the sustainability of agricultural ecosystems. Its accumulation is influenced by a multitude of factors, including soil type, climate, and land management practices. Among these farming activities, agroforestry, a land use system that integrates trees and shrubs with crops and/or livestock, has garnered significant attention for its potential to enhance SOC levels. The synergistic effects of soil type, climate, and agroforestry practices present a complex interplay that can substantially influence SOC build-up, ultimately affecting ecosystem services, agricultural productivity, and carbon sequestration efforts (Box 4.1). The cultivation practices may be of following two classes:

(i) **Practices enhancing SOC** content such as mulching, cover cropping, and reduced tillage can enhance SOC by improving soil structure and moisture retention. The effectiveness of these practices can vary significantly depending on the soil type and climatic conditions.
(ii) **Soil amendments** (e.g., biochar, compost) can also be tailored to specific soil types and climates to boost SOC levels.

4.7.5 Monitoring and Assessment of SOC Built-up Under Agroforestry

Soil organic carbon (SOC) is a critical component of soil health and plays a vital role in the global carbon cycle. It significantly influences soil fertility, water retention, and overall ecosystem functioning. Understanding the dynamics of SOC is particularly important in the context of agroforestry, a land use system that integrates trees and shrubs with crops and/or livestock. This practice not only enhances biodiversity and agricultural productivity but also offers a promising pathway for carbon sequestration, mitigating climate change effects.

The assessment and monitoring of SOC levels are essential for establishing a baseline understanding of soil health in agroforestry systems. Baseline SOC levels provide crucial information on the starting point from which changes can be measured. This understanding is pivotal as it allows researchers, land managers, and policymakers to gage the effectiveness of agroforestry practices in enhancing SOC stocks over time. Changes in SOC can result from various factors, including management practices, climatic conditions, and soil types. Thus, regular monitoring can reveal trends and fluctuations in SOC levels, facilitating informed decision-making regarding land management and agroforestry strategies.

Implementing a systematic monitoring program is vital for capturing the nuanced responses of SOC to different agroforestry practices. This involves employing a range of methodologies, from soil sampling and laboratory analysis to remote sensing and modeling techniques. Such comprehensive assessments enable stakeholders to identify which agroforestry practices are most effective in promoting SOC sequestration and under what conditions. Moreover, continuous monitoring helps in adapting management strategies to optimize the benefits of agroforestry systems, ensuring that they contribute not only to sustainable agricultural productivity but also to broader environmental goals.

So, monitoring and assessing SOC within agroforestry systems is essential for understanding baseline conditions and potential changes over time. Regular assessment not only informs adaptive management strategies but also enhances our capacity to leverage agroforestry as a viable approach to sequester carbon and improve soil health. This integrative approach can ultimately lead to more sustainable land management practices that align agricultural production with ecological integrity, thus fostering resilience in the face of climate change and environmental degradation.

4.8 Concluding Remarks

The integration of agroforestry systems into agricultural landscapes offers a compelling strategy for enhancing sustainability through the accumulation of soil organic carbon (SOC). By fostering microbial activity and nutrient cycling, agroforestry reduces reliance on chemical fertilizers and pesticides, yielding both economic savings and significant environmental benefits. This multifaceted approach not only addresses immediate agricultural productivity concerns but also lays the groundwork for a more resilient agricultural future that harmonizes productivity with ecological health.

Enhancing soil health and microbial activity: Agroforestry stands out as a powerful method for boosting soil microbial activity, which is crucial for soil health and crop productivity. The presence of trees and shrubs enhances organic matter inputs through leaf litter, root biomass, and decaying wood, leading to increased SOC levels. This organic matter serves as a food source for soil microbes, which in turn facilitate nutrient cycling and improve soil structure. Healthy microbial communities are essential for breaking down organic materials, making nutrients more available to crops, and supporting overall soil fertility.

Mitigating soil erosion: Soil erosion represents a significant challenge for sustainable agriculture, leading to loss of arable land and reduced productivity. Agroforestry mitigates erosion by stabilizing the soil with tree roots, which not only hold the soil in place but also improve its structure and water retention. The canopy cover provided by trees reduces the impact of raindrops on the soil surface, further protecting it from erosion. As a result, agroforestry systems can enhance both soil conservation and agricultural productivity, contributing to a more resilient agricultural landscape.

Climate adaptation and carbon sequestration: The interplay of soil type, climate, and agroforestry practices critically influences SOC dynamics. Agroforestry systems are versatile and can be adapted to various climatic and ecological contexts, making them effective tools for carbon sequestration. As global challenges like climate change intensify, leveraging agroforestry for its carbon storage potential becomes increasingly vital. By integrating trees into farming systems, farmers can enhance their carbon footprint while contributing to broader climate mitigation goals.

Overcoming challenges and barriers: Despite the clear benefits of agroforestry, several challenges hinder its widespread adoption. Issues related to market access, economic viability, and inadequate policy-support pose significant barriers. Farmers may lack information on the long-term benefits of agroforestry or face difficulties in accessing the necessary financial resources. Therefore, targeted interventions, such as improving infrastructure, creating supportive policies, and providing education, are essential to encourage the adoption of these sustainable practices.

The role of research and community engagement: Future research should focus on quantifying the diverse benefits of agroforestry across various systems and climates. Understanding the interactions between different tree and crop species and their specific contributions to soil health will enhance the effectiveness of agroforestry practices. Additionally, engaging local communities and farmers is crucial. Education and training programs can empower farmers with the knowledge needed to implement agroforestry successfully, thus fostering a culture of sustainability.

As the global agricultural landscape grapples with issues like soil degradation, food security, and climate change, agroforestry emerges as a promising solution that can restore soil health, enhance nutrient cycling, and promote biodiversity. By embracing agroforestry practices, we can create agricultural systems that are not only productive but also ecologically sound. Policymakers, practitioners, and farmers must collaborate to promote and adopt these sustainable land use practices, ensuring a resilient agricultural future that supports both human and ecological health. Ultimately, agroforestry presents a vital strategy for achieving sustainable development goals, particularly in the context of enhancing soil organic carbon and fostering environmental sustainability.

References

Acharya AK, Kafle N (2009) Land degradation issues in Nepal and its management through agroforestry. J Agric Environ 140:230–240

Ajayi OC, Akinnifesi FK, Sileshi G, Chakeredza S, Mn'gomba S, Ajayi O, Nyoka I and Chineke T (2009). Local solutions to global problems: the potential of agroforestry for climate change adaptation and mitigation in southern Africa. Invited paper presented at the tropical forests and climate change adaptation (TroFCCA) regional meeting "knowledge and action on forests for climate change adaptation in Africa", November 18–20, Accra, Ghana. https://www.researchgate.net/publication/275830270

Akoto DS, Partey ST, Denich M et al (2020) Towards bamboo agroforestry development in Ghana: evaluation of crop performance, soil properties and economic benefit. Agrofor Syst 94:1759–1780. https://doi.org/10.1007/s10457-020-00493-7

Akpalu SE, Dawoe EK, Abunyewa AA (2020) Effects of Faidherbia albida on some important soil fertility indicators on agroforestry parklands in the semi-arid zone of Ghana. Afr J Agric Res 2020(15):256–268. https://doi.org/10.5897/AJAR2019.14617

Al Mamun A, Muhammad S, Hossain MA, Hosssain MK, Kabir MA (2014) Eucalyptus agroforestry system in Chittagong: socio-economic potentiality amid controversies. Int J Agron

Agri Res 4(1):32–39. https://innspub.net/eucalyptus-agroforestry-system-in-chittagong-socio-economic-potentiality-amid-controversies/
Allen SC, Jose S, Nair PKR, Brecke BJ, Nkedi-Kizza P, Ramsey CL (2004) Safety-net role of tree roots: evidence from a pecan (*Carya illinoensis* K. Koch)–cotton (*Gossypium hirsutum* L.) alley cropping system in the southern United States. For Ecol Manag 192(2–3):395–407. ISSN 0378-1127. https://www.sciencedirect.com/science/article/pii/S0378112704001033. https://doi.org/10.1016/j.foreco.2004.02.009
Allen J, Pascual KS, Romasanta RR, Van Trinh M, Van Thach T, Van Hung N, Sander BO, Chivenge P (2020) Rice straw management effects on greenhouse gas emissions and mitigation options. Sustainable rice straw management, pp145–159. https://scholar.google.com/scholar_lookup?title=Rice%20Straw%20Management%20Effects%20on%20Greenhouse%20Gas%20Emissions%20and%20Mitigation%20Options&publication_year=2020&author=J.%20Allen&author=K.S.%20
Amare T, Amede T, Abewa A, Woubet A, Agegnehu G, Gumma M, Schulz S (2022) Remediation of acid soils and soil property amelioration via Acacia decurrensbased agroforestry system. Agrofor Syst 96:329–342
An Z, Pokharel P, Plante AF, Bork EW, Carlyle CN, Williams EK, Chang SX (2023) Soil organic matter stability in forest and cropland components of two agroforestry systems in western Canada. Geoderma 433(2023):116463. https://doi.org/10.1016/j.geoderma.2023.116463. ISSN 0016-7061. https://www.sciencedirect.com/science/article/pii/S0016706123001404
Atangana A, Khasa D, Chang S, Degrande A (2014) Phytoremediation in tropical agroforestry. In: Tropical agroforestry. Springer, Dordrecht. https://doi.org/10.1007/978-94-007-7723-1_19
Augustine CMJ, Vogt KA, Harrison RB, Hunsaker HM (2007) Nitrogen-fixing trees in small-scale agriculture of mountainous Southeast Guatemala: effects on soil quality and erosion control. J Sustain For 23(4):61–80. https://doi.org/10.1300/J091v23n04_04
Ayala-Montejo D, Valdés-Velarde E, Benedicto-Valdés GS, Escamilla-Prado E, Sánchez-Hernández R, Gallardo JF, Martínez-Zurimendi P (2022) Soil biological activity, carbon and nitrogen dynamics in modified coffee agroforestry systems in Mexico. Agronomy 12(8):1794. https://doi.org/10.3390/agronomy12081794
Bado VB, Bationo A (2018) Integrated management of soil fertility and land resources in Sub-Saharan Africa: involving local communities. In: Sparks DL (ed) Advances in agronomy, vol 150. Academic Press, pp 1–33., ISSN 0065-2113, ISBN 9780128151754, https://www.sciencedirect.com/science/article/pii/S006521131830018X. https://doi.org/10.1016/bs.agron.2018.02.001
Bado BV, Anthony Whitbread A, Manzo MLS (2021) Improving agricultural productivity using agroforestry systems: performance of millet, cowpea, and ziziphus-based cropping systems in West Africa Sahel. Agric Ecosyst Environ 305:107175. https://doi.org/10.1016/j.agee.2020.107175. ISSN 0167-8809. https://www.sciencedirect.com/science/article/pii/S0167880920303613
Bai W, Sun Z, Zheng J, Du G, Feng L, Cai Q, Yang N, Feng C, Zhang Z, Evers JB, van der Werf W, Zhang L (2016) Mixing trees and crops increases land and water use efficiencies in a semi-arid area. Agric Water Manag 178:281–290. https://doi.org/10.1016/j.agwat.2016.10.007
Banerjee S, Baah-Acheamfour M, Carlyle CN, Bissett A, Richardson AE, Siddique T, Bork EW, Chang SX (2016) Determinants of bacterial communities in Canadian agroforestry systems. Environ Microbiol 18(6):1805–1816. https://doi.org/10.1111/1462-2920.12986. Epub 2015 Aug 11
Béliveau A, Lucotte M, Davidson R, Paquet S, Mertens F, Passos CJ, Romana CA (2017) Reduction of soil erosion and mercury losses in agroforestry systems compared to forests and cultivated fields in the Brazilian Amazon. J Environ Manag 203(Pt 1):522–532. https://doi.org/10.1016/j.jenvman.2017.07.037. Epub 2017 Aug 30
Beule L, Karlovsky P (2021) Tree rows in temperate agroforestry croplands alter the composition of soil bacterial communities. PLoS One 16(2):e0246919. https://doi.org/10.1371/journal.pone.0246919

Beule L, Vaupel A, Moran-Rodas VE (2022) Abundance, diversity, and function of soil microorganisms in temperate alley-cropping agroforestry systems: a review. Microorganisms 10:616

Beuschel R, Piepho HP, Joergensen RG et al (2019) Similar spatial patterns of soil quality indicators in three poplar-based silvo-arable alley cropping systems in Germany. Biol Fertil Soils 55:1–14. https://doi.org/10.1007/s00374-018-1324-3

Bisht N, Sah VK, Satyawali K, Tiwari S (2017) Comparison of wheat yield and soil properties under open and poplar based agroforestry system. J Appl Nat Sci 2017(9):1540–1543

Borin M, Passoni M, Mara Thiene M, Tempesta T (2010) Multiple functions of buffer strips in farming areas. Eur J Agron 32(1):103–111. https://doi.org/10.1016/j.eja.2009.05.003. ISSN 1161-0301. https://www.sciencedirect.com/science/article/pii/S1161030109000422

Bracken P, Burgess PJ, Girkin NT (2023) Opportunities for enhancing the climate resilience of coffee production through improved crop, soil and water management. Agroecol Sustain Food Syst 47:1125–1157

Brathikan VM (2024) Sustainable agriculture in India: carbon credits and organic farming. https://www.researchgate.net/publication/381295245

Cao L, Wang S, Peng T, Cheng Q, Zhang L, Zhang Z, Yue F, Fryer AE (2020) Monitoring of suspended sediment load and transport in an agroforestry watershed on a karst plateau, Southwest China. Agric Ecosyst Environ 299:106976. https://doi.org/10.1016/j.agee.2020.106976

Cardinael R, Tiphaine C, Cambou A, Béral C, Barthès BG, Dupraz C, Durand C, Kouakoua E, Chenu C (2017) Increased soil organic carbon stocks under agroforestry: a survey of six different sites in France. Agric Ecosyst Environ 236(2017):243–255., ISSN 0167-8809. https://doi.org/10.1016/j.agee.2016.12.011

Chatterjee N, Nair PKR, Chakraborty S, Nair VD (2018) Changes in soil carbon stocks across the forest-agroforest-agriculture/pasture continuum in various agroecological regions: a meta-analysis, agriculture. Ecosyst Environ 266:55–67. https://doi.org/10.1016/j.agee.2018.07.014. ISSN 0167-8809. https://www.sciencedirect.com/science/article/pii/S0167880918302913

Cherubin MR, Chavarro-Bermeo JP, Silva-Olaya AM (2019) Agroforestry systems improve soil physical quality in northwestern Colombian Amazon. Agrofor Syst 93:1741–1753

Choudhary A, Rijhwani S (2022) Microbial diversity in selected agroforestry systems of Central Rajasthan: life sciences-botany for medicinal science. Int J Life Sci Pharma Res 10(5):65–73. https://doi.org/10.22376/ijpbs/lpr.2020.10.5.L65-73

Chowdhury FI, Barua I, Chowdhury AI, de Dios VR, Alam MS (2022) Agroforestry shows higher potential than reforestation for soil restoration after slash-and-burn: a case study from Bangladesh. Geol Ecol Landscap 6(1):48–54. https://doi.org/10.1080/24749508.2020.1743528

Das AK, Rahman MA, Keya SS et al (2020) Malta-based agroforestry system: an emerging option for improving productivity, profitability and land use efficiency. Environ Sustain 3:521–532. https://doi.org/10.1007/s42398-020-00139-5

Dawson IK, Leakey R, Clement CR, Weber JC, Cornelius JP, Roshetko JM, Vinceti B, Kalinganire A, Tchoundjeu Z, Masters E, Jamnadass R (2014) The management of tree genetic resources and the livelihoods of rural communities in the tropics: non-timber forest products, smallholder agroforestry practices and tree commodity crops. For Ecol Manag 333(2014):9–21. https://doi.org/10.1016/j.foreco.2014.01.021. ISSN 0378-1127. https://www.sciencedirect.com/science/article/pii/S0378112714000231

De Souza HN, De Goede RG, Brussaard L, Cardoso IM, Duarte EM, Fernandes RB, Gomes LC, Pulleman MM (2011) Protective shade, tree diversity and soil properties in coffee agroforestry systems in the Atlantic rainforest biome. Agric Ecosyst Environ 146(1):179–196. https://doi.org/10.1016/j.agee.2011.11.007

De Stefano A, Jacobson MG (2018) Soil carbon sequestration in agroforestry systems: a meta-analysis. Agroforest Syst 92:285–299. https://doi.org/10.1007/s10457-017-0147-9

De'Longe MS, Miles A, Carlisle L (2016) Investing in the transition to sustainable agriculture. Environ Sci Pol 2016(55):266–273. https://core.ac.uk/download/pdf/82089420.pdf

Dev I, Ram A, Ahlawat SP et al (2020) Bamboo-based agroforestry system (Dendrocalamus strictus + sesame–chickpea) for enhancing productivity in semi-arid tropics of Central India. Agrofor Syst 94:1725–1739. https://doi.org/10.1007/s10457-020-00492-8

Dhaliwal J, Kukal SS, Sharma S (2017) Soil organic carbon stock in relation to aggregate size and stability under tree-based cropping systems in Typic Ustochrepts. Agroforest Syst 92:275–284. (2018). https://doi.org/10.1007/s10457-017-0103-8

Diallo MB, Akponikpè PBI, Fatondji D, Abasse T, Agbossou EK (2019) Long-term differential effects of tree species on soil nutrients and fertility improvement in agroforestry parklands of the Sahelian Niger. For Trees Livelihoods. https://doi.org/10.1080/14728028.2019.1643792

Diel J, Vogel H-J, Schlüter S (2019) Impact of wetting and drying cycles on soil structure dynamics. Geoderma 345:63–71. https://doi.org/10.1016/j.geoderma.2019.03.018

Do VH, La N, Mulia R, Bergkvist G, Dahlin AS, Nguyen VT, Pham HT, Öborn I (2020) Fruit tree-based agroforestry systems for smallholder farmers in Northwest Vietnam—a quantitative and qualitative assessment. Land 9(11):451. https://doi.org/10.3390/land9110451

Don A, Seidel F, Leifeld J, Kätterer T, Martin M, Pellerin S, Emde D, Seitz D, Chenu C (2024) Carbon sequestration in soils and climate change mitigation—definitions and pitfalls. Glob Chang Biol 30:e16983. https://doi.org/10.1111/gcb.16983

Dossa EL, Khouma M, Diedhiou I, Sene M, Kizito F, Badiane AN, Samba SAN, Dick RP (2009) Carbon, nitrogen and phosphorus mineralization potential of semiarid Sahelian soils amended with native shrub residues. Geoderma 148(3–4):251–260., ISSN 0016-7061. https://www.sciencedirect.com/science/article/pii/S0016706108002838. https://doi.org/10.1016/j.geoderma.2008.10.009

Douglas G, Mackay A, Vibart R, Dodd M, McIvor I, McKenzie C (2020) Soil carbon stocks under grazed pasture and pasture-tree systems. Sci Total Environ 715:136910., ISSN 0048-9697. https://www.sciencedirect.com/science/article/pii/S0048969720304204. https://doi.org/10.1016/j.scitotenv.2020.136910

Du X, Jian J, Du C, Stewart RD (2022) Conservation management decreases surface runoff and soil erosion. Int Soil Water Conserv Res 10(2):188–196. ISSN 2095-6339. https://www.sciencedirect.com/science/article/pii/S2095633921000757. https://doi.org/10.1016/j.iswcr.2021.08.001

Eddy WC, Yang WH (2022) Improvements in soil health and soil carbon sequestration by an agroforestry for food production system. Agric Ecosyst Environ 333:Article 107945. https://doi.org/10.1016/j.agee.2022.107945

Elias E (2002) Farmers' perceptions of soil fertility change and management, SOS Sahel & Institute for sustainable development, Doctoral Thesis, Addis Ababa, Ethiopia, and 252p. http://hdl.handle.net/123456789/2162. http://publication.eiar.gov.et:8080/xmlui/handle/123456789/2162?show=full

Eyasu E (2002) Farmers' Perceptions of Soil Fertility Change and Management, SOS SAHEL & Institute for Sustainable Development, Doctoral Thesis, Addis Ababa, Ethiopia, and 252p

Fadl KEM (2013) Influence of Acacia Senegal agroforestry system on growth and yield of sorghum, sesame, roselle and gum in north Kordofan State, Sudan. J For Res 24:173–177. https://doi.org/10.1007/s11676-012-0319-4

Fagerholm N, Torralba M, Burgess PJ, Plieninger T (2016) A systematic map of ecosystem services assessments around European agroforestry. Ecologi Indicat 62:47–65. https://doi.org/10.1016/j.ecolind.2015.11.016. ISSN 1470-160X

Fahad S, Chavan SB, Chichaghare AR, Uthappa AR, Kumar M, Kakade V, Pradhan A, Jinger D, Rawale G, Yadav DK et al (2022) Agroforestry systems for soil health improvement and maintenance. Sustain For 14(22):14877. https://doi.org/10.3390/su142214877

Fahmi MKM, Dafa-Alla D-AM, Kanninen M, Luukkanen O (2018) Impact of agroforestry parklands on crop yield and income generation: case study of rainfed farming in the semi-arid zone of Sudan. Agroforest Syst 92:785–800. https://doi.org/10.1007/s10457-016-0048-3

Fatima S, Abbas S, Rebi A, Ying Z (2024) Sustainable forestry and environmental impacts: assessing the economic, environmental, and social benefits of adopting sustainable agricultural practices. Ecologi Front. ISSN 2950-5097 2024. https://doi.org/10.1016/j.ecofro.2024.05.009

Gebru H (2015) A review on the comparative advantage of intercropping systems. J Biol Agric Healthc. www.iiste.org, ISSN 2224-3208 (Paper) ISSN 2225-093X (Online) 2015.5(7):28–38

Gikas GD, Tsihrintzis VA, Sykas D (2016) Effect of trees on the reduction of nutrient concentrations in the soils of cultivated areas. Environ Monit Assess 188(6):327. https://doi.org/10.1007/s10661-016-5325-0. Epub 2016 May 5

Greipsson S (2011) Phytoremediation. Nat Educ 3. https://www.nature.com/scitable/knowledge/library/phytoremediation-17359669/#:~:text=Phytoremediation%20basically%20refers%20to%20thecost%2Deffective%20environmental%20restoration%20technology

Guillot E, Bertrand I, Rumpel C, Gomez C, Arnal D, Abadie J, Hinsinger P (2021) Spatial heterogeneity of soil quality within a Mediterranean alley cropping agroforestry system: comparison with a monocropping system. Eur J Soil Biol 105:103330

Guo J, Wang B, Wang G, Wu Y, Cao F (2020) Afforestation and agroforestry enhance soil nutrient status and carbon sequestration capacity in eastern China. Land Degrad Dev 31:392–403

Guo K, Yang J, Yu N, Luo L, Wang E (2023) Biological nitrogen fixation in cereal crops: progress, strategies, and perspectives. Plant Commun 4(2):100499., ISSN 2590-3462. https://doi.org/10.1016/j.xplc.2022.100499

Gupta H, Panchol R, Yadav R, Vasure N, Choudhary S, Singh MN, Rastogi M (2023) The role of agroforestry systems in enhancing climate resilience and sustainability—a review. Int J Environ Clim Change 13:4342–4353. https://doi.org/10.9734/IJECC/2023/v13i113615

Gusli S, Sumeni S, Sabodin R, Muqfi IH, Nur M, Hairiah K, Useng D, Van Noordwijk M (2020) Soil organic matter, mitigation of and adaptation to climate change in cocoa–based agroforestry systems. Land 9(9):323. https://doi.org/10.3390/land9090323

Haglund E, Ndjeunga J, Snook L, Pasternak D (2011) Dry land tree management for improved household livelihoods: farmer managed natural regeneration in Niger. J Environ Manag 92(7):1696–1705. https://doi.org/10.1016/j.jenvman.2011.01.027. Epub 2011 Apr 1

Handa AK, Dev I, Rizvi R H, Kumar N, Ram A, Kumar D, Kumar A, Bhaskar S, Dhyani SK, Rizvi J (eds) (2019) Successful agroforestry models for different agro-ecological regions in India. Jointly published by the Central Agroforestry Research Institute (CAFRI), Jhansi, and the South Asia Regional Programme (based in New Delhi) of World Agroforestry (ICRAF)

Hussain Z, Rasheed F, Tanvir MA, Zafar Z, Rafay M, Mohsin M et al (2020) Increased antioxidative enzyme activity mediates the phytoaccumulation potential of Pb in four agroforestry tree species: a case study under municipal and industrial wastewater irrigation. Int J Phytoremediation 23(7):704–714. https://doi.org/10.1080/15226514.2020.1849016

Ibrahim A, Abaidoo RC, Fatondji D, Opoku A (2015) Hill placement of manure and fertilizer micro-dosing improves yield and water use efficiency in the Sahelian low input millet-based cropping system. Field Crop Res 2015(180):29–36

ICAI (2009) Clean development mechanism and carbon credits—a primer. Celebrating the 60 year of excellence. In: Professional Development Committee, the Institute of Chartered Accountants of India (ICAI, set up by an Act of Parliament), New Delhi. ISBN: 978-81-8441-204-8. Website: www.icai.org

Jones SK, Bergamini N, Beggi F et al (2022) Research strategies to catalyze agroecological transitions in low- and middle-income countries. Sustain Sci 17:2557–2577. https://doi.org/10.1007/s11625-022-01163-6

Juo ASR (2004) Chap 3: Technologies for sustainable management of steep lands in Asia: harmonizing economic and ecological sustainability. In: Partap T (ed) Sustainable farming systems in upland areas ©APO 2004, ISBN: 92-833-7031-7. (STM 09-00) Report of the APO study meeting on sustainable farming systems in upland areas, held in New Delhi, India, 15–19 January 2001. Asian Productivity Organization, Chiyoda-ku, Tokyo, Japan, pp 48–58. URL: www.apo-tokyo.org

Kamau S, Barrios E, Karanja NK, Ayuke FO, Lehmann F (2017) Soil macrofauna abundance under dominant tree species increases along a soil degradation gradient. Soil Biol Biochem 112(2017):35–46. ISSN 0038-0717. https://www.sciencedirect.com/science/article/pii/S003807171630459X. https://doi.org/10.1016/j.soilbio.2017.04.016

Kandji ST, Verchot L, Mackensen J (2006) Climate change and variability in the southern Africa: impacts and adaptation in the agricultural sector. ICRAF, Nairobi, Kenya. https://apps.worldagroforestry.org/downloads/Publications/PDFS/B14549.pdf

Kannur S, Patil S, Inamati S (2020) Dynamics of soil properties as influenced by rubber based agroforestry system in hilly zone of Karnataka. Indian J Agrofor 22(1) https://epubs.icar.org.in/index.php/IJA/article/view/101716

Kessler JJ (1992) The influence of karite' (Viteillaria paradoxa) and ne're' (Parkia biglobosa) trees on sorghum production in Burkina Faso. Agrofor Syst 17:97–118

Kho RM, Yacouba B, Yayé M et al (2001) Separating the effects of trees on crops: the case of *Faidherbia albida* and millet in Niger. Agrofor Syst 52:219–238. https://doi.org/10.1023/A:1011820412140

Kinama JM, Stigter CJ, Ong CK, Ng'ang'a JK, Gichuki FN (2007) Contour hedgerows and grass strips in erosion and runoff control on sloping land in semi-arid Kenya. Arid Land Res Manag 21(1):1–19. https://doi.org/10.1080/15324980601074545

Kumar A, Dwivedi GK, Tewari S, Paul J, Anand R, Kumar N, Kumar P, Singh H, Rajesh Kaushal R (2020) Carbon mineralization and inorganic nitrogen pools under Terminalia chebula Retz.-based agroforestry system in Himalayan foothills, India. For Sci 66(5):634–643. https://doi.org/10.1093/forsci/fxaa012

Kuyah S, Whitney CW, Jonsson M et al (2019) Agroforestry delivers a win-win solution for ecosystem services in sub-Saharan Africa. A meta-analysis. Agron Sustain Dev 39:47. https://doi.org/10.1007/s13593-019-0589-8

Lacombe S, Bradley RL, Hamel C, Beaulieu C (2009) Do tree-based intercropping systems increase the diversity and stability of soil microbial communities? Agriculture, Ecosystems & Environment 131(1–2):25–31., ISSN 0167-8809. https://www.sciencedirect.com/science/article/pii/S0167880908002442. https://doi.org/10.1016/j.agee.2008.08.010

Lagerlof J, Adolfsson L, Boerjesson G, Ehlers K, Vinyoles GP, Sundh I (2014) Land-use intensification and agroforestry in the Kenyan highland: impacts on soil microbial community composition and functional capacity. Appl Soil Ecol 82:93–99

Lal R (2023) Carbon farming by recarbonization of agroecosystems. Pedosphere 33(5):676–679., ISSN 1002-0160. https://doi.org/10.1016/j.pedsph.2023.07.024

Li XA, Ge TD, Chen Z, Wang SM, Ou XK, Wu Y, Chen H, Wu JP (2020) Enhancement of soil carbon and nitrogen stocks by abiotic and microbial pathways in three rubber-based agroforestry systems in Southwest China. Land Degrad Dev 2020(31):2507–2515. https://doi.org/10.1002/ldr.3625

Liang WL (2020) Effects of stemflow on soil water dynamics in forest stands. In: Levia DF, Carlyle-Moses DE, Iida S, Michalzik B, Nanko K, Tischer A (eds) Forest-water interactions. Ecological studies, vol 240. Springer, Cham. https://doi.org/10.1007/978-3-030-26086-6_15

Licht LA and Isebrands JG (2005). Linking phytoremediated pollutant removal to biomass economic opportunities, Biom Bioener, 28, Issue 2, 2005, pages 203–218. ISSN 0961-9534. https://www.sciencedirect.com/science/article/pii/S096195340400162X. doi:https://doi.org/10.1016/j.biombioe.2004.08.015

Llorente M, Moreno G (2019) Factors determining distribution and temporal changes in soil carbon in Iberian Dehesa. Geophysical Research Abstracts Vol. 21, EGU2019-1962, 2019 EGU General Assembly 2019. (Geophysical Research Abstracts Vol. 21, EGU2019-1962, 2019. EGU General Assembly 2019)

Liu C, Jin Y, Hu Y, Tang J, Xiong Q, Xu M, Bibi F and Beng KC (2019) Drivers of soil bacterial community structure and diversity in tropical agroforestry systems. Agriculture, Ecosystems & Environment, 278, 24–34. https://doi.org/10.1016/j.agee.2019.03.015

Lorenz K, Lal R (2014) Soil organic carbon sequestration in agroforestry systems. A review. Agron Sustain Develop 34(2):443–454. https://doi.org/10.1007/s13593-014-0212-yff.ffhal01234833f

Mahmud K, Makaju S, Ibrahim R, Missaoui A (2020) Current progress in nitrogen fixing plants and microbiome research. Plants (Basel). 9(1):97. https://doi.org/10.3390/plants9010097. PMID: 31940996; PMCID: PMC7020401

Manson S, Nekaris KAI, Rendell A, Budiadi B, Imron MA, Campera M (2022) Agrochemicals and shade complexity affect soil quality in coffee home gardens. Earth 3(3):853–865. https://doi.org/10.3390/earth3030049

Mantino A, Volpi I, Micci M, Pecchioni G, Bosco S, Dragoni F, Mele M, Ragaglini G (2020) Effect of tree presence and soil characteristics on soybean yield and quality in an innovative alley-cropping system. Agronomy. 10(1):52. https://doi.org/10.3390/agronomy10010052

Marsden C, Martin-Chave A, Cortet J et al (2020) How agroforestry systems influence soil fauna and their functions—a review. Plant Soil 453:29–44. https://doi.org/10.1007/s11104-019-04322-4
Masebo N, Menamo M (2016) A review paper on: the role of agroforestry for rehabilitation of degraded soil. J Biol Agric Healthc 6(5):1–6. http://www.iiste.org
Matos PS, Cherubin MR, Damian JM et al (2022a) Short-term effects of agroforestry systems on soil health in southeastern Brazil. Agrofor Syst 96:897–908. https://doi.org/10.1007/s10457-022-00749-4
Matos PS, da Silva CF, Pereira MG, da Silva EMR, Tarré RM, Custódio ALC, Franco EZ (2022b) Short-term modifications of mycorrhizal fungi, glomalin and soil attributes in a tropical agroforestry. Acta Oecologica 114:103815. https://doi.org/10.1016/j.actao.2022.103815. ISSN 1146-609X. https://www.sciencedirect.com/science/article/pii/S1146609X22000054
Mayer S, Wiesmeier M, Sakamoto E, Hübner R, Cardinael R, Kühnel A, Kögel-Knabner I (2022) Soil organic carbon sequestration in temperate agroforestry systems—a meta-analysis, agriculture. Ecosyst Environ 323, 107689., ISSN 0167-8809. https://doi.org/10.1016/j.agee.2021.107689
Mesfin S, Haileselassie H (2022) Evaluation of soil physicochemical properties as affected by canopies of scattered agroforestry trees on croplands. South Afr J Plant Soil 39:153–162
Minoli S, Montaño M, Castro J P, Zabala A, Climate Focus, Hodgetts (VCMI) G, Ortega Pacheco DV, Witkowski (IICA) K (2023) Carbon market opportunities in the agriculture sector in Latin America and the Caribbean. Bakhtary H and Streck C (Coordinating authors) Climate Focus: Report. https://vcmintegrity.org/wp-content/uploads/2023/12/BVE23119401i-1.pdf
Mohamed AG (2005) Improvement of traditional Acacia senegal agroforestry: Ecophysiological characteristics as indicators for tree-crop interaction on sandy soil in western Sudan. Ph.D. Thesis, University of Helsinki, Helsinki, Finland, 2005. TradiAcaciaSenegalAgroforestryImprovement.pdf
Molina Cuasapaz GP (2020) Erosion control and carbon sequestration by bench terraces and agroforestry in the west Usambara mountains, Tanzania. Master Thesis, Utrecht University, the Netherlands. https://studenttheses.uu.nl/handle/20.500.12932/36982
Montagnini F (ed) (2019) Integrating landscapes: agroforestry for biodiversity conservation and food sovereignty. Softcover ISBN 978-3-319-88765-4. https://doi.org/10.1007/978-3-319-69371-2
Mooney HA, Vitousek PM, Matson PA (1987) Exchange of materials between terrestrial ecosystems and the atmosphere. Science 238(4829):926–932. https://doi.org/10.1126/science.238.4829.926
Muchane MN, Sileshi GW, Gripenberg S, Jonsson M, Pumariño L, Barrios E (2020) Agroforestry boosts soil health in the humid and sub-humid tropics: a meta-analysis. Agric Ecosyst Environ 295:106899., ISSN 0167-8809. https://doi.org/10.1016/j.agee.2020.106899
Mukhlis I, Muhammad SR, Hidayah I (2022) Understanding socio-economic and environmental impacts of agroforestry on rural communities. Forests 13(4):556. https://doi.org/10.3390/f13040556
Murthy IK, Dutta S, Varghese V, Joshi PP, Kumar P (2016) Impact of agroforestry systems on ecological and socio-economic systems: a review. Glob J Sci Front Res: H Environ Earth Science 16. Issue 5 Version 1.0 Year 2016. Online ISSN: 2249-4626 & Print ISSN: 0975-5896. https://www.researchgate.net/publication/315668550_Impact_of_Agroforestry_Systems_on_Ecological_and_Socio-Economic_Systems_A_Review
Muthuri CW, Kuyah S, Njenga M, Kuria A, Öborn I, van Noordwijk M (2023) Agroforestry's contribution to livelihoods and carbon sequestration in East Africa: a systematic review. Trees For People 14:100432., ISSN 2666-7193. https://doi.org/10.1016/j.tfp.2023.100432
Nair PKR (1993) An introduction to agroforestry. Kluwer Academic publisher with cooperation ICRAF, Dordrecht/London
Nair PKR (2008) Agroecosystem management in the 21st century: it is time for a paradigm shift. Journal of Tropical Agriculture 46:1–12
Narendra BH (2008) Teak (Tectona grandis LF) growth planted using agroforestry model on pumice-mined land. In Proceedings of international seminar on 'research on plantation Forest management: challenges and opportunities', Centre for Plantation Forest Research and Development, Bogor, Indonesia, 2008

Nath AJ, Lal R, Das AK (2015) Ethnopedology and soil properties in bamboo (Bambusa sp.) based agroforestry system in North East India. CATENA 135:92–99., ISSN 0341-8162. https://www.sciencedirect.com/science/article/pii/S0341816215300564. https://doi.org/10.1016/j.catena.2015.07.001

Navarro-Pedreño J, Almendro-Candel MB, Zorpas AA (2021) The increase of soil organic matter reduces global warming, myth or reality? Sci 2021(3):18. https://doi.org/10.3390/sci3010018

Ndoli A, Baudron F, Schut AGT, Mukuralinda A, Giller KE (2017) Disentangling the positive and negative effects of trees on maize performance in smallholdings of northern Rwanda. Field Crop Res 213(2017):1–11. ISSN 0378-4290. https://www.sciencedirect.com/science/article/pii/S0378429017306883. https://doi.org/10.1016/j.fcr.2017.07.020

Neufeldt H, Wilkes A, Zomer RJ, Xu J, Nang'ole E, Munster C, Place F (2009) Trees on farms: tackling the triple challenges of mitigation, adaptation and food security, World Agroforestry Centre Policy Brief 07. World Agroforestry Centre, Nairobi

Ngaba MJY, Mgelwa AS, Gurmesa GA, Uwiragiye Y, Zhu F, Qiu Q, Fang Y, Hu B, Rennenberg H (2024) Meta-analysis unveils differential effects of agroforestry on soil properties in different zonobiomes. Plant Soil 2024(496):589–607. https://doi.org/10.1007/s11104-023-06385-w

Okigbo BN, Lal R (1985) Soil conservation and management in developing countries working paper no. 6; role of cover crops in soil and water conservation, FAO, United Nations, viadelle Termedi Caracalla, Rome, Italy

Padovan MDP, Brook RM, Barrios M, Cruz-Castillo JB, Vilchez-Mendoza SJ, Costa AN, Rapidel B (2018) Water loss by transpiration and soil evaporation in coffee shaded by Tabebuia rosea Bertol and Simarouba glauca dc compared to unshaded coffee in sub-optimal environmental conditions. Agric For Meteor 248:1–14. https://doi.org/10.1016/j.agrformet.2017.08.036

Panda S, Dhara PK, Sarkar S, Das NC (2018) Chapter 35: Agroforestry—a multipurpose multistoreyed renewable plant treasure with medicinal values giving multireturn for livelihood and benefits over climate change. In book: In: Pandey C B, Gaur M K and Goyal R K (Ed.) Climate change and agroforestry adaptation, mitigation and livelihood security 1 : pp. 627 to 637. Publisher: nipa New India Publishing Agency, New Delhi . (https://www.researchgate.net/publication/324947113_Agroforestry_-_a_multipurpose_multistoreyed_renewable_plant_treasure_with_medicinal_values_giving_multireturn_for_livelihood_and_benefits_over_climate_change)

Pardon P, Mertens J, Reubens B et al (2020) Juglans regia (walnut) in temperate arable agroforestry systems: effects on soil characteristics, arthropod diversity and crop yield. Renew Agric Food Syst 35(5):533–549. https://doi.org/10.1017/S1742170519000176

Parwada C, Chipomho J, Mapope N, Masama E, Simango K (2022) Role of agroforestry on farmland productivity in semi-arid farming regions of Zimbabwe. Res World Agric Econ 3(2):39–47. https://doi.org/10.36956/rwae.v3i2.515

Pavlidis G, Tsihrintzis V (2017) Pollution control by agroforestry systems: a short review. Eur Water 2017(59):297–301. https://ewra.net/ew/pdf/EW_2017_59_40.pdf

Pavlidis G, Tsihrintzis VA, Karasali H, Alexakis D (2018). Tree uptake of excess nutrients and herbicides in a maize-olive tree cultivation system. J Environ Sci Health A Tox Hazard Subst Environ Eng 2018;53(1):1–12. doi: https://doi.org/10.1080/10934529.2017.1366241. Epub 2017 Oct 20

Pavlidis G, Karasali H, Tsihrintzis VA (2020) Pesticide and fertilizer pollution reduction in two alley cropping agroforestry cultivating systems. Wat Air Soil Pollut 231(5):241. https://doi.org/10.1007/s11270-020-04590-2

Pavlidis G, Karasali H, Tsihrintzis VA (2021) Dynamics of changes in the concentrations of herbicides and nutrients in the soils of a combined wheat-poplar tree cultivation: a field experimental model during the growing season. Agrofor Syst 95(2):321–338. https://doi.org/10.1007/s10457-020-00586-3

Purwaningsih R, Sartohadi J, Setiawan MA (2020) Trees and crops arrangement in the agroforestry system based on slope units to control landslide reactivation on volcanic foot slopes in Java, Indonesia. Land. 9(9):327. https://doi.org/10.3390/land9090327

Quandt A, Neufeldt H, Gorman K (2023) Climate change adaptation through agroforestry: opportunities and gaps. Curr Opin Environ Sustain 60:101244., ISSN 1877-3435. https://doi.org/10.1016/j.cosust.2022.101244

Ramirez MAJP, Roberto G, Visco RG, Predo CD, Galang MA (2022) Assessment of soil condition using soil quality index of different land use types in Liliw, Laguna, Philippines. Phil J Sci 151(3):1157–1169. ISSN 0031–7683. https://doi.org/10.56899/151.03.29

Rao MR, Nair PK, Ong CK (1998) Biophysical interactions in tropical agroforestry systems. Agrofor Syst 38(1):3–50. https://doi.org/10.1023/A:1005971525590

Ravi P, Barrios E, Bayala J, Diby L, Donovan J, Gyau A, Graudal L, Jamnadass R, Kahia J, Kehlenbeck K (2015) Agroforestry: realizing the promise of an agroecological approach. Agroecology for food security and nutrition, proceedings of the FAO international symposium, 18–19 September 2014, Rome, Italy. Food and Agriculture Organization of the United Nations (FAO) https://www.researchgate.net/publication/284609506_Agroforestry_realizing_the_promise_of_an_agroecological_approach

Rivest D, Martin-Guay MO, Cossette C (2022) Willows rapidly affect microclimatic conditions and forage yield in two temperate short-rotation agroforestry systems. Agrofor Syst 96(7):1–13. https://doi.org/10.1007/s10457-022-00758-3

Riyadh ZA, Rahman MA, Saha SR, Hossain MI (2018) Soil properties under jackfruit based agroforestry systems in Madhupur tract of Narsingdi district. J Sylhet Agril Univ 5(2):173–179. ISSN: 2308-1597. https://jsau.sau.ac.bd/wp-content/uploads/2021/01/10.-JSAU-V2N2-JD-078-Zabid2-11.pdf

Rizwan M, Rauf A, Rahmawaty, Akub EN (2020) Changes in soil chemical properties in agroforestry system at USU Arboretum, Kwala Bekala. In IOP Conference Series: Earth and Environmental Science 425 (2020) 012054. The 1st International Conference on Agriculture and Bioindustry 2019: IOP Publishing Ltd. https://doi.org/10.1088/1755-1315/425/1/012054

Rodríguez L, Suárez JC, Rodriguez W, Artunduaga KJ, Lavelle P (2021) Agroforestry systems impact soil macroaggregation and enhance carbon storage in Colombian deforested Amazonia. Geoderma 384:114810., ISSN 0016-7061. https://www.sciencedirect.com/science/article/pii/S0016706120325659. https://doi.org/10.1016/j.geoderma.2020.114810

Rolo V, Rivest D, Maillard É, Moreno G (2023) Agroforestry potential for adaptation to climate change: a soil-based perspective. Soil Use Manag 39(3):1006–1032. https://doi.org/10.1111/sum.12932

Rosati A, Borek R, Canali S (2021) Agroforestry and organic agriculture. Agrofor Syst 95:805–821

Sahoo G, Wani A M, Sharma A (2020) Int J Curr Microbiol App Sci. ISSN: 2319-7706. Special Issue-11: 2001–2020. Journal homepage: http://www.ijcmas.com

Salceda M, Udawatta RP, Nelson KA, Mendis SS, Bardhan S (2021) Spatial and temporal variability of soil organic carbon on a corn–soybean watershed with 23 years of agroforestry. Agron J 1–12:440. https://doi.org/10.1002/agj2.20948

Salve A, Bhardwaj DR (2020) Soil carbon stock and nutrient study in different agroforestry systems at Kinnaur district. Himachal Pradesh Plant Arch 2020(20):4251–4260

Saputra DD, Sari RR, Hairiah K et al (2020) Can cocoa agroforestry restore degraded soil structure following conversion from forest to agricultural use? Agrofor Syst 94:2261–2276. https://doi.org/10.1007/s10457-020-00548-9

Sarmiento-Soler A, Vaast P, Hoffmann MP, Rötter RP, Jassogne L, Van Asten PJ, Graefe S (2019) Water use of Coffea arabica in open versus shaded systems under smallholder's farm conditions in eastern Uganda. Agric For Meteor 266-267(2019):231–242. https://doi.org/10.1016/j.agrformet.2018.12.006

Sarveswaran S, Johar V, Singh V, Raghunathan C (2023) Agroforestry: a way forward for sustainable development. Ecol Environ Conserv 29 (August Suppl. Issue) 023:S300–S309. https://doi.org/10.53550/eec.2023.v29i04s.046

Schroth G, Burkhard J (2003) Chap 9: Nutrient exchange with the atmosphere. In: Schroth G, Sinclair FL (eds) Trees, crops and soil fertility: concepts and research methods. CABI Publishing, Wallingford, pp 181–189

Seiter S, Ingham EI, Ray DW (1999) Dynamics of soil fungal and bacterial biomass in a temperate climate alley cropping system. App Soil Ecol 12(2):139–147., ISSN 0929-1393. https://www.sciencedirect.com/science/article/pii/S0929139398001632. https://doi.org/10.1016/S0929-1393(98)00163-2

Sekhar M, Lallawmkimi M, Doppalapudi R, Upadhyay L, Tiwari P, Verma U, Rajput A (2024) Agroforestry practices and their impact on soil health and fertility: a review. J Exp Agric Int 46:511–528. https://doi.org/10.9734/jeai/2024/v46i92849

Sh F, Chavan S, Chichaghare A, Uthappa AR, Kumar M, Kakade V, Pradhan A, Jinger D, Rawale G, Yadav D, Kumar V, Farooq T, Ali B, Sawant AV, Saud S, Chen S, Poczai P (2022) Agroforestry systems for soil health improvement and maintenance. Sustain For 14. https://doi.org/10.3390/su142214877

Shanmughavel P, Peddappaiah RS, Muthukumar T (2000) Litter production and nutrient return in Bambusa bambos plantation. J Sustain For 11(3):71–82. https://doi.org/10.1300/J091v11n03_04

Shi L, Feng W, Xu J, Kuzyakov Y (2018) Agroforestry systems: meta-analysis of soil carbon stocks, sequestration processes, and future potentials. Land Degrad Dev 29(11):3886–3897. https://doi.org/10.1002/ldr.3136

Shoga'a Aldeen AH, Majid NM, Azani AM, Ghani ANA, Mohamed S (2013) Agroforestry impacts on soil fertility in the Rima'a Valley, Yemen. J Sustain For 32(3):286–309. https://doi.org/10.1080/10549811.2012.654723

Sileshi GW (2016) The magnitude and spatial extent of influence of Faidherbia albida trees on soil properties and primary productivity in drylands. J Arid Environ 132:1–14. https://doi.org/10.1016/j.jaridenv.2016.03.002. ISSN 0140-1963. https://www.sciencedirect.com/science/article/pii/S0140196316300283

Sileshi G W, Mafongoya P L, Akinnifesi F K, Phiri E, Chirwa P, Beedy T, Makumba W, Nyamadzawo G, Njoloma J, Wuta M, Nyamugafata P, and Jiri O (2014) Agroforestry: fertilizer trees. In: Neal Van Alfen, editor-in-chief. Encyclopedia of agriculture and food systems, Vol. 1. Elsevier, San Diego, pp 222–234

Sirohi C, Bangarwa KS (2017) Effect of different Spacings of poplar-based agroforestry system on soil chemical properties and nutrient status in Haryana, India. Curr Sci 113:1403–1407. https://doi.org/10.18520/cs/v113/i07/1403-1407

Sirohi C, Bangarwa KS, Dhillon RS, Chavan SB, Handa AK (2022) Productivity of wheat (*Triticum aestivum* L.) and soil fertility with poplar (*Populus deltoides*) agroforestry system in the semi-arid ecosystem of Haryana, India. Curr Sci 122:1072

Smith J, Pearce BD, Wolfe MS (2012) A European perspective for developing modern multifunctional agroforestry systems for sustainable intensification. Renew Agric Food Syst 27:323–332

Sonmez O, Turan V, Kaya C (2016) The effects of sulfur, cattle, and poultry manure addition on soil phosphorus. Turk J Agric For 2016(40):536–541. https://doi.org/10.3906/tar-1601-41. https://journals.tubitak.gov.tr/agriculture/vol40/iss4/6/

Steppler HA, Nair PKR (1987) Agroforestry a decade of development. International Council for Research in Agroforestry (ICRAF), Nairobi. ISBN 92 9059 036 X

Sudomo A, Leksono B, Tata HL, Rahayu AAD, Umroni A, Rianawati H, Asmaliyah K, Setyayudi A, Utomo MMB et al (2023) Can agroforestry contribute to food and livelihood security for Indonesia's smallholders in the climate change era? Agriculture. 13(10):1896. https://doi.org/10.3390/agriculture13101896

Suprayogo D, van Noordwijk M, Hairiah K, Meilasari N, Rabbani AL, Ishaq RM, Widianto W (2020) Infiltration-friendly agroforestry land uses on volcanic slopes in the Rejoso Watershed, East Java, Indonesia. Land. 9(8):240. https://doi.org/10.3390/land9080240

Surki AA, Nazari M, Fallah S et al (2020) The competitive effect of almond trees on light and nutrients absorption, crop growth rate, and the yield in almond–cereal agroforestry systems in semi-arid regions. Agrofor Syst 94:1111–1122. https://doi.org/10.1007/s10457-019-00469-2

Swamila M, Philip D, Akyoo AM, Sieber S, Bekunda M, Kimaro AA (2020) Gliricidia agroforestry technology adoption potential in slected dryland areas of Dodoma Region, Tanzania. Agriculture. 10(7):306. https://doi.org/10.3390/agriculture10070306

Temitope OO, Thonda OA (2016) Comparative study of the antibacterial and antifungal spectrum, phytochemical screening and antioxidant potentials of Alchornea laxiofolia and Piliostigma reticulatum leaf on pathogenic isolates. Pharmaceut Chem J 3(2):1–11. ISSN: 2349-7092. Online: www.tpcj.org

Thiesmeier A, Zander P (2023) Can agroforestry compete? A scoping review of the economic performance of agroforestry practices in Europe and North America. For Policy Econ 150:102939., ISSN 1389-9341. https://doi.org/10.1016/j.forpol.2023.102939

Torralba M, Fagerholm N, Burgess PJ, Moreno G, Plieninger T (2016) Do European agroforestry systems enhance biodiversity and ecosystem services? A meta-analysis. Agric Ecosyst Environ 230:150–161

Tsonkova P, Quinkenstein A, Böhm C, Freese D, Schaller E (2014) Ecosystem services assessment tool for agroforestry (ESAT-A): an approach to assess selected ecosystem services provided by alley cropping systems. Ecologi Indicat 2014:285–299., ISSN 1470-160X. https://www.sciencedirect.com/science/article/pii/S1470160X14001745. https://doi.org/10.1016/j.ecolind.2014.04.024

Tsufac AR, Awazi NP, Yerima BPK (2021a) Characterization of agroforestry systems and their effectiveness in soil fertility enhancement in the south-west region of Cameroon. Curr Res Environ Sustain 3:100024., ISSN 2666-0490. https://www.sciencedirect.com/science/article/pii/S2666049020300372. https://doi.org/10.1016/j.crsust.2020.100024

Tsufac AR, Awazi NP, Yerima BPK (2021b) Inorganic fertilizer use among agroforestry and non-agroforestry practicing farmers in South West Cameroon: a comparative assessment and policy ramifications. Afr J Agric Res 17(1):137–146. https://doi.org/10.5897/AJAR2020.15262. https://academicjournals.org/journal/AJAR/article-abstract/FB5138265849

Tsufac AR, Awazi NP, Tchamba MN (2022) Agroforestry as an agro-ecological pathway to phase out chemical fertilizer application in smallholder farms in Cameroon: state-of-the-art and policy implications. Effects Curr World Environ 17(1). https://doi.org/10.12944/CWE.17.1.9

Udawatta RP, Krstansky JJ, Henderson GS, Garrett HE (2002) Agroforestry practices, runoff, and nutrient loss: a paired watershed comparison. J Environ Qual 31:1214–1225

Udawatta RP, Garrett HE, Kallenbach RL (2010) Agroforestry and grass buffer effects on water quality in grazed pastures. Agroforest Syst 79:81–87. https://doi.org/10.1007/s10457-010-9288-9

Uthappa AR, Chavan S, Handa AK, Newaj R, Kumar D, Sridhar KB, Chaturvedi OP (2017) Chap 1: Agroforestry—a sustainable solution to address climate change challenges. In: Gupta SK, Panwar P, Kaushal R (eds) Agroforestry for increased production and livelihood security. ISBN 978-93-85516-76-4. New India Publishing Agency (NIPA), New Delhi, pp 1–22

van Noordwijk M, Ekadinata A, Leimona B, Catacutan D, Martini E, Tata HL, Öborn I, Hairiah K, Wangpakapattanawong P, Mulia R, Dewi S, Rahayu S, Zulkarnain MT (2020) Agroforestry options for degraded landscapes in Southeast Asia. Chapter 11. In: Dagar JC, Teketay D (eds) Agroforestry for degraded landscapes: recent advances and emerging challenges. Springer, Singapore. ISBN:978-981-15-4135-3. https://cdn.forru.org/publication-files/forru-0000164-0001-en.pdf

Van Stan JT, Lewis ES, Hildebrandt A, Rebmann C, Friesen J (2016) Impact of interacting bark structure and rainfall conditions on stemflow variability in a temperate beech-oak forest, Central Germany. Hydrol Sci J 61(11):2071–2083. https://doi.org/10.1080/02626667.2015.1083104

Vandermeer JH (1989) The ecology of intercropping. Cambridge University Press, Cambridge. 237 p. Online ISBN: 9780511623523. https://doi.org/10.1017/CBO9780511623523

Vignozzi N, Agnelli AE, Brandi G, Gagnarli E, Goggioli D, Lagomarsino A, Pellegrini S, Simoncini S, Simoni S, Valboa G, Caruso G, Gucci R (2019) Soil ecosystem functions in a high-density olive orchard managed by different soil conservation practices. Appl Soil Ecol 134(2019):64–76. ISSN 0929-1393. https://www.sciencedirect.com/science/article/pii/S0929139318303834. https://doi.org/10.1016/j.apsoil.2018.10.014

Wang Y, Zhang B, Lin L, Zepp H (2011) Agroforestry system reduces subsurface lateral flow and nitrate loss in Jiangxi Province, China. Agric Ecosyst Environ 140(3–4):441–453. ISSN 0167-8809. https://www.sciencedirect.com/science/article/pii/S0167880911000089. https://doi.org/10.1016/j.agee.2011.01.007

Wartenberg AC, Blaser WJ, Roshetko JM et al (2020) Soil fertility and Theobroma cacao growth and productivity under commonly intercropped shade-tree species in Sulawesi, Indonesia. Plant Soil 453:87–104. https://doi.org/10.1007/s11104-018-03921-x

Wilson MH, Lovell ST (2016) Agroforestry—the next step in sustainable and resilient agriculture. Sustainability 8(6):574. https://doi.org/10.3390/su8060574

Xiao T, Li P, Fei W, Wang J (2024) Effects of vegetation roots on the structure and hydraulic properties of soils: a perspective review. Sci Total Environ 906:167524. https://doi.org/10.1016/j.scitotenv.2023.167524. ISSN 0048-9697. https://www.sciencedirect.com/science/article/pii/S004896972306151X

Yadav RS, Yadav BL, Chhipa BR, Dhyani SK, Ram M (2011) Soil biological properties under different tree based traditional agroforestry systems in a semi-arid region of Rajasthan. India Agrofor Syst 2011(81):195–202. https://doi.org/10.1007/s10457-010-9277-z

Yengwe J, Amalia O, Lungu OI, De Neve S (2018) Quantifying nutrient deposition and yield levels of maize (Zea mays) under Faidherbia albida agroforestry system in Zambia. Eur J Agron 2018(99):148–155. https://doi.org/10.1016/j.eja.2018.07.004

Young A (1989) Agroforestry for soil management. CAB international with cooperation ICRAF, Nairobi

Zake J, Pietsch S, Friedel J, Zechmeister-Boltenstern S (2015) Can agroforestry improve soil fertility and carbon storage in smallholder banana farming systems? J Plant Nutr Soil Sci 178(2):237–249. https://doi.org/10.1002/jpln.201400281

Zhang D, Du G, Sun Z, Bai W, Wang Q, Feng L, Zheng J, Zhang Z, Liu Y, Yang S, Yang N, Feng C, Cai Q, Evers JB, van der Werf W, Zhang L (2018) Agroforestry enables high efficiency of light capture, photosynthesis and dry matter production in a semi-arid climate. Eur J Agron 94:1–11., ISSN 1161-0301. https://www.sciencedirect.com/science/article/pii/S1161030118300017. https://doi.org/10.1016/j.eja.2018.01.001

Zhang X, Gao G, Wu Z, Wen X, Zhong H, Zhong Z, Bian F, Gai X (2019) Agroforestry alters the rhizosphere soil bacterial and fungal communities of moso bamboo plantations in subtropical China. Appl Soil Ecol 143(2019):192–200., ISSN 0929-1393. https://www.sciencedirect.com/science/article/pii/S0929139319303178. https://doi.org/10.1016/j.apsoil.2019.07.019

Zhu X, Chen C, Wu J, Yang J, Zhang W, Zou X, Liu W, Jiang X (2019) Can intercrops improve soil water infiltrability and preferential flow in rubber-based agroforestry system? Soil Tillage Res 191:327–339

Zhu M, Cao X, Guo Y, Shi S, Wang W, Wang H (2022) Soil P components and soil fungi community traits in poplar shelterbelts and neighbouring farmlands in northeastern China: total alterations and complex associations. Catena 2022(218):106531. https://doi.org/10.1007/s10457-020-00548-9

Zhu Q, Yan K, Dong Y et al (2023a) Rhizosphere bacterial communities and soil nutrient conditions reveal sexual dimorphism of Populus deltoides. J For Res 34:761–771. https://doi.org/10.1007/s11676-022-01517-x

Zhu X, Yuan X, Lu E, Yang B, Wang H, Du Y, Singh AK, Liu W (2023b, 2023) Soil splash erosion: an overlooked issue for sustainable rubber plantation in the tropical region of China. Int Soil Wat Conserv Res 30-42(11, 1). https://doi.org/10.1016/j.iswcr.2022.05.005. ISSN 2095-6339. https://www.sciencedirect.com/science/article/pii/S2095633922000399

Zinngrebe Y, Borasino E, Chiputwa B et al (2020) Agroforestry governance for operationalising the landscape approach: connecting conservation and farming actors. Sustain Sci 15:1417–1434. https://doi.org/10.1007/s11625-020-00840-8

Zuazo VHD, Pleguezuelo CRR, Tavira SC, Martinez JRF (2014) Linking soil organic carbon stocks to land-use types in a Mediterranean agroforestry landscape. J Agric Sci Technol 16:667–679

Agroforestry: Improving Soil Health and Healthy Food Production

5

Abstract

Agroforestry, by integrating trees with crops and livestock, enhances soil health, improves ecosystem functions, and helps as buffer against extreme weather events like droughts and floods. It boosts soil properties like water retention, nutrient cycling, and erosion resistance, leading to stable yields, increased biodiversity, and carbon sequestration. Wide adaptability of agroforestry makes it a versatile solution for diverse climates, promoting both environmental conservation and food security. However, research gaps, also in high-income countries, remain the bottleneck in valuing the socio-economic impacts of agroforestry. To fully realize its potential, large scale studies and standardized monitoring are needed, alongside stronger collaboration between governments, policymakers, and farmers. Agroforestry offers a powerful tool to address soil degradation, biodiversity loss, and food insecurity amid growing climate challenges.

Keywords

Agroforestry · Trees · Crops · Livestock · Epedaphic biota · Euedaphic biota · Molt · Socio-economic value · Agrochemicals · Soil health · Human health · Healthy food · Food production · Soil as living skin · Ecosystem functions · Extreme weather · Droughts · Floods · Soil properties · Water cycle · Water retention · Nutrient cycling · Eutrophication · Erosion resistance · Particulate matter · Yields · Soil life · Soil biodiversity · Soil organisms · Soil foodweb · Soil microbes · Disease · Soil borne pathogens · Anthrax · Helminths · Allergies · Antibiotics · Antihelminth resistance · Fungi in stress tolerance · Biological control · Human health · Climate change · Biodiversity · Trophic levels · Land degradation · Agricultural landscapes · Carbon sequestration · Environmental conservation · Food security · High income countries · Governments · Policymakers · Farmers · Soil degradation · Biodiversity loss · Air quality · Water quality · Climate challenges

S. Panda, *Agroforestry*, https://doi.org/10.1007/978-981-96-6855-7_5

Abbreviations

AF: Agroforestry
AFS: Agroforestry system
CBD: Convention on Biological Diversity
CEC: Cation exchange capacity
CO_2: Carbon dioxide
DNA: Deoxyribonucleic acid
FAO: Food and Agriculture Organization
HICs: High income countries
ITPS: Intergovernmental Technical Panel on Soils
PM: Particulate matter
SOM: Soil organic matter
SWSR: Status of the World's Soil Resources
WHO: World Health Organization

5.1 Role of Agroforestry on Soil Health and Healthy Food Production: An Introduction

Soil health is often overlooked, despite its crucial role in sustaining life on Earth. Soil, often described as the planet's "living skin," is not just a passive medium for plant growth but an active, dynamic system that sustains ecosystems, regulates the climate, and supports human food production. The increasing challenges posed by climate change, biodiversity loss, and land degradation have underscored the importance of soil health, as it is deeply intertwined with global efforts to mitigate environmental degradation, enhance biodiversity, and promote sustainable food systems. In this context, agroforestry, an integrated approach that combines trees, crops, and livestock, has emerged as a powerful strategy for improving soil health and supporting healthy food production. The prime cause of this is the intricate relationship between soil health, carbon sequestration, and the broader suite of ecosystem services that soil provides. Healthy soils are foundational base not only for agricultural productivity but also for regulating water cycles, supporting biodiversity, and storing carbon. Agroforestry practices, which blend tree cultivation with crop and livestock production, can enhance soil structure, fertility, and carbon storage, while also contributing to biodiversity conservation and resilience in farming systems. By incorporating trees into agricultural landscapes, agroforestry provides multiple benefits that improve both the ecological health of the soil and the sustainability of food production systems.

Soil health is determined by a combination of factors, including its texture, mineral composition, and organic matter content. These factors influence key soil functions such as nutrient cycling, water retention, erosion control, and carbon sequestration. Healthy soils are teeming with life, spanning from microorganisms to earthworms, which contribute to nutrient cycling and organic matter decomposition, ensuring that crops receive the nutrients they need for growth. Moreover, soils rich in organic matter have better structure, promoting water infiltration and retention, which helps prevent soil erosion and improves resilience to droughts and floods.

Agroforestry practices contribute to soil health by increasing organic matter input and creating a more diverse, resilient soil ecosystem. The deep-rooted trees in agroforestry systems help improve soil structure and prevent erosion, while their leaflitter provides organic material that enriches the soil. This, in turn, supports the growth of crops and enhances soil fertility. In addition, trees have been shown to promote better nutrient cycling and enhance the soil's water-holding capacity, further contributing to its overall health.

In the fight against climate change, the role of soil carbon sequestration has gained significant attention. Soils, particularly those rich in organic matter, act as important carbon sinks, storing carbon dioxide (CO_2) from the atmosphere. Through the process of carbon sequestration, soil captures and locks away carbon in the form of soil organic matter (SOM), potentially for decades to centuries. Agroforestry systems are uniquely positioned to enhance this process. Trees, by virtue of their deep roots and high biomass production, capture more carbon than conventional agricultural systems, and their presence in farming landscapes increases the soil's capacity to store carbon.

The synergies between agroforestry and soil carbon sequestration are profound. Agroforestry systems not only increase soil organic carbon but also promote the growth of soil organisms that help break down organic material, further enriching the soil and creating a positive feedback loop for carbon storage. This dual benefit, improving soil health and increasing carbon sequestration, makes agroforestry a key strategy for addressing climate change. Additionally, practices such as agroforestry contribute to soil conservation by reducing erosion, which is particularly important in areas vulnerable to desertification and land degradation.

Beyond carbon sequestration, agroforestry systems provide a suite of ecosystem services that are critical to the health of the land and the people who depend on it. These services include nutrient cycling, water filtration, biodiversity support, and pest regulation. Trees in agroforestry systems help maintain biodiversity by creating habitat for a variety of species, from soil microorganisms to pollinators and wildlife. By fostering a more diverse ecosystem, agroforestry helps ensure the long-term sustainability of agricultural systems and enhances the resilience of farming communities to climate related shocks.

Agroforestry also promotes food security by diversifying the types of crops grown and by increasing overall farm productivity. The combination of trees, crops, and livestock can lead to more stable yields and reduce vulnerability to price fluctuations or crop failure, thus providing farmers with more reliable sources of food and income. Furthermore, the integration of diverse plant species and animals in agroforestry systems reduces the risk of pest outbreaks and plant diseases, thereby promoting more sustainable food production without the need for excessive use of chemical fertilizers or pesticides.

The interconnectedness between soil health, agroforestry, and climate resilience is critical to understanding how agroforestry can contribute to sustainable land management. As climate change intensifies, the need for resilient agricultural systems that can withstand extreme weather events, such as droughts, floods, and heatwaves, has never been more urgent. Healthy soils, enriched by agroforestry practices, are better equipped to handle such stresses. By improving water retention, enhancing biodiversity, and storing carbon, agroforestry helps buffer the impacts of climate variability and promotes long-term ecological stability.

Moreover, agroforestry offers a practical and adaptable approach to land management that can be tailored to different regions, climates, and agricultural practices. Whether in tropical, temperate, or arid zones, agroforestry systems can be designed to meet local needs and conditions, making them a versatile tool for both farmers and conservationists.

The integration of agroforestry into agricultural landscapes holds immense potential for improving soil health, enhancing food security, and combating climate change. Agroforestry offers a holistic approach to land management that acknowledges the interconnectedness of soil, trees, crops, and ecosystems. By improving soil health, sequestering carbon, enhancing biodiversity, and promoting sustainable food production, agroforestry systems can help address some of the most pressing environmental and societal challenges of our time.

However, realizing the full potential of agroforestry will require a concerted effort to promote sustainable land management practices, strengthen agricultural policies, and invest in research and education. As the global community works toward building more resilient and sustainable food systems, agroforestry stands out as a promising solution that supports the health of the soil, the planet, and future generations (Qaiser et al. 2016, Shukla et al. 2018).

5.2 Soil Health: The Basis of Agroforestry and Sustainable Agriculture

Resilience to climate change: It refers to the ability of soil to function effectively in supporting plant growth, regulating water, cycling nutrients, and maintaining ecosystem services. Healthy soils are teeming with biodiversity, supporting an intricate web of organisms, from bacteria and fungi to earthworms and larger organisms, that contribute to nutrient cycling, organic matter decomposition, and overall soil fertility. The concept of soil health is not just about maintaining fertility but also about creating a resilient, sustainable ecosystem that benefits both agriculture and the broader environment.

5.2.1 Factors Determining Soil Health

Soil health is influenced by several key factors, each of which plays a role in the soil's ability to support plant growth and maintain ecological balance:

1. **Soil texture:** The relative proportions of sand, silt, and clay in the soil determine its texture. Soils with an ideal balance, often called loam, are typically the best

for growing crops, as they allow for adequate water infiltration and root penetration. Soils that are too sandy may drain too quickly, while those that are too clayey may become compacted and waterlogged.

2. **Mineral composition:** The mineral content of the soil, including essential macro and micronutrients (such as nitrogen, phosphorus, potassium, calcium, and magnesium), is vital for plant growth. Soil minerals also affect its pH, which influences nutrient availability to plants. For example, soils with too much acidity or alkalinity may limit nutrient uptake.
3. **Soil organic matter (SOM):** SOM, decomposed plant and animal residues, enhances soil structure, water retention, and nutrient availability. It promotes the activity of soil organisms like fungi, bacteria, and earthworms, all of which contribute to nutrient cycling and organic matter decomposition. Organic matter also improves cation exchange capacity (CEC) of soil, which is essential for retaining and releasing nutrients to plants.

5.2.2 Key Soil Functions for Agricultural Productivity

Healthy soils play a fundamental role in supporting the sustainability of agriculture through several essential functions:

1. **Nutrient Cycling:** Soil organisms decompose organic matter, releasing nutrients into the soil in a form that plants can absorb. This continuous cycling of nutrients helps maintain soil fertility over time, reducing the need for synthetic fertilizers and minimizing nutrient loss.
2. **Water retention and infiltration:** Healthy soils are better able to retain water, which is crucial for crops, especially during periods of drought. Soils rich in organic matter improve water infiltration and reduce runoff, allowing rainwater to percolate deeper into the soil and recharge groundwater supplies.
3. **Erosion control:** Soil structure plays a critical role in preventing erosion. Well aggregated soils are less prone to erosion by wind and water. Healthy soils, with high organic matter content, are more stable and able to resist physical disruption, thus protecting the land from degradation.
4. **Carbon sequestration:** Soils are an important carbon sink, storing more carbon than the atmosphere and vegetation combined. Organic matter in the soil can sequester carbon for long periods, helping mitigate climate change by reducing the amount of carbon dioxide in the atmosphere.

5.2.3 The Role of Agroforestry in Soil Health

Agroforestry, the practice of integrating trees and shrubs into agricultural systems, plays a pivotal role in enhancing soil health and promoting sustainable farming. By combining trees with crops and/or livestock, agroforestry creates a more diverse, resilient, and productive farming ecosystem. Generally, six roles of AF are identified as:

1. **Improvement of soil structure:** Trees in agroforestry systems have deep and extensive root systems that help break up compacted layers of soil, enhancing soil aeration and root penetration. This improves soil structure, making it easier for crops to grow and for water to infiltrate.
2. **Increased organic matter:** The leaf litter and organic material deposited by trees provide valuable inputs of organic matter into the soil. As these organic materials decompose, they enhance soil fertility, structure, and moisture retention. Over time, this input of organic matter helps maintain healthy soil by replenishing nutrients and supporting soil organisms.
3. **Erosion prevention:** The roots of trees anchor the soil, preventing erosion by stabilizing the topsoil. In agroforestry systems, trees act as a natural barrier to wind and water erosion, which is particularly important on slopes or in areas prone to heavy rainfall.
4. **Enhanced nutrient cycling:** Trees in agroforestry systems contribute to nutrient cycling by fixing nitrogen (in the case of nitrogen-fixing species like legumes) and enhancing soil nutrient availability through leaf litter and root decay. This creates a more balanced and sustainable nutrient cycle, benefiting both trees and crops.
5. **Improved water holding capacity:** Trees can increase the water holding capacity of the soil by reducing surface runoff and increasing infiltration. The deeper roots of trees can also tap into water sources unavailable to shallow-rooted crops, providing a buffer during dry periods and improving the resilience of the entire farming system to droughts.
6. **Biodiversity support:** Agroforestry systems foster biodiversity by providing habitats for a variety of plants, animals, and beneficial microorganisms. A diverse soil ecosystem is key to maintaining soil health, as different species contribute to nutrient cycling, pest control, and organic matter decomposition.

5.2.4 The Synergy Between Agroforestry and Sustainable Agriculture

Agroforestry contributes to the overall sustainability of agricultural systems by improving soil health in ways that conventional monoculture farming cannot. Monoculture farming often depletes soil health due to a lack of crop diversity, soil erosion, and overreliance on chemical fertilizers and pesticides. In contrast, agroforestry systems encourage natural processes that maintain soil fertility, enhance biodiversity, and protect against environmental degradation. By promoting healthier soils, agroforestry can reduce the need for external inputs like chemical fertilizers and pesticides, making agricultural systems more economically viable and environmentally friendly. Furthermore, agroforestry systems can help farmers adapt to climate change by improving soil resilience to extreme weather events, such as droughts and floods.

5.3 Ecosystem Services of Agroforestry in Enhancing Biodiversity and Food Security

Agroforestry, the integration of trees, crops, and/or livestock in a single farming system, offers far reaching benefits beyond carbon sequestration. By supporting a range of ecosystem services, agroforestry plays a critical role in enhancing biodiversity, improving soil health, and promoting food security. This is important to note the key ecosystem services provided by agroforestry, their contribution to biodiversity conservation, and their potential in ensuring food security in agricultural landscapes (Table 5.1).

Key ecosystem services of agroforestry: Generally, three identified ecosystem services of agroforestry are (1) Nutrient cycling, (2) Water filtration and regulation, and (3) Biodiversity support.

Table 5.1 Deep root systems and nutrient uptake by agroforestry trees in nutrient cycling[a]

Observations	Details
Tree root depth	Agroforestry trees can have deep root systems that penetrate several meters into the soil, which helps in nutrient cycling and stabilization
Nutrient uptake	Deep root systems allow agroforestry trees to access nutrients from deeper soil layers, including nitrogen, phosphorus, potassium, and micronutrients
Nutrient cycling	Agroforestry trees contribute to nutrient cycling by bringing up nutrients from deeper layers into the topsoil where crops can access them
Improvement of soil fertility	Deep-rooted trees can improve soil fertility through the deposition of organic matter (leaf litter, root turnover), which enriches the soil with nutrient.
Soil organic matter (SOM)	Agroforestry trees enhance SOM, which improves soil structure, water retention, and nutrient availability
Ecosystem services	Deep-rooted trees contribute to soil erosion control, water retention, and biodiversity, promoting long-term soil health and productivity
Soil nutrient redistribution	Agroforestry trees can help redistribute nutrients that are otherwise out of reach for shallow-rooted crops, contributing to more balanced soil nutrient availability
Examples of agroforestry trees	Trees like *Gliricidia sepium*, *Acacia mangium*, and *Leucaena leucocephala*, etc. have deep roots that effectively contribute to nutrient cycling
Impact on soil pH	Nutrient cycling by deep-rooted trees can help buffer soil pH, making soil more conducive for crop growth and biodiversity
Soil erosion mitigation	Deep roots of agroforestry trees prevent soil erosion by anchoring the soil and stabilizing the land

[a](Ref: FAO and ITPS 2015)

5.3.1 Nutrient Cycling in Agroforestry

One of the key environmental services provided by agroforestry is nutrient cycling. This process plays a critical role in enhancing soil fertility, maintaining ecosystem health, and improving agricultural productivity. Nutrient cycling in agroforestry is a dynamic process that involves the movement of nutrients through various biotic and abiotic components of the system, with trees and decaying organic matter playing crucial roles as key contributors to soil fertility and health. Trees help cycle nutrients from deep soil layers to the surface, reducing the need for synthetic fertilizers, while decaying plant debris enhances soil structure, moisture retention, and microbial activity. Together, these processes contribute to sustainable farming practices, improve soil health, and support long-term agricultural productivity. The continued adoption and integration of agroforestry practices can be an important strategy for addressing soil degradation and improving food security (FAO and ITPS 2015). Thus, role of agroforestry in nutrient cycling is mainly of two ways: (1) Role of agroforestry trees in nutrient cycling: Agroforestry systems enhance soil fertility by facilitating nutrient cycling. Trees, through their root systems, access deep soil layers and bring nutrients to the surface, which are then recycled back into the soil as organic matter. This natural fertilization process reduces the need for synthetic fertilizers and supports long-term soil health; and (2) Role of decaying plant and tree debris from agroforestry: Decaying plant and tree matter contributes organic carbon to the soil, enhancing soil structure, moisture retention, and microbial activity, which, in turn, improves the soil's nutrient holding capacity.

5.3.1.1 Role of Agroforestry Trees in Nutrient Cycling

Agroforestry trees contribute significantly to nutrient cycling by acting as a conduit for the uptake, cycling, and redistribution of nutrients in the soil. The role of trees in nutrient cycling can be linked with several key functions.

5.3.1.1.1 Deep Root Systems and Nutrient Uptake

One of the primary ways in which trees enhance nutrient cycling is through their deep root systems. Trees can reach deep soil layers that are inaccessible to shallow rooted crops. By accessing deep soil horizons, trees can take up nutrients such as potassium, phosphorus, calcium, and other minerals that are typically not available to surface-dwelling crops. These nutrients are then transported up through the tree and either used in the tree growth or shed back into the soil via leaf litter and root exudates.

5.3.1.1.2 Nutrient Redistribution

After trees absorb nutrients from deep soil layers, they release these nutrients back into the surface soil through leaf litter, root turnover, and other organic matter. This process enhances the nutrient availability in the upper soil layers where crops can access them. The continual cycling of nutrients from deep to shallow soils helps maintain or even increase the fertility of the topsoil over time.

5.3.1.1.3 Nitrogen Fixation

Certain species of agroforestry trees, particularly legumes, can fix atmospheric nitrogen through symbiotic relationships with nitrogen-fixing bacteria (e.g., *Rhizobium* species). These trees enrich the soil with nitrogen, a key nutrient for plant growth. When these trees shed leaves or are pruned, the nitrogen is incorporated into the soil, benefiting the surrounding crops and other vegetation.

5.3.1.1.4 Reduction of Synthetic Fertilizer Dependence

By facilitating natural nutrient cycling, agroforestry systems reduce the need for synthetic fertilizers. The nutrient inputs from trees, coupled with improved soil organic matter, can sustain soil fertility over the long term, decreasing the reliance on external chemical fertilizers. This not only reduces input costs for farmers but also minimizes the environmental pollution associated with fertilizer runoff, such as eutrophication of water bodies.

5.3.1.1.5 Enhancement of Soil Organic Matter

As trees shed leaves, branches, and roots, they contribute organic material to the soil, which is decomposed by soil microorganisms. The breakdown of organic matter releases essential nutrients like nitrogen, phosphorus, and potassium, which are then available for uptake by plants. Over time, this process enhances the soil organic matter content, improving its structure, water-holding capacity, and overall fertility.

5.3.2 Role of Decaying Plant and Tree Debris from Agroforestry

The decaying plant and tree matter in agroforestry systems plays a pivotal role in enhancing soil health and nutrient cycling. As trees and other plants shed leaves, branches, and other organic materials, this debris becomes a critical source of nutrients and organic carbon for the soil (Fig. 5.1, Table 5.2).

5.3.2.1 Contribution to Organic Carbon

Decaying plant material, including tree leaves, twigs, and roots, contributes substantial amounts of organic carbon to the soil. Organic carbon is a key component of soil organic matter (SOM), which improves soil structure, increases water holding capacity, and promotes better soil aeration. SOM is also the primary energy source for soil microorganisms, which break down organic material and recycle nutrients back into the soil in the forms that are available for plants to use.

5.3.2.2 Improvement of Soil Structure and Moisture Retention

The decomposition of tree and plant debris results in the formation of humus, a stable form of organic matter that helps to bind soil particles together. This enhances soil structure, creating a well-aerated and porous environment that facilitates root growth and water infiltration. Humus also helps retain moisture in the soil, which is especially beneficial in drought-prone areas. Better moisture retention ensures that nutrients are more readily available to plants during dry periods.

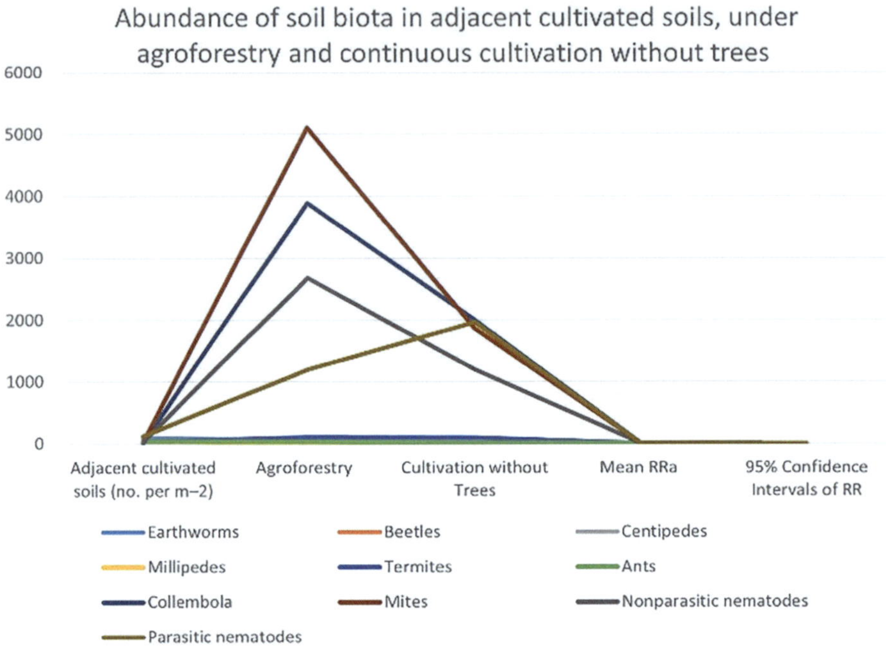

Fig. 5.1 Abundance of soil biota under agroforestry compared to cultivated soil without trees. (Ref: Sileshi and Mafongoya 2006a, b; Tian et al. 1997; Tian et al. 2000; Barrios et al. 2012; Fonte et al. 2010; Rahman et al. 2012; León et al. 2006; Hairiah et al. 2006; Geissen et al. 2009; Adejuyigbe et al. 1999; Kang et al. 1999; Kandji et al. 2001, 2002; Desaeger and Rao 2000, 2003; Banful et al. 2000, Musokwa and Mafongoya 2021)

5.3.2.3 Increased Microbial Activity

The organic matter from decaying plant and tree debris serves as a food source for soil microorganisms, including bacteria, fungi, and earthworms. These microorganisms are essential for the breakdown of organic material into smaller, nutrient-rich components. The increase in microbial activity in the soil promotes nutrient mineralization, a process in which nutrients are converted from organic forms into inorganic forms that are easily taken up by plants.

5.3.2.4 Nutrient Holding Capacity

As plant and tree debris decomposes, it adds to the soil's pool of organic matter, which enhances the soil's ability to retain and supply nutrients. Organic matter has a high cation exchange capacity (CEC), which allows it to hold onto positively charged nutrients like calcium, magnesium, and potassium, preventing them from leaching out of the soil with rainfall or irrigation. This increases nutrient retention and reduces nutrient loss, improving overall soil fertility.

Table 5.2 Soil health benefits from the functioning of trees in agroforestry

Soil health function	Benefit	Impact	Source
1. Organic inputs (above and below ground)	Contribution of organic matter through pruning and root systems	Up to 20 t dry matter ha^{-1} $yr.^{-1}$ from pruning; 358 kg N, 28 kg P, 232 kg K, 144 kg Ca, 60 kg Mg from pruning per year	Palm (1995)
2. Biological nitrogen fixation (BNF)	Nitrogen input to soil through nitrogen fixation by leguminous trees	56 to 675 kg N ha^{-1} yr^{-1} depending on climate, tree species, and management system	Nygren et al. (2012)
3. Deep uptake and recycling of nutrients	Trees take up water and nutrients from below crop root zone, reducing competition for water and nutrients	50% of water uptake by trees like *Vitellaria paradoxa* from below crop root zone during dry season	Bayala and Prieto (2020)
4. Water filtration and erosion reduction	Trees act as barriers to overland flow and help absorb water, reducing erosion and improving water filtration	50% reduction in soil erosion rates, 2.8 times increase in water infiltration rates	Muchane et al. (2020) and Ilstedt et al. (2007, 2016)

5.3.3 Agroforestry and Its Role in Improving Healthy Food Production

Agroforestry, the integration of trees and shrubs into agricultural landscapes, can play a pivotal role in improving food production while also supporting soil health and biodiversity. The concept discussed in the text emphasizes the critical connection between soil biodiversity and human health. It highlights that soil is not just a physical medium for growing food but a dynamic, living system that houses a vast array of organisms, including microbes, animals, and plants, which collectively contribute to the ecosystem's functionality. The long-term health of both soil and food systems depends on maintaining soil biodiversity through sustainable land management practices. Agroforestry offers a promising model in this context.

Soil is described as a dynamic reservoir of biodiversity, where complex interactions between microbes, plants, and animals provide critical services such as disease control, nutrient cycling, and the regulation of water and air quality. Healthy soils with high biodiversity contribute to the quantity and quality of the food we consume. In agroforestry systems, the inclusion of trees and perennial plants alongside annual crops enhances soil structure and promotes greater biodiversity compared to monoculture farming systems. The root systems of trees, for example, help prevent soil erosion, promote water infiltration, and improve nutrient cycling, all of which contribute to healthier soil and, by extension, healthier food production (Fig. 5.2).

It is well accepted that sustainable use of soil is essential for maintaining its longterm health and the services it provides. Human activities such as deforestation, agriculture, urbanization, and land degradation can disrupt soil biodiversity, leading

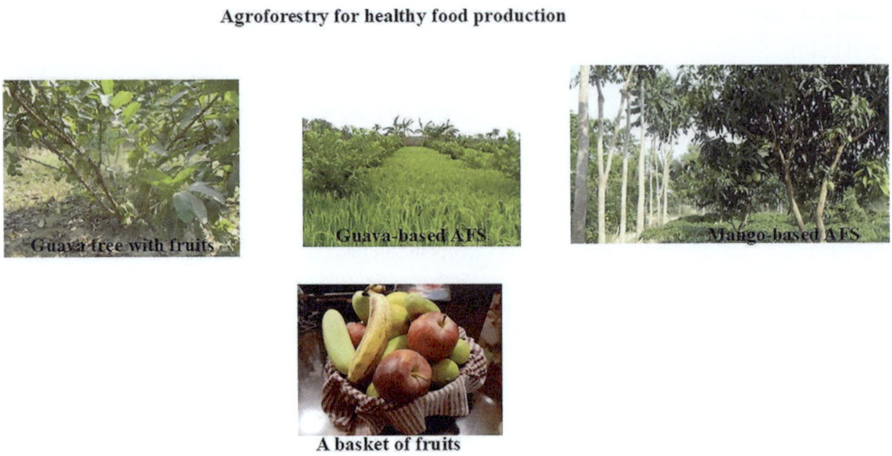

Fig. 5.2 Agroforestry for healthy food production

to negative consequences like increased soil erosion, loss of nutrients, and reduced water retention. Agroforestry, by integrating trees into agricultural systems, mitigates some of these negative effects. The diversity of plant species and the presence of trees can improve the resilience of agricultural landscapes, reduce erosion, and maintain or even enhance soil fertility, all of which are essential for sustainable food production.

Agroforestry systems also promote the careful management of soil by reducing reliance on synthetic fertilizers and pesticides. The symbiotic relationships between plants, trees, and soil organisms can naturally suppress pests and diseases, reducing the need for chemical interventions. This approach not only maintains soil health but also supports the production of healthier, more nutritious food.

5.3.4 Role of Biodiversity Under Agroforestry in Plant Pest and Disease Control

In the present day concern, it is a key point on the role of soil biodiversity that is required to be thoroughly examined in disease control. Healthy soils support a diverse microbial community that can help control soil-borne pathogens and pests, reducing the incidence of crop diseases. Agroforestry enhances this aspect of soil health by increasing the complexity of the plant and microbial communities. For instance, certain tree species can act as barriers to pests or host beneficial insects that naturally control pest populations. By fostering a more biodiverse ecosystem, agroforestry systems reduce the need for chemical pesticides and herbicides, which can have negative effects on both human health and the environment.

Moreover, it is of equal concern that disruptions to soil biodiversity, such as those caused by land use changes like deforestation or desertification, can lead to an increase in harmful pathogens and pests. By promoting agroforestry practices, which maintain or restore soil biodiversity, it is possible to minimize these risks and promote healthier food systems.

5.4 Human Health and Food Security

The connection between soil health and human health is emphasized by the recognition that the quality of the food we eat, the air we breathe, and the water we drink are all influenced by soil biodiversity. Soil is the foundation of food production, and its ability to support healthy crops, fishes, and livestock is intrinsically linked to the health of the broader ecosystem. Agroforestry systems, which integrate a variety of plants, fishes and animals, can increase the diversity of food sources, improve the resilience of agricultural systems to climate change, and help ensure food security in the long term.

The presence of trees and perennial plants in agroforestry systems also contributes to the overall health of the landscape. Trees sequester carbon, improve air quality, and provide habitat for wildlife, all of which contribute to ecological stability and human wellbeing. Additionally, the diversification of crops in agroforestry systems can provide more nutritious food, which is essential for improving human health, especially in areas that are vulnerable to food insecurity and malnutrition (Fig. 5.2, Table 5.3).

5.5 Agroforestry as a Solution to Global Challenges

The present environment inked health concern needs thorough review on the broader environmental changes driven by human activities, such as global warming, land use changes, and urbanization, which threaten soil health and biodiversity. These changes can disrupt the balance of ecosystems and contribute to the spread of diseases. Agroforestry, by promoting sustainable land management and protecting soil biodiversity, can help mitigate some of these global challenges. For example, agroforestry systems have been shown to enhance soil carbon sequestration, reduce greenhouse gas emissions, and improve water retention, all of which contribute to climate change mitigation and adaptation.

Furthermore, agroforestry offers a pathway to restoring degraded lands. By incorporating trees and other perennial plants into agricultural systems, farmers can restore the productivity of degraded soils, reduce erosion, and improve water management. This restoration of soil health not only benefits food production but also provides broader ecological and health benefits.

Table 5.3 Agroforestry and soil biodiversity interactions: impact on soil-borne pathogens, anthrax, helminths, allergies, antibiotics, antihelminth resistance, biological control, and health

Interacting factors	Key findings
1. Soil biodiversity and soil-borne pathogens (Berg et al. 2005; Pepper et al. 2009; Bultman et al. 2013; Brevik and Burgess 2014; Brevik and Sauer 2015; Ferris and Tuomisto 2015; Ling et al. 2015; Oliver and Gregory 2015)	– Soil biodiversity is largely beneficial, contributing to the control of soil-borne pathogens and pests
	– Most soil organisms do not pose risks to human health but can be instrumental in disease and pest control
	– Some pathogens, like *Pseudomonas* and *Enterobacter*. Control plant root pathogens and promote plant growth
	– Soil-borne pathogens that cause human diseases often require specific soil conditions to persist, such as soil type and moisture
2. Soil and anthrax (Smith et al. 2000; Ganz et al. 2014)	– *Bacillus anthracis* spores can survive in soils for decades and are triggered by heavy rains to infect grazing animals
	– Soil types with high calcium content (Chernozem soils) and neutral to alkaline favor anthrax outbreaks
	– Burning vegetation and maintaining soil cover can help reduce anthrax prevalence by limiting spore spread and airborne transmission
3. Soil and helminths (de Silva et al. 2003; Schär et al. 2013; Khieu et al. 2014)	– Soil-transmitted helminths, like Strongyloides and hookworms, are prevalent in tropical and subtropical regions
	– The risk of helminth infection is higher in soils with low organic carbon, especially in areas where forests have been converted to cropland
	– Increasing soil organic carbon may reduce helminth infections, linking soil health to human health outcomes

(continued)

Table 5.3 (continued)

Interacting factors	Key findings
4. Soil and allergies (Kay 2000; Matricardi and Bonini 2000; Rook 2010; Hanski et al. 2012; Brevik and Sauer 2015; Ruokolainen et al. 2015; WHO and CBD 2015)	– Exposure to soil microorganisms helps develop immune tolerance, potentially reducing allergies
	– Urban populations, with less exposure to diverse soil organisms, may have a higher prevalence of allergic diseases
	– Access to natural areas or small livestock could help urban populations gain immune benefits from soil microorganisms
5. Soil, antibiotics, and antihelminth resistance (Prichard 2009; Corbett et al. 2014; Ling et al. 2015)	– Soil is a potential source of novel antibiotics, yet many soil bacteria remain uncultured
	– Antibiotic resistance and antihelminth resistance are increasing, but proper land management can help control parasitic infections, e.g., rotating pastures and managing grass growth
6. Soil biological control (Gray 1983; Andersen 1987; Gehring et al. 2002; Weller et al. 2002; Charles et al. 2005; Keesing et al. 2006; Sánchez-Moreno and Ferris 2007; Suzán et al. 2009; Searle et al. 2011; Stirling 2011; Costa et al. 2012; Ferris et al. 2012; Johnson et al. 2012; van Elsas et al. 2012; Johnson et al. 2013; Penton et al. 2014; Stirling 2014)	– Soil biodiversity plays a key role in biological control of pests, with soil predators and parasites reducing pathogen and pest populations
	– Indigenous soil organisms like entomopathogenic nematodes can naturally control pests like *Diaprepes abbreviatus*, reducing the need for harmful chemicals
	– Overall soil biodiversity, rather than individual species, is crucial for effective disease suppression
7. Maintaining soil biodiversity for health (Wu et al. 2011; Bates et al. 2013; Lauber et al. 2013; Tedersoo et al. 2014; Ramirez et al. 2014)	– Soil biodiversity varies across regions and is crucial for maintaining ecosystem health, which indirectly benefits human health
	– Protecting rare and region-specific species is essential, as different soil organisms contribute to disease suppression, nutrient cycling, and other ecosystem functions
	– Agroforestry practices that maintain or enhance soil biodiversity can promote resilience against pests and diseases, benefiting both plant health and human health

5.6 Global Implications of Agroforestry and Nutrient Cycling

Agroforestry, being the powerful approach that supports healthy food production by maintaining soil biodiversity and promoting sustainable land management practices. The integration of trees and shrubs into agricultural systems enhances soil health, reduces the need for, chemical inputs, and helps manage pests and diseases naturally. By improving soil biodiversity and fostering ecological balance, agroforestry systems can contribute to better food security, greater resilience to climate change, and, improved human health. As the global population grows and environmental challenges intensify, agroforestry will play an increasingly key role in ensuring a sustainable and healthy food supply for future generations.

In this connection, it is obvious to note that the role of agroforestry in nutrient cycling is crucial for the long-term sustainability of global agricultural systems. According to the FAO and ITPS (2015) report, "Status of the World's Soil Resources," soil degradation is a critical concern globally, with nutrient depletion being one of the main drivers of soil degradation. Agroforestry systems, by enhancing nutrient cycling, can mitigate soil degradation, improve soil fertility, and promote sustainable land management practices.

In regions where conventional agricultural systems rely heavily on synthetic fertilizers and monocropping, agroforestry provides a natural, cost-effective alternative. The integration of trees into farming landscapes can reduce soil erosion, enhance biodiversity, and improve water retention, all while maintaining or increasing agricultural productivity. This is particularly important in the face of climate change, where soil health and water conservation are becoming increasingly critical.

Nutrient cycling through agroforestry is one of the supporting types of ecosystem services. This involves mainly two categories of actions, viz. (1) transformations of organic materials of soil organisms and (2) retention and release of nutrients on charged surfaces. As the plant communities influence soil water, labile carbon intake, pH, and nutrient cycling, they have a significant impact on microbial activity and organic matter input into soil, making them the primary governing factors of soil carbon storage processes. Particularly, production of food and fiber depends on soil biodiversity. It also play a significant role in controlling other essential soil functions, such as water purification, greenhouse gas emission control, and nutrient cycling. Additionally, soil biodiversity stocks are acknowledged as a significant biological and genetic resource for biotechnological use (Wall et al. 2012; Brevik and Sauer 2015; Table 5.4).

5.6.1 Water Filtration and Regulation

Agroforestry also involves two important ecosystem services as follows:

1. Agroforestry practices help regulate water quality and availability by stabilizing water flow and reducing soil erosion. The deep-rooted trees improve water infiltration, reduce surface runoff, and prevent soil degradation.

Table 5.4 Agroforestry and soil biodiversity interactions: impact on food, fiber, biomass production

Key findings	Description	References
Soil-based crop production	All terrestrial crop production is soil-based, emphasizing the importance of sustainable soil management for long-term food security	Pepper et al. (2009)
Soil-borne pathogens and crop productivity	Soil-borne pathogens can disrupt nutrient flow, reduce plant biomass (above- and belowground), and reduce food quality and quantity	Mujtar et al. (2019) and Cappelli et al. (2022)
Soil organisms enhancing productivity	Earthworms increase productivity in humid and cooler climates, while termites improve yields in warmer, drier climates	Evans et al. (2011) and van Groenigen et al. (2014)
Soil life and nutrient efficiency	Enriched soil life improves nutrient-use efficiency, enhances plant nutrient uptake, and increases crop yields	Bender and van der Heijden (2015)
Soil foodweb resilience	A diverse soil foodweb enhances ecosystem resilience and resistance, benefiting agricultural sustainability	Wall et al. (2010), Ferris et al. (2012) and de Vries et al. (2012, 2013)
Symbiotic soil microbes	Symbiotic soil microbes are crucial for nutrient cycling and biofortification of plants, especially for micronutrients like zinc	Rillig and Mummey (2006) and Oliver and Gregory (2015)
Role of fungi in stress tolerance	Fungi, particularly endophytes, help plants tolerate stress, potentially enhancing crop resilience	Rodriguez et al. (2008)
Impact of agricultural intensification	Practices like ploughing, agrochemical use, and excessive fertilization reduce soil biodiversity. Negatively impacting longterm agricultural sustainability	Wall et al. (2012) and Tsiafouli et al. (2015)
Balanced use of agrochemicals	While agrochemicals are important, their use should be optimized in terms of timing, rate, and location to prevent detrimental effects on soil biodiversity	Meena et al. (2020) and Mishra et al. (2024)

2. Trees also act as natural filters, trapping sediment and absorbing excess nutrients and pollutants from runoff, thus improving water quality in nearby rivers, lakes, or groundwater systems. The increased presence of trees in agroforestry systems can mitigate flooding and reduce the impact of droughts by maintaining more consistent water levels.

5.6.2 Biodiversity Support Services of Agroforestry

One of the most important ecosystem services of agroforestry is the support it provides for biodiversity. Agroforestry systems create a mosaic of habitats, offering shelter and food sources for various species, from microorganisms in the soil to larger wildlife.

The integration of trees in agricultural landscapes increases habitat complexity, which supports a wide range of species, including pollinators (bees, butterflies) and beneficial predators (birds, insects) that help control pest populations.

A diverse range of plants and trees in agroforestry systems promotes genetic diversity within the farming ecosystem, which is vital for the resilience of both the agricultural system and the surrounding natural ecosystems (Table 5.5).

Decline in soil biodiversity greatly impacts numerous ecosystem processes, including interaction within plant diversity, decomposition, nutrient retention, and nutrient cycling. Relationships between aboveground and below-ground biodiverse communities imply that soil organisms may be impacted by the same processes that affect aboveground extinction (Wardle et al. 2004; De Deyn and van der Putten 2005; Bardgett and van der Putten 2014; Wagg et al. 2014; Wall et al. 2015; Schittko et al. 2022). Agricultural intensification may diminish soil biodiversity, which would result in fewer functional groupings and a simpler food web. Crops and plants, fertilizers, pH, tillage techniques, crop residue retention, pesticides, herbicides, and pollution are other factors that impact biodiversity in agricultural soils. Converting natural habitat to agricultural production changes the biological and physical qualities of the soil (such as temperature, pH, and water-holding capacity) as well as the microhabitat. The distribution and diversity of soil mesofauna may change in response to changes in various soil characteristics. In agriculture, organisms that have evolved and adapted to high levels of physical disturbance thrive, leading to the decreasing the diversity and richness of soil fauna (Crossley et al. 1992; Bardgett and van der Putten 2014; Tsiafouli et al. 2015; Table 5.6 and 5.7).

5.6.3 Pest and Disease Regulation

1. The biodiversity fostered by agroforestry helps regulate pest populations and reduces the spread of plant diseases. A diverse environment can host natural predators and parasitoids that keep pest species in check, thereby reducing the need for chemical pesticides.
2. Tree cover also disrupts the habitat of pest species, further limiting their ability to spread and damage crops. By decreasing pest outbreaks, agroforestry supports more sustainable farming practices and reduces the risks associated with pest resistance to chemicals (Table 5.7).

5.6.4 Contribution to Food Security

5.6.4.1 Diversification of Agricultural Products

Agroforestry systems increase the diversity of products available to farmers, which in turn enhances food security. By growing a variety of crops alongside trees, farmers can harvest multiple products, including fruits, nuts, timber, medicinal plants, and fodder, in addition to traditional staple crops.

Table 5.5 Agroforestry and soil biodiversity interactions: impact on air quality

Key findings	Description
Impact of land-use change on air quality	Land-use changes, especially intensive agriculture, are linked to increase dust storms, greenhouse gas emissions, and volatile organic compounds (Garrison et al. 2003). These changes contribute to particulate matter (PM 10) formation, which negatively affects air quality and human health
Soil organisms in air	Soil organisms like bacteria, fungi, nematodes, and mites can be transported hundreds to thousands of kilometers by wind (Brown and Hovmøller 2002; Nkem et al. 2006), impacting air quality by introducing biological particles that contribute to dust storms and pollutants
Dust and health risks	Intensive agricultural practices, such as over-ploughing, burning biomass, and leading fields bare, increase dust and particulate matter (PM10). PM10 is associated with respiratory problems, lung damage, and diseases like lung cancer (Schenker 2000; Garrison et al. 2003). Polluted soils, containing heavy metals and pathogens, exacerbate these health risks when disturbed
Valley fever example	In the southwestern USA, agricultural practices disturb soil and release spores of *Coccidioides immitis*, a fungus that can cause valley fever (lung disease). This fungus, which normally helps stabilize the soil, becomes airborne when the soil is disturbed, highlighting the health risks of disrupting soil biodiversity (Tabor et al. 2011; Nguyen et al. 2013)
Role of soil biodiversity in dust formation	Soil organisms like arbuscular mycorrhizae, saprotrophic fungi, and earthworms play a key role in soil stabilization. When disturbed by intensive farming, these organisms are harmed, increasing the likelihood of dust formation and degradation of air quality (Frey et al. 1999; Six et al. 2006)
Agroforestry as a mitigation strategy	Agroforestry and agroecological management practices can reduce dust and particulate matter formation. Techniques like reduced tillage, planting windbreaks, adding manure, incorporating cover crops, and retaining crop residues help maintain soil stability, protect soil biodiversity, and improve air quality (Nordstrom and Hotta 2004; Madden et al. 2008)
Impact of reduced tillage on air quality	Reduced tillage has been shown to lower the formation of PM10, thereby reducing the risk of respiratory diseases and heart conditions linked to air pollution (Madden et al. 2008; Anderson et al. 2012). This practice helps conserve soil structure and biodiversity while mitigating airborne pollutants
Consequence of soil pollution on airborne pathogens	Soils contaminated with heavy metals, pathogens, and antibiotic-resistant organisms can release harmful particles into the air. These can contribute to respiratory diseases and exacerbate environmental health risks when disturbed by agricultural activities (Madden et al. 2008; Sprigg et al. 2014)
Link between agricultural intensification and human health	The intensification of agriculture disrupts soil biodiversity, leading to the destabilization of soil and increased dust formation. This not only damages the environment but also poses significant risks to human health, underlining the need for more sustainable farming practices (Garrison et al. 2003; Six et al. 2006)

Table 5.6 Agroforestry and soil biodiversity interactions impact on water quality

Key finding	Implications for water quality	Reference
Pollution from agrochemicals and poor sanitation compromises water quality (e.g., fecal contamination)	Increases contamination of drinking water, highlighting the need for natural systems to help filter contaminants	Alavanja et al. (2013) and Oliver and Gregory (2015)
Soil biodiversity enhances soil structure, promoting water infiltration and percolation	Improves water-use efficiency by crops and reduces agricultural runoff: helping to manage water quality in adjacent areas	Abraham and Silambarasan (2013) and Rayu et al. (2012)
Soil organisms can degrade harmful pollutants, such as selenium in agricultural drainage water	Microbial activity in soils can remediate pollutants, reducing the impacts of contaminants like selenium on water systems	Frankenberger and Arshad (2001)
Agroforestry practices support soil biodiversity, enhancing the filtration of contaminants	Reduces the flow of agrochemical runoff and pathogens into surface waters, helping to maintain water quality and ecosystem health	Rayu et al. (2012) and Tiemann et al. (2015)
Land-use changes, such as urbanization, affect water runoff and infiltration, impacting local water bodies	Altered hydrological cycles can exacerbate water quality issues in downstream areas, requiring mitigation through land management	Alavanja et al. (2013), Oliver and Gregory (2015)
Interventions like reduced irrigation and rotational land fallowing reduce contaminants such as selenium	Mitigates the build-up of harmful contaminants in water sources, improving practices	Tiemann et al. (2015)

This diversification reduces dependency on a single crop, thereby mitigating the risk of total crop failure due to pests, diseases, or climate extremes. It ensures a more balanced and stable food supply for local communities.

5.6.4.2 Enhanced Farm Productivity

The synergistic effects of trees, crops, and livestock lead to higher overall farm productivity. Trees provide shade, windbreaks, and organic matter, which help improve microclimate and reduce water stress on crops. These factors increase crop resilience, leading to more stable yields (Fig. 5.3).

Livestock integrated into agroforestry systems benefit from the nutritional diversity provided by tree leaves and other. Plant materials, while their manure can be used to fertilize crops, closing the nutrient cycle on the farm.

5.6.4.3 Climate Resilience

Agroforestry systems provide a buffer against climate-related shock such as droughts, floods, and temperature extremes. Trees in agroforestry systems help regulate local climates by moderating temperature fluctuations, improving moisture retention in the soil, and protecting crops from wind damage.

The diverse ecosystem created by agroforestry enhances resilience by ensuring that even if certain crops fail due to climate stressors, others may still thrive. This

Table 5.7 Interactions of agroforestry, soil biodiversity and land use

Key findings	Summary	References
Losses in soil biodiversity affect multiple ecosystem functions	Declines in soil biodiversity impact plant diversity, decomposition, nutrient retention, and cycling	Wagg et al. (2014)
Links between aboveground and below-ground communities	Factors affecting above-ground extinction may also affect soil organisms, emphasizing the connection between aboveground and below-ground biodiversity	Wardle et al. (2004), De Deyn and van der Putten (2005) and Bardgett and van der Putten (2014)
Agricultural intensification reduces soil biodiversity	Intensive agricultural practices reduce soil biodiversity, decreasing food-web complexity and the number of functional groups	Tsiafouli et al. (2015)
Factors influencing soil biodiversity in agricultural soils	Soil biodiversity is influenced by factors such as crop type, fertilizers, pH, tillage practices, crop residue retention, pesticides, herbicides, and pollution	Bardgett and van der Putten (2014)
Changes in soil properties due to land-use conversion affect soil biodiversity	Conversion of natural habitat to agriculture alters soil properties like temperature, pH, and water-holding capacity, which in turn impacts soil mesofauna diversity	Crossley et al. (1992) and Bardgett and van der Putten (2014)
Agricultural management practices degrade soil communities	Intensive agricultural practices (e.g., monocultures, tillage, and chemical inputs) degrade the soil food web, resulting in reduced biodiversity and increased pest incidence	Decaëns and Jiménez (2002) and Eggleton et al. (2002)
Soil fauna contribute to organic matter decomposition, influenced by temperature and moisture	Soil fauna play a crucial role in organic matter decomposition, with temperature and moisture conditions significantly affecting their contribution	Wall et al. (2008)
Distance from the equator and precipitation affect fungal biodiversity	A global study found that fungal species richness is influenced by proximity to the equator and annual precipitation, providing benchmarks for assessing human impacts on soil biodiversity	Tedersoo et al. (2014)
Soil management influences biodiversity, with pressures affecting 56% of EU soils	Soil biodiversity in the EU is under threat from factors like intensive exploitation, reduced organic matter, habitat disturbance pollution, and climate change	Gardi et al. (2013)
Thresholds in soil biodiversity may lead to functional decline	Experimental evidence suggests that below a certain threshold of soil biodiversity, ecosystem functions may decline, although such levels are rarely seen in natural systems	van der Heijden et al. (1998, 2008), Liiri et al. (2002) and Setälä and McLean (2004)

(continued)

Table 5.7 (continued)

Key findings	Summary	References
Functional redundancy in soil communities	Soil communities show functional redundancy for general processes (e.g., carbon cycling) but changes may impact specialized processes. High biodiversity increases resilience to environmental perturbations	Nielsen et al. (2011) and Jiang et al. (2024)
Soil biodiversity enhances ecosystem resilience	More diverse soil communities are more resilient to environmental changes, as they can compensate for the loss or compromise of certain components	Kibblewhite et al. (2008)

Fig. 5.3 Agroforestry associated components for healthy food production

resilience helps farming communities adapt to changing climatic conditions and ensures a more stable food supply.

5.6.4.4 Reduction in Dependency on External Inputs

Agroforestry systems reduce the need for external inputs like chemical fertilizers, pesticides, and herbicides. By maintaining healthy soils, promoting biodiversity, and reducing pest pressures, agroforestry systems can produce more food with fewer external inputs. This not only lowers costs for farmers but also decreases environmental pollution and the depletion of natural resources.

The integration of trees and other plant species helps maintain the natural balance of nutrients, further promoting sustainable farming practices without compromising productivity.

5.7 Nurturing Soil Life and Human Health: On Global Scale

Agroforestry systems that combine trees and crops able to coexist through positive interactions, e.g., symbiosis, will generate organic inputs of different quality that can preserve soil cover and increase the diversity and persistence of active soil biota. Thus, agroforestry, being the integration of trees with crops and/or livestock, offers a promising approach to bridging the gap between agriculture and forestry by addressing both environmental and socio-economic objectives. A comprehensive global assessment of below-ground biodiversity is still lacking, despite recent efforts such as the Global Soil Biodiversity Atlas (Orgiazzi et al. 2016) and the Global Soil Biodiversity Initiative. While progress is being made, especially through DNA-based methods, no global benchmark values exist to quantify changes or losses in soil biodiversity due to natural or anthropogenic factors. This absence of standardized methodologies and accessible, comprehensive datasets, particularly for morphological data, limits our ability to fully assess and monitor soil biodiversity across regions and trophic levels. To address these gaps, there is a pressing need for more coordinated, large scale studies to understand the impacts of soil biodiversity loss on ecosystem functioning and to develop a common, publicly accessible database that integrates both molecular and morphological data (FAO and ITPS 2015). Empirical research indicates that agroforestry can provide multiple benefits, including the prevention of environmental degradation, enhancing agricultural productivity, increasing carbon sequestration, and supporting soil and ecosystem health, while also offering stable incomes, human health, and other social welfare benefits (Pauli et al. 2011; Kamau et al. 2017; Kamau et al. 2020; Dierks et al. 2021). A study by Castle et al. (2022) aims to address the existing research gap by collating and reviewing evidence on the impacts of agroforestry on ecosystem services and human wellbeing in high income countries (HICs). It examines the outcomes of agroforestry practices in relation to environmental sustainability, agricultural productivity, and human welfare, while also identifying key evidence gaps and research biases. An extensive search through 31,852 articles across five primary databases and 24 organizational websites, covering peer reviewed and gray literature from January 1990 to June 2020, this study compiled a dataset of 585 primary articles, 6 ongoing studies, and 41 systematic reviews. The findings from Castle et al. (2022) show that research on agroforestry in HICs has concentrated primarily on linear boundary plantings such as hedgerows, shelterbelts, and riparian buffers, and silvopasture systems. The most studied outcome is the regulation and maintenance of ecosystem conditions, such as soil and water quality, carbon storage, and biodiversity, followed by agricultural yield. Despite this, there remains a significant lack of studies examining the impacts of agroforestry on human wellbeing, particularly on aspects like health, housing, food security, and nutrition. Notably, cultural and subjective well-being outcomes, such as aesthetic and cultural value, were more prominent in the HIC literature, suggesting that cultural drivers, rather than purely functional benefits, play a significant role in the adoption of agroforestry practices in these regions. Moreover, while the environmental benefits of agroforestry, especially related to carbon sequestration, have been well documented, there is a surprising paucity of research on agroforestry's role in climate change adaptation and mitigation in HICs. This gap may be linked to the relatively low adoption rates of agroforestry systems, which limit opportunities for large scale studies examining the resilience of agroforestry systems under extreme climate events in those countries. The study by Castle et al. (2022) highlights critical research gaps, particularly the need for more comprehensive studies on the effects of agroforestry practices at the landscape scale as well as more evidence on policy interventions aimed at promoting agroforestry. The findings underscore the importance of expanding the evidence base on agroforestry in high income countries to inform future research and policy decisions, ultimately supporting the development of integrated systems that can deliver both environmental and socio-economic benefits.

5.8 Concluding Remarks

The interconnectedness between soil health, agroforestry, and climate resilience plays a pivotal role in sustainable land management, especially as the world grapples with the intensifying effects of climate change. The urgency of building agricultural systems that can endure extreme weather events—such as droughts, floods, and heatwaves — has never been clearer. Agroforestry, which integrates trees with crops and livestock, offers a powerful solution by enhancing soil health and improving ecosystem functions. Healthy soils, enriched through agroforestry practices, are better equipped to absorb water, retain nutrients, and resist erosion, providing a natural buffer against climate variability. This not only stabilizes yields but also boosts biodiversity and helps sequester carbon, addressing both the ecological and climatic challenges we face today.

Adaptability of agroforestry is another key advantage. By tailoring systems to suit specific climates and regions, whether tropical, temperate, or arid, agroforestry supports local agricultural practices while simultaneously promoting environmental conservation. This makes it a versatile tool for both farmers and conservationists, offering solutions for diverse ecological and socio-economic contexts. Importantly, agroforestry strengthens the resilience of farming communities to environmental shocks and economic stresses by diversifying agricultural production, enhancing food security, and reducing dependency on external inputs.

At the core of these benefits is soil health. Healthy soils are the foundation of productive agricultural systems, supporting not only crop yields but also crucial ecosystem functions like nutrient cycling and water regulation. By fostering soil fertility and biodiversity, agroforestry systems enhance these essential processes, creating a virtuous cycle of ecological health and agricultural productivity. This, in turn, supports long-term food security by stabilizing agricultural outputs and reducing vulnerability to climate induced disruptions.

However, while benefits of agroforestry for biodiversity, soil health, and climate resilience are well documented, still there remains significant research gaps. For example, studies on the role of agroforestry in climate change adaptation and mitigation, particularly in high-income countries, are still limited. While environmental benefits like carbon sequestration are widely acknowledged, less attention has been given to the broader socio-economic impacts, including human well-being, health, and cultural values. There is a pressing need for more comprehensive, large scale studies those assess potential of agroforestry at both local and landscape scales, as well as the development of standardized methodologies for monitoring soil biodiversity.

As global populations grow and the impacts of climate change deepen, agroforestry presents an invaluable opportunity to promote sustainable land management. By integrating trees into agricultural systems, we can address multiple pressing challenges, such as soil degradation, biodiversity loss, water scarcity, and food insecurity, while enhancing the resilience of farming communities. To fully realize opportunities from agroforestry, governments, policymakers, and farmers must work together to prioritize research, expand adoption, and create policies that support the integration of agroforestry into modern agricultural landscapes. This collaborative approach will help ensure that agroforestry becomes a cornerstone of sustainable agricultural systems, capable of securing both our environmental and food futures.

References

Abraham J, Silambarasan S (2013) Biodegradation of chlorpyrifos and its hydrolyzing metabolite 3,5,6-trichloro-2-pyridinol by Sphingobacterium sp. JAS3. Process Biochem 48(10):1559–1564. https://doi.org/10.1016/j.procbio.2013.06.034. https://www.sciencedirect.com/science/article/pii/S1359511313003875. ISSN 1359-5113

Adejuyigbe CO, Tian G, Adeoye GO (1999) Soil microarthropod populations under natural and planted fallows in southwestern Nigeria. Agrofor Syst 47:263–272. https://biblio.iita.org/documents/adejuyigbe-soil-1999.pdf-74b9106a80f04b407015bcb18e19ac1e.pdf

Alavanja MC, Ross MK, Bonner MR (2013) Increased cancer burden among pesticide applicators and others due to pesticide exposure. CA Cancer J Clin 63(2):120–142. https://doi.org/10.3322/caac.21170. Epub 2013 Jan 15. PMID: 23322675. https://pubmed.ncbi.nlm.nih.gov/23322675/

Andersen DC (1987) Below-ground herbivory in natural communities: a review emphasizing fossorial animals. Q Rev Biol 62(3):261–286. https://doi.org/10.1086/415512

Anderson JO, Thundiyil JG, Stolbach A (2012) Clearing the air: a review of the effects of particulate matter air pollution on human health. J Med Toxicol 8(2):166–175. https://doi.org/10.1007/s13181-011-0203-1. https://pubmed.ncbi.nlm.nih.gov/22194192/. PMID: 22194192; PMCID: PMC3550231

Banful B, Dzietror A, Ofori I et al (2000) Yield of plantain alley cropped with Leucaena leucocephala and Flemingia macrophylla in Kumasi, Ghana. Agrofor Syst 49:189–199. https://doi.org/10.1023/A:1006335710243

Bardgett R, van der Putten W (2014) Belowground biodiversity and ecosystem functioning. Nature 515:505–511. https://doi.org/10.1038/nature13855

Barrios E, Sileshi GW, Shepherd K, Sinclair F (2012) Agroforestry and soil health: linking trees, soil biota, and ecosystem services. In: Wall DH et al (eds) Soil ecology and ecosystem services. Chapter 5.2, 1st edn. OUP, Oxford, pp 316–330. https://apps.worldagroforestry.org/downloads/Publications/PDFS/BC12072.pdf

Bates ST, Clemente JC, Flores GE, Walters WA, Parfrey LW, Knight R, Fierer N (2013) Global biogeography of highly diverse protistan communities in soil. ISME J 7(3):652–659. https://

doi.org/10.1038/ismej.2012.147. https://pubmed.ncbi.nlm.nih.gov/23235291/. Epub 2012 Dec 13. PMID: 23235291; PMCID: PMC3578557

Bayala J, Prieto I (2020) Water acquisition, sharing and redistribution by roots: applications to agroforestry systems. Plant and Soil 453:17–28. https://doi.org/10.1007/s11104-019-04173-z

Bender SF, van der Heijden MGA (2015) Soil biota enhance agricultural sustainability by improving crop yield, nutrient uptake and reducing nitrogen leaching losses. J Appl Ecol 51(1):228–239. https://doi.org/10.1111/1365-2664.12351

Berg G, Eberl L, Hartmann A (2005) The rhizosphere as a reservoir for opportunistic human pathogenic bacteria. Environ Microbiol 7(11):1673–1685. https://doi.org/10.1111/j.1462-2920.2005.00891.x. PMID: 16232283. https://pubmed.ncbi.nlm.nih.gov/16232283/

Brevik EC, Burgess LC (2014) The influence of soils on human health. Nat Educ Knowl 5(12):1. https://www.nature.com/scitable/knowledge/library/the-influence-of-soils-on-human-health-127878980/

Brevik EC, Sauer TJ (2015) The past, present, and future of soils and human health studies. Soil 1:35–46. https://doi.org/10.5194/soil-1-35-2015

Brown JK, Hovmøller MS (2002) Aerial dispersal of pathogens on the global and continental scales and its impact on plant disease. Science 297(5581):537–541. https://doi.org/10.1126/science.1072678. PMID: 12142520. https://pubmed.ncbi.nlm.nih.gov/12142520/

Bultman MW, Fisher FS, Pappagianis D (2013) An overview of the ecology of soil-borne human pathogens. In: Selinus O (ed) Essentials of medical geology, section II. Pathways and exposures, Ch. 19, 1st edn. Academic Press, New York. eBook ISBN: 9780080454191. https://shop.elsevier.com/books/essentials-of-medical-geology/selinus/978-0-08-045419-1

Cappelli SL, Domeignoz-Horta LA, Loaiza V, Laine AL (2022) Plant biodiversity promotes sustainable agriculture directly and via belowground effects. Trends Plant Sci 27(7):674–687. https://doi.org/10.1016/j.tplants.2022.02.003. ISSN 1360-1385. https://www.sciencedirect.com/science/article/pii/S1360138522000322

Castle SE, Miller DC, Merten N et al (2022) Evidence for the impacts of agroforestry on ecosystem services and human well-being in high-income countries: a systematic map. Environ Evid 11:10. https://doi.org/10.1186/s13750-022-00260-4. https://rdcu.be/d0hyN

Charles L, Carbone I, Davies KG, Bird D, Burke M, Kerry BR, Opperman CH (2005) Phylogenetic analysis of Pasteuria penetrans by use of multiple genetic loci. J Bacteriol 187(16):5700–5708. https://doi.org/10.1128/JB.187.16.5700-5708.2005. PMID: 16077116; PMCID: PMC1196054. https://pmc.ncbi.nlm.nih.gov/articles/PMC1196054/

Corbett CJ, Love S, Moore A, Burden FA, Matthews JB, Denwood MJ (2014) The effectiveness of faecal removal methods of pasture management to control the cyathostomin burden of donkeys. Parasit Vectors 7:48. https://doi.org/10.1186/1756-3305-7-48. PMID: 24460700; PMCID: PMC3904009. https://pubmed.ncbi.nlm.nih.gov/24460700/

Costa SR, Kerry BR, Bardgett RD, Davies KG (2012) Interactions between nematodes and their microbial enemies in coastal sand dunes. Oecologia 170(4):1053–1066. https://doi.org/10.1007/s00442-012-2359-z. Epub 2012 May 24. PMID: 22622874. https://pubmed.ncbi.nlm.nih.gov/22622874/

Crossley DA, Mueller BR, Perdue JC (1992) Biodiversity of microarthropods in agricultural soils: relations to processes. Agric Ecosyst Environ 40(1–4):37–46. https://doi.org/10.1016/0167-8809(92)90082-M. ISSN 0167-8809. https://www.sciencedirect.com/science/article/pii/016788099290082M

De Deyn GB, Van der Putten WH (2005) Linking aboveground and belowground diversity. Trends Ecol Evol 20(11):625–633. https://doi.org/10.1016/j.tree.2005.08.009. Epub 2005 Sep 8. PMID: 16701446. https://pubmed.ncbi.nlm.nih.gov/16701446/

de Silva NR, Brooker S, Hotez PJ, Montresor A, Engels D, Savioli L (2003) Soil-transmitted helminth infections: updating the global picture. Trends Parasitol 19(12):547–551. https://doi.org/10.1016/j.pt.2003.10.002. PMID: 14642761. https://pubmed.ncbi.nlm.nih.gov/14642761/

de Vries F, Liiri M, Bjørnlund L et al (2012) Land use alters the resistance and resilience of soil food webs to drought. Nat Clim Chang 2:276–280. https://doi.org/10.1038/nclimate1368. https://www.nature.com/articles/nclimate1368#citeas

de Vries FT, Thébault E, Liiri M et al (2013) Soil food web properties explain ecosystem services across European land use systems. Proc Natl Acad Sci U S A 110(35):14296–14301. https://doi.org/10.1073/pnas.1305198110. Epub 2013 Aug 12. PMID: 23940339; PMCID: PMC3761618. https://pubmed.ncbi.nlm.nih.gov/23940339/

Decaëns T, Jiménez J (2002) Earthworm communities under an agricultural intensification gradient in Colombia. Plant and Soil 240:133–143. https://doi.org/10.1023/A:1015844610604

Desaeger J, Rao MR (2000) Parasitic nematode populations in natural fallows and improved cover crops and their effects on subsequent crops in Kenya. Field Crop Res 65:41–56. https://doi.org/10.1016/S0378-4290(99)00071-4. ISSN 0378-4290. https://www.sciencedirect.com/science/article/pii/S0378429099000714

Desaeger J, Rao MR (2003) Significance of lesion and spiral nematodes in crotalaria-maize rotation in Western Kenya. Nematropica 33(1):27–39. https://www.researchgate.net/publication/286951649_Significance_of_lesion_and_spiral_nematodes_in_Crotalaria-maize_rotation_in_Western_Kenya

Dierks J, Blaser-Hart WJ, Gamper HA, Nyoka IB, Barrios E, Six J (2021) Trees enhance abundance of arbuscular mycorrhizal fungi, soil structure, and nutrient retention in low-input maize cropping systems. Agric Ecosyst Environ 318:107487. https://doi.org/10.1016/j.agee.2021.107487. ISSN 0167-8809. https://www.sciencedirect.com/science/article/pii/S0167880921001912

Eggleton P, Bignell DE, Hauser S, Dibog L, Norgrove L, Madong B (2002) Termite diversity across an anthropogenic disturbance gradient in the humid forest zone of West Africa. Agric Ecosyst Environ 90(2):189–202. https://doi.org/10.1016/S0167-8809(01)00206-7. ISSN 0167-8809. https://www.sciencedirect.com/science/article/pii/S0167880901002067

Evans TA, Dawes TZ, Ward PR, Lo N (2011) Ants and termites increase crop yield in a dry climate. Nat Commun 2:262. https://doi.org/10.1038/ncomms1257. PMID: 21448161; PMCID: PMC3072065. https://pubmed.ncbi.nlm.nih.gov/21448161/

FAO and ITPS (2015) Status of the world's soil resources (SWSR)—Main report. FAO of the UN and Intergovernmental Technical Panel on Soils, Rome. ISBN 978-92-5-109004-6. https://openknowledge.fao.org/server/api/core/bitstreams/6ec24d75-19bd-4f1f-b1c5-5becf50d0871/content

Ferris H, Tuomisto H (2015) Unearthing the role of biological diversity in soil health. Soil Biol Biochem 85(2015):101–109. https://doi.org/10.1016/j.soilbio.2015.02.037. ISSN 0038-0717. https://www.sciencedirect.com/science/article/pii/S0038071715000917

Ferris H, Pocasangre LE, Serrano E, Muñoz J, Garcia S, Perichi G, Martinez G (2012) Diversity and complexity complement apparent competition: nematode assemblages in banana plantations. Acta Oecol 40:11–18. https://doi.org/10.1016/j.actao.2012.02.003. ISSN 1146-609X. https://www.sciencedirect.com/science/article/pii/S1146609X12000161

Fonte SJ, Barrios E, Six J (2010) Earthworms, soil fertility and aggregate-associated soil organic matter dynamics in the Quesungual agroforestry system. Geoderma 155(3–4):320–328. https://doi.org/10.1016/j.geoderma.2009.12.016. ISSN 0016-7061. https://www.sciencedirect.com/science/article/pii/S0016706109004212

Frankenberger WT Jr, Arshad M (2001) Bioremediation of selenium-contaminated sediments and water. BioFactors 14(1–4): 241–254. https://doi.org/10.1002/biof.5520140130

Frey SD, Elliott ET, Paustian K (1999) Bacterial and fungal abundance and biomass in conventional and no-tillage agroecosystems along two climatic gradients. Soil Biol Biochem 31(4):573–585. https://doi.org/10.1016/S0038-0717(98)00161-8. ISSN 0038-0717. https://www.sciencedirect.com/science/article/pii/S0038071798001618

Ganz HH, Turner WC, Brodie EL, Kusters M, Shi Y et al (2014) Interactions between *Bacillus anthracis* and plants may promote anthrax transmission. PLoS Negl Trop Dis 8(6):e2903. https://doi.org/10.1371/journal.pntd.0002903

Gardi C, Jeffery S, Saltelli A (2013) An estimate of potential threats levels to soil biodiversity in EU. Glob Chang Biol 19(5):1538–1548. https://doi.org/10.1111/gcb.12159

Garrison VH, Shinn EA, Foreman WT et al (2003) African and Asian dust: from desert soils to coral reefs. BioScience 53(5):469–480. https://doi.org/10.1641/0006-3568(2003)053[0469:AAADFD]2.0.CO;2. https://academic.oup.com/bioscience/article/53/5/469/241414

Gehring CA, Wolf JE, Theimer TC (2002) Terrestrial vertebrates promote arbuscular mycorrhizal fungal diversity and inoculum potential in a rain forest soil. Ecol Lett 5(4):540–548. https://doi.org/10.1046/j.1461-0248.2002.00353.x

Geissen V, Peña-Peña K, Huerta E (2009) Effects of different land use on soil chemical properties, decomposition rate and earthworm communities in tropical Mexico. Pedobiologia 53(1):75–86. https://doi.org/10.1016/j.pedobi.2009.03.004. ISSN 0031-4056. https://www.sciencedirect.com/science/article/pii/S0031405609000365

Gray NF (1983) Ecology of nematophagous fungi: Panagrellus redivivus as the target organism. Plant and Soil 73:293–297. https://doi.org/10.1007/BF02197725

Hairiah K, Sulistyani H, Suprayogo D, Widianto PP, Widodo RH, Van Noordwijk M (2006) Litter layer residence time in forest and coffee agroforestry systems in Sumberjaya, West Lampung. For Ecol Manage 224(1–2):45–57. https://doi.org/10.1016/j.foreco.2005.12.007. ISSN 0378-1127. https://www.sciencedirect.com/science/article/pii/S0378112705007371

Hanski I, von Hertzen L, Fyhrquist N, Koskinen K, Torppa K, Laatikainen T, Karisola P, Auvinen P, Paulin L, Mäkelä MJ, Vartiainen E, Kosunen TU, Alenius H, Haahtela T (2012) Environmental biodiversity, human microbiota, and allergy are interrelated. Proc Natl Acad Sci U S A 109(21):8334–8339. https://doi.org/10.1073/pnas.1205624109. Epub 2012 May 7. PMID: 22566627; PMCID: PMC3361383. https://pubmed.ncbi.nlm.nih.gov/22566627/

Ilstedt U, Malmer A, Verbeeten E, Murdiyarso D (2007) The effect of afforestation on water infiltration in the tropics: a systematic review and meta-analysis. For Ecol Manag 251(1–2):45–51. https://doi.org/10.1016/j.foreco.2007.06.014. ISSN 0378-1127. https://www.sciencedirect.com/science/article/pii/S0378112707004665

Ilstedt U, Tobella BA, Bazié HR, Bayala J, Verbeeten E, Nyberg G, Sanou J, Benegas L, Murdiyarso D, Laudon H et al (2016) Intermediate tree cover can maximize groundwater recharge in the seasonally dry tropics. Sci Rep 6:1–12. https://www.nature.com/articles/srep21930

Jiang L, Ushio M, Imai N et al (2024) Changes of bacterial versus fungal community composition along a forest degradation gradient of logged-over tropical rain forests, and their consequences on soil enzyme activities in Malaysian Borneo. Plant Soil. https://doi.org/10.1007/s11104-024-06939-6

Johnson PT, Preston DL, Hoverman JT, Henderson JS, Paull SH, Richgels KL, Redmond MD (2012) Species diversity reduces parasite infection through cross-generational effects on host abundance. Ecology 93(1):56–64. https://doi.org/10.1890/11-0636.1. PMID: 22486087. https://pubmed.ncbi.nlm.nih.gov/22486087/

Johnson PT, Preston DL, Hoverman JT, Richgels KL (2013) Biodiversity decreases disease through predictable changes in host community competence. Nature 494(7436):230–233. https://doi.org/10.1038/nature11883. PMID: 23407539. https://pubmed.ncbi.nlm.nih.gov/23407539/

Kamau S, Barrios E, Karanja NK, Ayuke FO, Lehmann J (2017) Spatial variation of soil macrofauna and nutrients in tropical agricultural systems influenced by historical charcoal production in south Nandi, Kenya. Appl Soil Ecol 119:286–293. https://doi.org/10.1016/j.apsoil.2017.07.007. ISSN 0929-1393. https://www.sciencedirect.com/science/article/pii/S0929139317303293

Kamau S, Barrios E, Karanja NK, Ayuke FO, Lehmann J (2020) Dominant tree species and earthworms affect soil aggregation and carbon content along a soil degradation gradient in an agricultural landscape. Geoderma 359:113983. https://doi.org/10.1016/j.geoderma.2019.113983. ISSN 0016-7061. https://www.sciencedirect.com/science/article/pii/S0016706118322419

Kandji ST, Ogol CKPO, Albrecht A (2001) Diversity of plant-parasitic nematodes and their relationships with some soil physico-chemical characteristics in improved fallows in western Kenya. Appl Soil Ecol 18(2):143–157. https://doi.org/10.1016/S0929-1393(01)00157-3. ISSN 0929-1393. https://www.sciencedirect.com/science/article/pii/S0929139301001573

Kandji ST, Ogol CKPO, Albrecht A (2002) Influence of some agroforestry practices on the temporal structures of nematodes in western Kenya. Eur J Soil Biol 38(2):197–203. https://doi.org/10.1016/S1164-5563(02)01144-5. ISSN 1164-5563. https://www.sciencedirect.com/science/article/pii/S1164556302011445

Kang BT, Caveness FE, Tian G, Kolawole G (1999) Longterm alley cropping with four hedgerow species on an Alfisol in southwestern Nigeria–effect on crop performance, soil chemical properties and nematode population. Nutr Cycl Agroecosyst 54:145–155. https://hdl.handle.net/10568/98012

Kay AB (2000) Overview of 'allergy and allergic diseases: with a view to the future'. Br Med Bull 56(4):843–864. https://doi.org/10.1258/0007142001903481. PMID: 11359624. https://pubmed.ncbi.nlm.nih.gov/11359624/

Keesing F, Holt RD, Ostfeld RS (2006) Effects of species diversity on disease risk. Ecol Lett 9(4):485–498. https://doi.org/10.1111/j.1461-0248.2006.00885.x. PMID: 16623733. https://pubmed.ncbi.nlm.nih.gov/16623733/

Khieu V, Schär F, Forrer A, Hattendorf J, Marti H, Duong S, Vounatsou P, Muth S, Odermatt P (2014) High prevalence and spatial distribution of Strongyloides stercoralis in rural Cambodia. PLoS Negl Trop Dis 8(6):e2854. https://doi.org/10.1371/journal.pntd.0002854. PMID: 24921627; PMCID: PMC4055527. https://pubmed.ncbi.nlm.nih.gov/24921627/

Kibblewhite MG, Ritz K, Swift MJ (2008) Soil health in agricultural systems. Philos Trans R Soc Lond B Biol Sci 363(1492):685–701. https://doi.org/10.1098/rstb.2007.2178. PMID: 17785275; PMCID: PMC2610104. https://pubmed.ncbi.nlm.nih.gov/17785275/

Lauber CL, Ramirez KS, Aanderud Z, Lennon J, Fierer N (2013) Temporal variability in soil microbial communities across land-use types. ISME J 7(8):1641–1650. https://doi.org/10.1038/ismej.2013.50. Epub 2013 Apr 4. PMID: 23552625; PMCID: PMC3721119. https://pubmed.ncbi.nlm.nih.gov/23552625/

León YS, de Melo E, Soto G, Johnson-Maynard J, Lugo-Pérez J (2006) Earthworm populations, microbial biomass and coffee production in different experimental agroforestry management systems in Costa Rica. Caribb J Sci 42(3):397–409. College of Arts and Sciences. University of Puerto Rico, Mayagüez. https://www.researchgate.net/publication/262069595_Earthworm_Populations_Microbial_Biomass_and_Coffee_Production_in_Different_Experimental_Agroforestry_Management_Systems_in_Costa_Rica

Liiri M, Setälä H, Haimi J, Pennanen T, Fritze H (2002) Soil processes are not influenced by the functional complexity of soil decomposer food webs under disturbance. Soil Biol Biochem 34(7):1009–1020. https://doi.org/10.1016/S0038-0717(02)00034-2. ISSN 0038-0717. https://www.sciencedirect.com/science/article/pii/S0038071702000342

Ling L, Schneider T, Peoples A et al (2015) A new antibiotic kills pathogens without detectable resistance. Nature 517:455–459. https://doi.org/10.1038/nature14098

Madden MM, Southard RJ, Mitchell JP (2008) Conservation tillage reduces PM10 emissions in dairy forage rotations. Atmos Environ 42(16):3795–3808. https://doi.org/10.1016/j.atmosenv.2007.12.058. https://www.sciencedirect.com/science/article/pii/S1352231008000095

Matricardi PM, Bonini S (2000) High microbial turnover rate preventing atopy: a solution to inconsistencies impinging on the hygiene hypothesis? Clin Exp Allergy 30(11):1506–1510. https://doi.org/10.1046/j.1365-2222.2000.00994.x. PMID: 11069557. https://pubmed.ncbi.nlm.nih.gov/11069557/

Meena RS, Kumar S, Datta R, Lal R, Vijayakumar V, Brtnicky M, Sharma MP, Yadav GS, Jhariya MK, Jangir CK et al (2020) Impact of agrochemicals on soil microbiota and management: a review. Land 9(2):34. https://doi.org/10.3390/land9020034

Mishra MK, Shailendra Kumar Mishra SK, Panday L, Pandey R (2024) Agrochemicals and their impact on environment. In: Recent trends in plant protection, 1st edn. NIPA. https://www.researchgate.net/publication/375485756_Agrochemicals_and_Their_Impact_on_Environment/citations

Muchane MN, Sileshi GW, Gripenberg S, Jonsson M, Pumariño L, Barrios E (2020) Agroforestry boosts soil health in the humid and sub-humid tropics: a meta-analysis. Agric Ecosyst Environ 295:106899. https://doi.org/10.1016/j.agee.2020.106899. ISSN 0167-8809

Mujtar VEL, Muñoz N, Mc Cormick BP, Pulleman M, Tittonell P (2019) Role and management of soil biodiversity for food security and nutrition; where do we stand? Glob Food Sec

20(2019):132–144. https://doi.org/10.1016/j.gfs.2019.01.007. ISSN 2211-9124. https://www.sciencedirect.com/science/article/pii/S2211912418300300

Musokwa M, Mafongoya PL (2021) Effects of improved pigeonpea fallows on biological and physical soil properties and their relationship with maize yield. Agrofor Syst 95:443–457. https://doi.org/10.1007/s10457-021-00598-7

Nguyen C, Barker BM, Hoover S, Nix DE, Ampel NM, Frelinger JA, Orbach MJ, Galgiani JN (2013) Recent advances in our understanding of the environmental, epidemiological, immunological, and clinical dimensions of coccidioidomycosis. Clin Microbiol Rev 26(3):505–525. https://doi.org/10.1128/CMR.00005-13. PMID: 23824371; PMCID: PMC3719491. https://pubmed.ncbi.nlm.nih.gov/23824371/

Nielsen UN, Ayres E, Wall DH, Bardgett RD (2011) Soil biodiversity and carbon cycling: a review and synthesis of studies examining diversity–function relationships. Eur J Soil Sci 62(1):105–116. https://doi.org/10.1111/j.1365-2389.2010.01314.x. Special Issue: Soil Organic Matters: February 2011

Nkem JN, Wall DH, Virginia RA et al (2006) Wind dispersal of soil invertebrates in the McMurdo dry valleys, Antarctica. Polar Biol 29:346–352. https://doi.org/10.1007/s00300-005-0061-x

Nordstrom KF, Shintaro Hotta S (2004) Wind erosion from cropland in the USA: a review of problems, solutions and prospects. Geoderma 121(3–4):157–167. https://doi.org/10.1016/j.geoderma.2003.11.012. ISSN 0016-7061. https://www.sciencedirect.com/science/article/pii/S0016706103003355

Nygren P, Fernández MP, Harmand JM et al (2012) Symbiotic dinitrogen fixation by trees: an underestimated resource in agroforestry systems? Nutr Cycl Agroecosyst 94:123–160. https://doi.org/10.1007/s10705-012-9542-9

Oliver MA, Gregory PJ (2015) Soil, food security and human health: a review. Eur J Soil Sci 66(2):257–276. https://doi.org/10.1111/ejss.12216

Orgiazzi A, Bardgett RD, Barrios E et al (eds) (2016) Global soil biodiversity atlas. European Commission, Publications Office of the European Union, Luxembourg. https://data.europa.eu/doi/10.2788/2613; https://publications.europa.eu/en/publication-detail/-/publication/c54ece8e-1e4d-11e6-ba9a-01aa75ed71a1; https://op.europa.eu/s/zYlL

Palm CA (1995) Contribution of agroforestry trees to nutrient requirements of intercropped plants. Agrofor Syst 30:105–124. https://doi.org/10.1007/BF00708916

Pauli N, Barrios E, Conacher AJ, Oberthür T (2011) Soil macrofauna in agricultural landscapes dominated by the Quesungual slash-and-mulch agroforestry system, western Honduras. Appl Soil Ecol 47(2):119–132. https://doi.org/10.1016/j.apsoil.2010.11.005. ISSN 0929-1393. https://www.sciencedirect.com/science/article/pii/S0929139310002155

Penton CR, Gupta VV, Tiedje JM, Neate SM, Ophel-Keller K, Gillings M, Harvey P, Pham A, Roget DK (2014) Fungal community structure in disease suppressive soils assessed by 28S LSU gene sequencing. PLoS One 9(4):e93893. https://doi.org/10.1371/journal.pone.0093893. PMID: 24699870; PMCID: PMC3974846. https://pubmed.ncbi.nlm.nih.gov/24699870/

Pepper IL, Gerba CP, Newby DT, Rice CW (2009) Soil: a public health threat or savior? Crit Rev Environ Sci Technol 39(5):416–432. https://doi.org/10.1080/10643380701664748

Prichard R (2009) Drug resistance in nematodes. In: Mayers DL (ed) Antimicrobial drug resistance. Infectious disease. Humana Press, Totowa, pp 621–628. https://doi.org/10.1007/978-1-59745-180-2_44

Qaiser KN, Dutt V, Mughal AH, Mugloo JA (2016) Apple based Horti-Agri-pasture system under temperate conditions of Kashmir valley. In: Chaturvedi OP, Sikka AK, Handa AK, Bajpai CK (eds) Agroforestry technologies for different agro-climatic zones of the country. All India Coordinated Research Project on Agroforestry, ICAR-Central Agroforestry Research Institute, Jhansi, pp 1–177

Rahman PM, Varma RV, Sileshi GW (2012) Abundance and diversity of soil invertebrates in annual crops, agroforestry and forest ecosystems in the Nilgiri biosphere reserve of Western Ghats, India. Agrofor Syst 85:165–177. https://doi.org/10.1007/s10457-011-9386-3

Ramirez KS, Leff JW, Barberán A, Bates ST, Betley J, Crowther TW, Kelly EF et al (2014) Biogeographic patterns in below-ground diversity in New York City's Central Park are similar to those observed globally. Proc Biol Sci 281(1795):20141988. https://doi.org/10.1098/rspb.2014.1988. PMID: 25274366; PMCID: PMC4213626. https://pubmed.ncbi.nlm.nih.gov/25274366/

Rayu S, Karpouzas DG, Singh BK (2012) Emerging technologies in bioremediation: constraints and opportunities. Biodegradation 23(6):917–926. https://doi.org/10.1007/s10532-012-9576-3. Epub 2012 Jul 27. PMID: 22836784. https://pubmed.ncbi.nlm.nih.gov/22836784/

Rillig MC, Mummey DL (2006) Mycorrhizas and soil structure. New Phytol 171(1):41–53. https://doi.org/10.1111/j.1469-8137.2006.01750.x. PMID: 16771981. https://pubmed.ncbi.nlm.nih.gov/16771981/

Rodriguez RJ, Henson J, Van Volkenburgh E, Hoy M, Wright L, Beckwith F, Kim YO, Redman RS (2008) Stress tolerance in plants via habitat-adapted symbiosis. ISME J 2(4):404–416. https://doi.org/10.1038/ismej.2007.106. Epub 2008 Feb 7. PMID: 18256707. https://pubmed.ncbi.nlm.nih.gov/18256707/

Rook GA (2010) 99th Dahlem conference on infection, inflammation and chronic inflammatory disorders: darwinian medicine and the 'hygiene' or 'old friends' hypothesis. Clin Exp Immunol 160(1):70–79. https://doi.org/10.1111/j.1365-2249.2010.04133.x. PMID: 20415854; PMCID: PMC2841838. https://pubmed.ncbi.nlm.nih.gov/20415854/

Ruokolainen L, von Hertzen L, Fyhrquist N, Laatikainen T, Lehtomäki J, Auvinen P, Karvonen AM, Hyvärinen A, Tillmann V, Niemelä O, Knip M, Haahtela T, Pekkanen J, Hanski I (2015) Green areas around homes reduce atopic sensitization in children. Allergy 70(2):195–202. https://doi.org/10.1111/all.12545. PMID: 25388016; PMCID: PMC4303942. https://pubmed.ncbi.nlm.nih.gov/25388016/

Sánchez-Moreno S, Ferris H (2007) Suppressive service of the soil food web: effects of environmental management. Agric Ecosyst Environ 119(1–2):75–87. https://doi.org/10.1016/j.agee.2006.06.012. ISSN 0167-8809. https://www.sciencedirect.com/science/article/pii/S0167880906002659

Schär F, Trostdorf U, Giardina F, Khieu V, Muth S, Marti H, Vounatsou P, Odermatt P (2013) Strongyloides stercoralis: global distribution and risk factors. PLoS Negl Trop Dis 7(7):e2288. https://doi.org/10.1371/journal.pntd.0002288. PMID: 23875033; PMCID: PMC3708837. https://pubmed.ncbi.nlm.nih.gov/23875033/

Schenker M (2000) Exposures and health effects from inorganic agricultural dusts. Environ Health Perspect 108 Suppl 4(Suppl 4):661–664. https://doi.org/10.1289/ehp.00108s4661. PMID: 10931784; PMCID: PMC1637665. https://pmc.ncbi.nlm.nih.gov/articles/PMC1637665/

Schittko C, Onandia G, Bernard-Verdier M, Heger T, Jeschke JM, Kowarik I, Maaß S, Joshi J (2022) Biodiversity maintains soil multifunctionality and soil organic carbon in novel urban ecosystems. J Ecol 110:916–934. https://doi.org/10.1111/1365-2745.13852

Searle CL, Biga LM, Spatafora JW, Blaustein AR (2011) A dilution effect in the emerging amphibian pathogen Batrachochytrium dendrobatidis. Proc Natl Acad Sci U S A 108(39):16322–16326. https://doi.org/10.1073/pnas.1108490108. Epub 2011 Sep 19. PMID: 21930900; PMCID: PMC3182747. https://pubmed.ncbi.nlm.nih.gov/21930900/

Setälä H, McLean MA (2004) Decomposition rate of organic substrates in relation to the species diversity of soil saprophytic fungi. Oecologia 139(1):98–107. https://doi.org/10.1007/s00442-003-1478-y. Epub 2004 Jan 22. PMID: 14740289. https://pubmed.ncbi.nlm.nih.gov/14740289/

Shukla AK, Kumar S, Ram SN (2018) Yield, soil health and moisture status in Bael (*Aegle marmelos*)-based Horti-Pasture System in semi-arid ecosystem of India. Agrofor Syst 93:1275–1283. https://doi.org/10.1007/s10457-018-0241-7

Sileshi G, Mafongoya PL (2006a) Variation in macrofaunal communities under contrasting land use systems in eastern Zambia. Appl Soil Ecol 33:49–60. https://apps.worldagroforestry.org/downloads/Publications/PDFS/ja06040.pdf

Sileshi G, Mafongoya PL (2006b) Long-term effects of improved legume fallows on soil invertebrate macrofauna and maize yield in eastern Zambia. Agric Ecosyst Environ 115:69–78. https://doi.org/10.1016/j.agee.2005.12.010. https://www.sciencedirect.com/science/article/pii/S0167880905005645

Six J, Frey SD, Thiet RK, Batten KM (2006) Bacterial and fungal contributions to carbon sequestration in agroecosystems. Soil Sci Soc Am J 70(2):555–569. https://doi.org/10.2136/sssaj2004.0347

Smith KL, DeVos V, Bryden H, Price LB, Hugh-Jones ME, Keim P (2000) Bacillus anthracis diversity in Kruger National Park. J Clin Microbiol 38(10):3780–3784. https://doi.org/10.1128/JCM.38.10.3780-3784.2000. PMID: 11015402; PMCID: PMC87475

Sprigg WA, Nickovic S, Galgiani JN et al (2014) Regional dust storm modeling for health services: the case of valley fever. Aeolian Res 14:53–73. https://doi.org/10.1016/j.aeolia.2014.03.001. https://www.sciencedirect.com/science/article/pii/S1875963714000202

Stirling GR (2011) Biological control of plant-parasitic nematodes: an ecological perspective, a review of progress and opportunities for further research. In: Davies K, Spiegel Y (eds) Biological control of plant-parasitic nematodes: progress in biological control, vol 11. Springer, Dordrecht. https://doi.org/10.1007/978-1-4020-9648-8_1

Stirling GR (2014) Ecosystem services and the concept of 'integrated soil biology management'. In: Stirling GR (ed) Biological control of plant-parasitic nematodes: soil ecosystem management in sustainable agriculture, 2nd edn, pp 3–11. https://doi.org/10.1079/9781780644158.0000

Suzán G, Marcé E, Giermakowski JT, Mills JN, Ceballos G, Ostfeld RS, Armién B, Pascale JM, Yates TL (2009) Experimental evidence for reduced rodent diversity causing increased hantavirus prevalence. PLoS One 4(5):e5461. https://doi.org/10.1371/journal.pone.0005461. Epub 2009 May 6. PMID: 19421313; PMCID: PMC2673579. https://pmc.ncbi.nlm.nih.gov/articles/PMC2673579/

Tabor JA, O'rourke MK, Lebowitz MD, Harris RB (2011) Landscape-epidemiological study design to investigate an environmentally based disease. J Expo Sci Environ Epidemiol 21(2):197–211. https://doi.org/10.1038/jes.2009.67. Epub 2010 Mar 3. PMID: 20197794. https://pubmed.ncbi.nlm.nih.gov/20197794/

Tedersoo L, Bahram M, Põlme S, Kõljalg U, Yorou NS, Wijesundera R, Villarreal Ruiz L et al (2014) Fungal biogeography. Global diversity and geography of soil fungi. Science 346(6213):1256688. https://doi.org/10.1126/science.1256688. PMID: 25430773. https://pubmed.ncbi.nlm.nih.gov/25430773/

Tian G, Kang BT, Brussaard L (1997) Effect of mulch quality on earthworm activity and nutrient supply in the humid tropics. Soil Biol Biochem 29:369–373. https://doi.org/10.1016/S0038-0717(96)00099-5. https://www.sciencedirect.com/science/article/pii/S0038071796000995

Tian G, Olimah JA, Adjeoye GO, Kang BT (2000) Regeneration of earthworm populations in a degraded soil by natural and planted fallows under humid tropical conditions. Soil Sci Soc Am J 64:222–228. https://doi.org/10.2136/sssaj2000.641222x

Tiemann LK, Grandy AS, Atkinson EE, Marin-Spiotta E, McDaniel MD (2015) Crop rotational diversity enhances belowground communities and functions in an agroecosystem. Ecol Lett 18(8):761–771. https://doi.org/10.1111/ele.12453. Epub 2015 May 25. PMID: 26011743. https://pubmed.ncbi.nlm.nih.gov/26011743/

Tsiafouli MA, Thébault E, Sgardelis SP et al (2015) Intensive agriculture reduces soil biodiversity across Europe. Glob Chang Biol 21(2):973–985. https://doi.org/10.1111/gcb.12752. Epub 2014 Nov 17. PMID: 25242445. https://pubmed.ncbi.nlm.nih.gov/25242445/

van der Heijden M, Klironomos J, Ursic M et al (1998) Mycorrhizal fungal diversity determines plant biodiversity, ecosystem variability and productivity. Nature 396:69–72. https://doi.org/10.1038/23932

van der Heijden MG, Bardgett RD, van Straalen NM (2008) The unseen majority: soil microbes as drivers of plant diversity and productivity in terrestrial ecosystems. Ecol Lett 11(3):296–310. https://doi.org/10.1111/j.1461-0248.2007.01139.x. Epub 2007 Nov 29. Erratum in: Ecol Lett.2008 Jun;11(6):651. PMID: 18047587. https://pubmed.ncbi.nlm.nih.gov/18047587/

van Elsas JD, Chiurazzi M, Mallon CA, Elhottova D, Kristufek V, Salles JF (2012) Microbial diversity determines the invasion of soil by a bacterial pathogen. Proc Natl Acad Sci U S A 109(4):1159–1164. https://doi.org/10.1073/pnas.1109326109. Epub 2012 Jan 9. PMID: 22232669; PMCID: PMC3268289. https://pubmed.ncbi.nlm.nih.gov/22232669/

van Groenigen JW, Lubbers IM, Vos HM, Brown GG, De Deyn GB, van Groenigen KJ (2014) Earthworms increase plant production: a meta-analysis. Sci Rep 4:6365. https://doi.org/10.1038/srep06365. PMID: 25219785; PMCID: PMC5376159. https://pmc.ncbi.nlm.nih.gov/articles/PMC5376159/

Wagg C, Bender SF, Widmer F, van der Heijden MG (2014) Soil biodiversity and soil community composition determine ecosystem multifunctionality. Proc Natl Acad Sci U S A 111(14):5266–5270. https://doi.org/10.1073/pnas.1320054111. Epub 2014 Mar 17. PMID: 24639507; PMCID: PMC3986181. https://pubmed.ncbi.nlm.nih.gov/24639507/

Wall DH, Bradford MA, St John MG et al (2008) Global decomposition experiment shows soil animal impacts on decomposition are climate-dependent. Glob Chang Biol 14(11):2661–2677. https://doi.org/10.1111/j.1365-2486.2008.01672.x. PMCID: PMC3597247. https://pmc.ncbi.nlm.nih.gov/articles/PMC3597247/

Wall DH, Bardgett R, Kelly E (2010) Biodiversity in the dark. Nat Geosci 3:297–298. https://doi.org/10.1038/ngeo860. https://research.manchester.ac.uk/en/publications/biodiversity-in-the-dark

Wall DH et al (eds) (2012) Soil ecology and ecosystem services. Academic, Oxford. https://doi.org/10.1093/acprof:oso/9780199575923.001.0001

Wall D, Nielsen U, Six J (2015) Soil biodiversity and human health. Nature 528:69–76. https://doi.org/10.1038/nature15744

Wardle DA, Bardgett RD, Klironomos JN, Setälä H, van der Putten WH, Wall DH (2004) Ecological linkages between aboveground and belowground biota. Science 304(5677):1629–1633. https://doi.org/10.1126/science.1094875. PMID: 15192218. https://pubmed.ncbi.nlm.nih.gov/15192218/

Weller DM, Raaijmakers JM, Gardener BB, Thomashow LS (2002) Microbial populations responsible for specific soil suppressiveness to plant pathogens. Annu Rev Phytopathol 40:309–348. https://doi.org/10.1146/annurev.phyto.40.030402.110010. Epub 2002 May 13. PMID: 12147763. https://pubmed.ncbi.nlm.nih.gov/12147763/

WHO and CBD (2015) Connecting global priorities: biodiversity and human health. A state of knowledge review. World Health Organization and Secretariat of the Convention on Biological Diversity, Geneva. https://www.cbd.int/health/SOK-biodiversity-en.pdf

Wu T, Ayres E, Bardgett RD, Wall DH, Garey JR (2011) Molecular study of worldwide distribution and diversity of soil animals. Proc Natl Acad Sci U S A 108(43):17720–17725. https://doi.org/10.1073/pnas.1103824108. Epub 2011 Oct 17. PMID: 22006309; PMCID: PMC3203765. https://pubmed.ncbi.nlm.nih.gov/22006309/

Agroforestry: Mitigating Climate Change Through Carbon Sequestration

6

Abstract

Agroforestry, a sustainable land use practice that integrates agriculture and forestry, has emerged as a key nature-based solution for combating climate change while enhancing environmental protection. This approach not only boosts agricultural productivity but also plays a critical role in carbon sequestration, a process that captures and stores atmospheric carbon dioxide (CO_2) to mitigate the greenhouse effect. Agroforestry systems, which combine trees, shrubs, crops, and livestock, facilitate significant carbon storage through photosynthesis, converting CO_2 into organic matter stored in the soil, biomass, and products. The carbon sequestration potential of agroforestry is driven by several factors: the presence of trees and shrubs with extensive root systems that enhance soil organic matter retention; improved soil quality that supports better carbon retention; diversified cropping systems that increase carbon storage; and reduced greenhouse gas emissions from reduced reliance on fossil fuel based inputs. In addition to environmental benefits, agroforestry contributes to enhanced agricultural productivity and supports rural livelihoods. This practice, therefore, represents a vital strategy in the global fight against climate change, offering a sustainable, resilient approach to land use for the future.

Keywords

Agroforestry · Agriculture · Forestry · Nature-based solution · Climate change · Radiative balance · Cement production · Environmental protection · Carbon sequestration · Emission intensity · Carbon footprint · Energy footprint · Carbon captures · Carbon retention · Carbon storage · *Kyoto* protocol · Atmospheric carbon dioxide (CO_2) · Carbon farming · Carbon yield · Quarter Girth formula · Greenhouse effect · Agroforestry systems · Trees · Shrubs · Crops · Livestock · Feed · Anti-methanogenic · Rumen · Enteric fermentation · Breeding animals · Ration balancing · Photosynthesis · Nutrient pumps · Weathering · Organic matter · Soil organic matter · Biomass · Root systems · Rhizosphere · Safety net ·

S. Panda, *Agroforestry*, https://doi.org/10.1007/978-981-96-6855-7_6

Root exudates · Improved soil quality · Diversified cropping systems · Greenhouse gas emissions · Fossil fuel based inputs · Environmental benefits · Agricultural productivity · Rural livelihoods · Sustainable land use · Resilient land use

Abbreviations

AF:	Agroforestry
CF:	Carbon footprint
CO_2:	Carbon dioxide
CO_2e:	Carbon dioxide equivalent
DBH:	Diameter at breast height
EUE:	Energy Use Efficiency
GHG:	Greenhouse gas
GWP:	Global warming potential
GWP100:	Global warming potential over a period of 100 years
ICLS:	Integrated crop-livestock systems
IPCC:	Intergovernmental Panel on Climate Change
IRESA:	Integrated Renewable Energy and Sustainable Agriculture
LIFE:	Lifestyle for Environment
LOF:	Liquid organic fertilizers
MoEFCC:	Ministry of Environment, Forest and Climate Change
MtCO_2e:	Metric tonne of carbon dioxide equivalent
nbs:	Nature based solution
NUE:	Nitrogen use efficiency
SOC:	Soil organic carbon
SOM:	Soil organic matter

6.1 Agroforestry: The Nature Based Solution for Climate Change Through Carbon Sequestration

Agroforestry, an integrated land use approach combining agricultural and forestry practices, has emerged as a powerful and sustainable solution for combating climate change while enhancing environmental protection. As highlighted in this chapter, agroforestry not only boosts agricultural productivity but also plays a pivotal role in sequestering carbon, making it an effective nature based strategy for mitigating global warming.

Carbon sequestration, the process of capturing and storing atmospheric carbon dioxide (CO_2), is vital process in reducing the concentration of greenhouse gases that contribute to climate change. Agroforestry systems, which blend trees, shrubs, crops, and livestock, provide an excellent mechanism for carbon capture. The trees and other vegetation in these systems absorb CO_2 through photosynthesis, storing it in their biomass, soil, and products, and thereby reducing the overall levels of carbon in the atmosphere.

6.1.1 Factors Contributing to the Success of Agroforestry in Enhancing Carbon Sequestration

(a) **Trees and shrubs**: The inclusion of trees and shrubs in agroforestry systems significantly increases the capacity for carbon storage. These woody plants have expansive root systems that help retain soil organic matter, creating a robust carbon sink. The foliage and branches of the trees also store carbon, contributing to higher overall sequestration rates.
(b) **Soil improvement**: Agroforestry practices improve soil health by boosting organic matter content, enhancing soil structure, and reducing erosion. Healthy soils are more effective at retaining carbon, making them essential in the long term sequestration of CO_2. Furthermore, improved soil quality supports more productive agricultural systems, enhancing food security in the process.
(c) **Diversified cropping systems**: By incorporating a variety of tree and crop species, agroforestry systems can increase carbon sequestration rates. Different species have varying abilities to capture and store carbon, and when combined, they create a more effective carbon sink than monoculture systems. The diversity of plant life also strengthens ecosystem resilience to changing climatic conditions.
(d) **Reduced emissions**: Agroforestry can help reduce greenhouse gas emissions from traditional farming practices. Trees and shrubs can reduce the need for synthetic fertilizers and pesticides, which are energy intensive to produce and apply. Additionally, agroforestry systems often rely more on natural processes, reducing the carbon footprint of agricultural activities.

Beyond its role in carbon sequestration, agroforestry provides a multitude of other environmental and socio-economic benefits. It supports biodiversity, improves water retention, and combats soil degradation, while also offering farmers a more diversified and resilient source of income. By incorporating trees into farming systems, agroforestry also promotes sustainable land use and climate adaptation strategies, essential for building resilience against climate change.

In this way, agroforestry represents a key opportunity to address both environmental degradation and the challenge of climate change. By fostering carbon sequestration, enhancing soil health, and promoting biodiversity, agroforestry not only helps mitigate climate change but also boosts agricultural productivity and supports rural livelihoods. As such, it holds significant promise in the global pursuit of a more sustainable and climate resilient future for all (Dixon et al. 1994; Pandey 2002: Albrecht and Kandji 2003; Lai 2004; Montagnini and Nair 2004; Borges et al. 2017; Noumi et al. 2018; Ramos et al. 2018; Doddabasawa and Murthy 2019; Zaro et al. 2019; Birhane et al. 2020; Rallings et al. 2020; Sarto et al. 2020; Goncalves et al. 2021).

Thus, the subsequent discussions necessarily deals with two aspects of AF systems related to carbon sequestration like (1) carbon farming, i.e., carbon storage through farming, and (2) minimizing carbon emission from farming, as depicted by the following relation:

$$AF \text{ carbon sequestration} = \text{carbon storage} - \text{carbon emission} \tag{6.1}$$

So, for a successful AF, relation (6.1) must hold good as per the following condition:

$$AF \text{ carbon sequestration} = +\left(\text{carbon storage} - \text{carbon emission}\right) \tag{6.2}$$

$$\text{provided, Carbon storage} > \text{carbon emission} \tag{6.3}$$

If + (carbon storage — carbon emission) is considered as Carbon yield, or carbon farming, It is obvious that

$$\text{Carbon yield} = \text{carbon farming} \tag{6.4}$$

6.2 Minimizing Carbon Emission Through Agroforestry

Agroforestry systems, which integrate trees into agricultural landscapes, have increasingly been recognized as a promising and effective land use strategy to enhance soil health and mitigate the impacts of climate change, particularly through increasing both aboveground and soil carbon stocks. These systems are vital not only for improving soil fertility but also for addressing the growing concern of greenhouse gas emissions, particularly carbon dioxide (CO_2), methane (CH_4), and nitrous oxide (N_2O), the latter being the third most potent greenhouse gas. Numerous scientific studies have demonstrated the potential of agroforestry to sequester carbon, thus playing a crucial role in reducing the atmospheric concentrations of greenhouse gases.

6.2.1 Minimizing Carbon Emission in Tree and Crop Cultivation

In the context of climate change, one of the key factors contributing to greenhouse gas emissions is the use of nitrogen-based fertilizers. These fertilizers are widely applied in modern agriculture to boost crop yields. However, they also spur the microbial activity in the soil that leads to the production of nitrous oxide, a gas that is approximately 298 times more effective at trapping heat in the atmosphere than carbon dioxide. Furthermore, nitrous oxide is detrimental to the ozone layer, contributing to its depletion. This process underscores the need for alternative agricultural practices that minimize the reliance on synthetic fertilizers and their associated environmental impacts.

Agroforestry systems, particularly those that incorporate leguminous trees, offer a sustainable solution. Leguminous trees, along with a few non leguminous species, have the ability to fix nitrogen from the atmosphere into the soil through a symbiotic relationship with nitrogen fixing bacteria in their root nodules. This natural process not only enriches the soil with nitrogen, reducing the need for synthetic nitrogen fertilizers, but also helps to maintain a more balanced and sustainable nutrient cycle in agricultural ecosystems. Nitrogen fixation is particularly beneficial in agroforestry systems, as it enhances soil fertility, improves crop productivity, and reduces the environmental footprint of farming.

Moreover, trees play a crucial role as nutrient pumps. Their roots can access deep layers of the soil that are otherwise beyond the reach of most crops. These deeper rooted trees take up essential nutrients that have been released through the weathering of rocks in the subsoil, which would otherwise remain unavailable to surface crops. This nutrient uptake is especially valuable in nutrient depleted soils, as trees help cycle nutrients and make them available to surrounding crops. In addition, trees also improve soil structure and fertility through the exudation of growth promoting substances in their rhizosphere, which benefits soil microbes and enhances nutrient availability.

The roots of trees also act as a "safety net" by trapping and recycling nutrients that would otherwise be lost to leaching, a common problem in intensive agricultural systems. Leaching occurs when excess water washes away essential nutrients from the topsoil, leading to nutrient depletion. However, the presence of nets of tree roots can prevent this loss by absorbing nutrients and, through their association with mycorrhizal fungi, recycling nutrients within the soil. Mycorrhizal networks, which form a symbiotic relationship with plant roots, further enhance the ability of trees to capture and recycle nutrients, ensuring that valuable soil nutrients are retained and made available to the entire ecosystem.

Empirical studies have demonstrated the significant contributions of leguminous trees in agroforestry systems to nutrient cycling. For example, in alley cropping systems, where rows of trees are planted between crop fields, leguminous trees have been found to contribute substantial amounts of nitrogen and other essential nutrients to the soil. In one study, leguminous trees were shown to provide up to 358 kg of nitrogen (N) per hectare per year, along with 28 kg of phosphorus (P), 232 kg of potassium (K), 144 kg of calcium (Ca), and 60 kg of magnesium (Mg) (Palm 1995). These nutrients are critical for crop growth and contribute to the overall fertility and productivity of agricultural systems, ensuring long term sustainability without the need for excessive chemical inputs.

Thus, agroforestry systems, particularly those incorporating nitrogen fixing trees, offer a multifaceted approach to sustainable agriculture. They help mitigate the negative environmental effects of synthetic fertilizers, enhance soil carbon sequestration, improve soil structure and fertility, and reduce the risk of nutrient leaching. Moreover, by improving the resilience of farming systems to climate change, agroforestry can contribute to both food security and environmental sustainability. As such, agroforestry represents a powerful tool in the fight against climate change, providing a model for integrated land management that supports both ecological health and agricultural productivity (Sathaye and Ravindranath 1998; Verchot et al. 2007; Verma et al. 2014; Arora et al. 2014; Goswami et al. 2013; Kanime et al. 2013; Box 6.1).

6.2.2 Minimizing Methane Emission in Livestock Management

The integration of improved animal nutrition, anti-methanogenic feed additives, and effective livestock management practices is pivotal in reducing methane emissions while enhancing livestock productivity. Programs like the Ration Balancing

Program focus on optimizing feed conversion efficiency, thereby minimizing methane production through more digestible and energy rich diets. By incorporating high-quality forages and essential feed additives, such as tannins and essential oils, the health and digestion of animals are improved, leading to both better animal welfare and lower emissions. Notably, the use of feed supplements like *Harit Dhara* of BAIF (Aware 2023), which reduces protozoa in the rumen, has been shown to lower methane emissions by up to 20%, while also boosting milk production and weight gain. In addition to improved feeding practices, selective breeding and enhanced management strategies are key to optimizing feed utilization and reducing emissions per unit of output. Breeding animals that naturally produce less methane and are more efficient in converting feed into milk or meat helps to meet consumer demand while simultaneously reducing the environmental impact (Box 6.2, Fig. 6.1).

Moreover, livestock waste management plays a critical role in emission reduction. Innovations required to offer a practical solution for converting animal dung into biogas for cooking, reducing methane emissions, for example, reducing by up to 4 tons of CO_2 equivalent per year per unit through Integrated Renewable Energy and Sustainable Agriculture (IRESA) Model based Biogas Unit, developed by BAIF (Aware 2023). The byproduct, digested slurry, can be used as organic fertilizer, improving soil quality and supporting sustainable farming practices.

By adopting a holistic approach that combines optimized nutrition, selective breeding, manure management, and the use of biogas technology, we can reduce methane emissions, improve agricultural productivity, and promote sustainable livelihoods, particularly in rural communities. This approach not only contributes to mitigating global warming but also fosters a more sustainable, efficient, and resilient agricultural system (Table 6.1, Box 6.3).

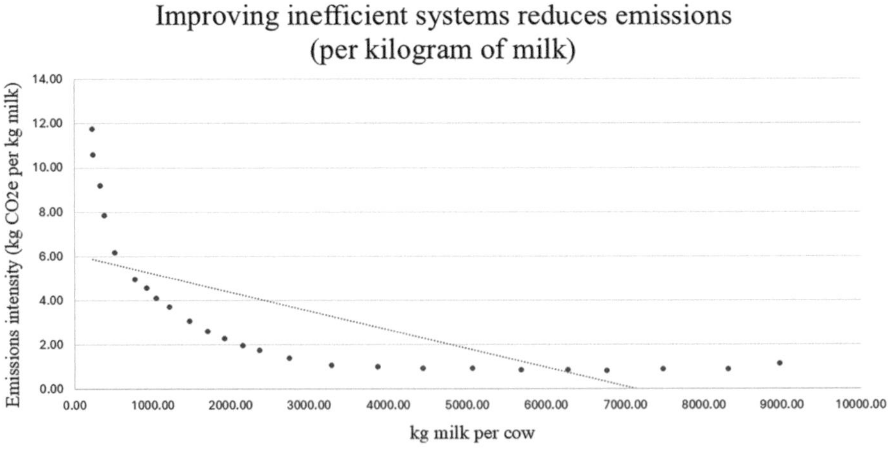

Fig. 6.1 Efficient livestock management reducing methane emission. (Based on Gerber et al. 2011 as referred in Fig. 23 in Gerber et al. 2013)

Box 6.1 GHGs, GWP, CO_2e measurement units

CO_2e (carbon dioxide equivalent) is a key metric used to quantify and compare the global warming potential of various greenhouse gases, helping track progress in reducing emissions and combating climate change. By measuring emissions in terms of CO_2e, we can assess the impact of gases like methane, nitrous oxide, and ozone, and determine whether efforts to transition to clean, renewable energy are effectively curbing global warming.

Greenhouse gases are essential for regulating the earth's temperature, but human activity has disrupted the natural carbon cycle, causing an overabundance of these gases in the atmosphere. This excess, particularly of carbon dioxide, methane, and nitrous oxide, is trapping too much energy from the sun, leading to global warming and climate change. Scientists use the CO_2e metric to quantify the impact of these gases, which collectively contribute to the ongoing rise in global temperatures.

The main gases measured using CO_2e include carbon dioxide (CO_2), methane (CH_4), nitrous oxide (N_2O), and several others, as the key contributors to global warming and climate change, with varying levels of harmful impact on the environment, but all of them play a role in raising global surface temperatures. Other gases are hydrofluorocarbons (HFCs), perfluorocarbons (PFCs), sulfur hexafluoride (SF_6), nitrogen trifluoride (NF_3).

CO_2e (carbon dioxide equivalent) provides a standardized metric that allows scientists to quantify and compare the impact of various greenhouse gases on global warming in a unified unit.

MtCO_2e (metric tonne of carbon dioxide equivalent) is a unit of measurement used to quantify the impact of different greenhouse gases on the environment, standardizing their effects in terms of carbon dioxide emissions to assess their contribution to global warming.

The key difference between carbon dioxide and carbon emissions is that carbon dioxide (CO_2) is a specific compound, whereas carbon emissions refer to a broader range of greenhouse gases, including methane (CH_4), nitrous oxide (N_2O), and others, which contribute to climate change. While CO_2 is naturally occurring, excess levels of CO_2, along with these other gases, can have harmful environmental impacts.

The greenhouse effect is a natural process that keeps earth warm enough to support life, but human activities like burning fossil fuels and deforestation are intensifying it, leading to rising temperatures, melting glaciers, and higher sea levels, disrupting the planet's carbon cycle and threatening ecosystems.

Global warming potential (GWP) is a metric used to compare the warming effect of different greenhouse gases over a specific time period, with carbon dioxide (CO_2) serving as the baseline at a GWP of 1. A gas with a higher GWP contributes more to global warming per ton of emissions than CO_2 over the same timeframe.

The global warming potential (GWP) of greenhouse gases varies significantly, with methane having a GWP of 28–36 over 100 years, nitrous oxide ranging from 265 to 298, and chlorofluorocarbons (CFCs) exhibiting an exceptionally high GWP due to their density and ability to trap significant amounts of heat.

Since the Industrial Revolution (spanning over the period of 1760–1840, the term coined by the English Economic Historian,1852–83), the rapid increase in CO_2 levels from burning fossil fuels has disrupted the Earth's carbon cycle, leading to a buildup of greenhouse gases that trap heat and disrupt the planet's climate regulation.

An overwhelming amount of carbon dioxide exceeds the natural capacity of earth to absorb it, leading to rising global temperatures, more intense storms, altered precipitation patterns, and rising sea levels, disrupting ecosystems, causing extreme weather events, and threatening human life and infrastructure.

Total greenhouse gas emissions, measured in kilotons (kt) of CO_2 equivalent, encompass carbon dioxide (CO_2) emissions, excluding those from short-cycle biomass burning (such as agricultural waste and savanna fires), but including emissions from other forms of biomass burning like forest fires, post-burn decay, peat fires, and decay of drained peatlands. Additionally, the total emissions account for all anthropogenic sources of methane (CH_4), nitrous oxide (N_2O), and fluorinated gases (F-gases), including hydrofluorocarbons (HFCs), perfluorocarbons (PFCs), and sulfur hexafluoride (SF_6).

The GHG totals are expressed in CO_2 equivalent using the GWP100 metric of the second assessment report of IPCC and include CO_2 (GWP100 = 1), CH_4 (GWP100 = 21), N_2O (GWP100 = 310) and F-gases (c-C_4F_8 GWP = 8700, C_2F_6 GWP = 9200, C_3F_8 GWP = 7000, C_4F_{10} GWP = 7000, C_5F_{12} GWP = 7500, C_6F_{14} GWP = 7400, C_7F_{16} GWP = 7820, CF_4 GWP = 6500, HFC-125 GWP = 2800, HFC-134a GWP = 1300, HFC-143a GWP = 3800, HFC-152a GWP = 140, HFC-227ea GWP = 2900, HFC-23 GWP = 11,700, HFC-236fa GWP = 6300, HFC-245fa GWP = 858, HFC-32 GWP = 650, HFC-365mfc GWP = 804, HFC-43-10-mee GWP = 1300, SF_6 GWP = 23,900).

The release of man-made greenhouse gases, particularly carbon dioxide (CO_2), into the atmosphere disrupts the earth's radiative balance, contributing to global warming and climate change. This is largely driven by human activities such as the burning of fossil fuels for energy, deforestation, and industrial processes like cement production. As a result, we are seeing rising global temperatures, melting ice caps, and rising sea levels, which, in turn, impact agriculture, ecosystems, and human societies. The term "emission intensity" helps to quantify the environmental impact of various activities or fuels by comparing the amount of a pollutant emitted relative to the intensity of the activity. While CO_2 is the most significant greenhouse gas in terms of global warming, other gases like methane (CH_4) and nitrous oxide (N_2O) also play important roles, especially in agricultural economies. To better assess the overall impact of a country or sector on climate change, it is important to consider the total emissions of all greenhouse gases, which are often expressed in CO_2 equivalents. Global efforts, such as the *Kyoto* protocol, aim to reduce these emissions and mitigate the adverse effects of climate change.

(continued)

(Ref: 1. https://www.inspirecleanenergy.com/blog/clean-energy-10-l/what-is-co2e#fu3
2. World Bank (2024). https://databank.worldbank.org/metadataglossary/world-development-indicators/series/EN.ATM.GHGT.KT.CE)

Box 6.2 Minimizing methane emissions from ruminants[a]

The challenge: methane emissions from ruminants

- **Ruminant livestock** (cattle, sheep, and goats) are responsible for about half of agricultural emissions.
- The largest source of these emissions is **enteric methane**, produced by microbes in the ruminant stomach.
- **Enteric methane** accounts for a significant portion of agricultural GHG emissions, contributing to climate change.

Key strategies to reduce methane emissions

1. **Improving feed quality**
   - Better feeds increase animal productivity (more meat and milk per kg of feed).
   - Improving feed quality in poorer countries (with lower-quality feeds) has the greatest potential to reduce methane emissions.
   - Efficiency gains: Moving from low- to average-quality feeds results in significant emissions reductions, especially in inefficient systems.
2. **Microbial manipulation**
   - Vaccines: Potential to alter stomach microbiota to reduce methane production.
   - Selective breeding: Identifying and breeding animals that naturally produce less methane.
   - Feeds and supplements: Special dietary interventions, drugs, or supplements aimed at reducing methane emissions.

Challenges:

Many attempts to manipulate the microbiome have not been successful (e.g., methane-inhibiting drugs quickly loose effectiveness).

Emerging technology: 3-NOP (3-Nitrooxypropan)

- Promising solution: A compound called 3-NOP has shown to reduce methane emissions by 30%.
- Additional benefits: It may increase animal growth rates, making it a dual-purpose solution.

Challenges for widespread adoption:

- Requires daily dosing to be effective, making it challenging for grazing operations.
- Further refinement needed for cost-effective application on large scales.

Policy recommendations for scaling solutions

1. **Incentivize private sector innovation:**
   - Offer incentives for the development and use of compounds like 3-NOP once they prove effective at a reasonable cost.
2. **Fund large-scale demonstration projects:**
   - Invest in demonstration projects to test and refine the application of 3-NOP and similar compounds at a commercial scale.
3. **Sustain public research:**
   - Maintain public research funding into new compounds and technologies to reduce methane emissions from enteric fermentation.

(continued)

Box 6.2 (continued)

The path forward

- Improving livestock productivity through better feed quality and novel compounds can reduce methane emissions significantly.
- Emerging technologies like 3-NOP offer substantial promise but require further research, development, and scaling.
- Government action is crucial to create the incentives and infrastructure necessary for these innovations to succeed.

Key takeaways

- Ruminants are major contributors to methane emissions.
- The greatest opportunities to reduce emissions lie in improving feed quality in poorer countries.
- 3-NOP shows promise in reducing methane emissions and increasing growth rates but requires further refinement.
- Governments should support R&D, large-scale demonstrations, and provide incentives for private sector innovation in this space.

[a](Ref. Gerber et al. 2013; Hristov et al. 2013, 2015; Martínez-Fernández et al. 2014; Reynolds et al. 2014; Romero-Perez et al. 2015; Aware 2023)

Box 6.3 Salient Features for Reducing Emissions from Manures Left on Pasture[a]

Nitrogen in manure *vs.* fertilizer emissions: Nitrogen in faeces and urine is converted into nitrous oxide at about twice the rate of nitrogen in fertilizer, according to IPCC emissions factors.

Productivity improvements in manure: By 2050, improvements in productivity have already been factored in, lowering emissions from manure on pasture by 25% compared to 2010. Further feed efficiency improvements could lead to additional, though modest, emission reductions.

Limited mitigation potential: Other studies suggest limited global potential for mitigating nitrous oxide emissions from manure. However, there is cautious optimism based on the development of two emerging technologies.

Technologies to inhibit nitrous oxide emissions:

(1) Chemical nitrification inhibitors: These have been effective in New Zealand when applied to pastures and, in a few experiments, when ingested by cows.

(2) Biological nitrification inhibition: Some varieties of *Brachiaria* grass in Latin America produce almost no nitrous oxide when manure is deposited on them. This trait could be bred into other grasses.

(3) Targeting emissions more efficiently: New research suggests that current emissions factors for unmanaged manure might be too high. Emissions rates are higher in wetter areas, allowing for more targeted and cost-effective application of these techniques to intensively grazed fields.

Call for increased research and funding: Research funding to reduce nitrification of manure is currently very limited. There is a strong call for governments and research agencies to substantially increase funding for these efforts.

Policy recommendations: Governments should commit to implementing regulatory or financial incentives to encourage private sector research and adoption of these techniques once they become viable.

[a](Ref. Galbally et al. 2010; Doole and Paragahawewa 2011; Barneze et al. 2014; Sordi et al. 2014; Mazzetto et al. 2014; Pelster et al. 2016; Ward et al. 2016; Byrnes et al. 2017; FAO 2022; Aware 2023)

Table 6.1 Improved manure management systems for managing emissions in agricultural practices[a]

Key points	Details
Managed *vs.* unmanaged manure	Manure is "managed" when animals are raised in confined settings and manure is removed and disposed of. Manure deposited directly on fields by ruminants is considered "unmanaged"
Emissions from managed manure	Managed manure generates methane and nitrous oxide
Contribution to emissions by animal type	– Pigs: −50% of emissions – Dairy cows: −33% of emissions
Dry *vs.* wet systems	– Dry systems manage 40% of total managed manure emissions despite having low-emissions rates – Wet systems can generate up to 20 times more emissions per ton of manure
Mitigation potential	Mitigation potential for dry systems is low due to high costs per ton of emissions
Separation of liquids and solids	Separating liquids from solids in manure has underappreciated potential to reduce emissions and can also reduce hauling costs and increase the value of manure as fertilizer
Cost of mitigation in wet systems	Existing wet systems are easier to mitigate, with an estimated $22 per ton of CO_2e mitigated for a sophisticated system for managing pig manure in North Carolina
Impact on retail price	Sophisticated systems for managing pig manure would add only 2% to the retail price of pork
Digesters	Digesters convert manure into methane for energy use, reducing emissions if manure is managed in wet form. Must have safeguards to prevent methane leakage
Broader benefits of manure management	Improving manure management can address environmental pollution, human health, and nuisance concerns. Mitigation can often be "free" from a socioeconomic perspective
Promising strategies for manure management	– Phased regulation of farms to encourage innovation – Government-funded competition to develop cost-effective technologies – Programs to detect and remediate digester leaks

[a](Ref. IPCC 2006, Table 10.17, Gerber et al. 2013; Mathews and Haley 2015; Schnepf 2015; Aware 2023)

6.3 Minimizing Carbon Emissions Through Agroforestry: A Policy Matter

In recent decades, carbon management has emerged as a crucial focal point in global efforts to combat climate change. With the intensifying threat posed by rising global temperatures, the identification and implementation of effective mitigation strategies to reduce carbon emissions have never been more urgent. The rapid increase in greenhouse gas (GHG) emissions has been a major contributor to climate change, with the global emission of carbon dioxide equivalent (CO_2e) escalating almost linearly from 746.5 Mt CO_2e in 1970 to a staggering 3375 Mt CO_2e in 2018. This steady rise in emissions is not only a reflection of industrialization and urbanization but also highlights the urgent need for systemic changes to reduce carbon footprints across sectors (Tables 6.2, 6.3, 6.4, 6.5, and 6.6).

Table 6.2 Fertilizer management for increasing nitrogen use efficiency (NUE) and reducing emissions[a]

Salient features	Details
Emissions from fertilizers (2010)	1.3 Gt CO_2e in 2010, mainly from the manufacture, transportation, and application of nitrogen
Projected emissions by 2050	Emissions expected to rise to 1.7 Gt CO_2e in the baseline scenario by 2050
Nitrogen Use Efficiency (NUE)	Globally, less than half of the nitrogen applied to farm fields is absorbed by crops. The rest is lost to the environment through runoff or as gases, including the potent heat-trapping gas nitrous oxide
Factors affecting NUE	Variation across countries and individual farms in nitrogen application rates and NUE
Mitigation strategies	Focus on changing agronomic practices, improving nitrogen application efficiency. High NUE possible but expensive compared to cheap nitrogen fertilizer
Innovations for improving NUE	Nitrification inhibitors, enhanced efficiency fertilizers, and biological nitrification inhibition show potential but are underused and underfunded
NUE improvement scenarios	Achieving 71% global NUE by 2050 could reduce emissions by 600 million tons, keeping nitrogen emissions close to 2010 levels
Government policy recommendations	– Implement flexible regulatory targets for fertilizer companies – Shift subsidies to support higher NUE – Fund research into biological nitrification inhibition – Support demonstration projects with high nitrogen-using farmers

[a](Ref. Zhang et al. 2015a, b; Aware 2023)

Table 6.3 Adoption of emission-reducing rice management and varieties[a]

Salient aspects	Details
Global impact of paddy rice	At least 10% of global agricultural GHG emissions in 2010, primarily methane
Technical potential for mitigation	High potential to mitigate rice emissions with economic benefits such as higher yields and reduced water use
Main mitigation options	1. Increase rice yields: High yields reduce the area needed, thus reducing emissions
	2. Remove rice straw before reflooding: Reduces methane production, straw can be used for production of mushrooms and bioenergy
	3. Reduce duration of flooding: Shortening flood duration reduces methane-producing bacteria
	4. Breed lower-methane rice: Some rice varieties emit less methane, but these traits are not widespread
Drawdown strategy (water level reduction)	Reducing flooding through "drawdowns" can reduce methane by up to 90%
Current practices	In China and Japan, drawdowns increase yields and reduce methane, but this is not observed in the USA
Benefits of reduced water levels	Saves irrigation water, especially at the fam scale
Obstacles	1. Dry planting increases weed growth
	2. Drawdowns are only feasible with reliable water replacement and flat fields
	3. Drawdowns may increase nitrous oxide emissions requiring nitrification inhibitors
Proposed strategies	1. Engineering analyses: Identify farmers with suitable irrigation systems for drawdowns and incentivize them
	2. Major breeding effort: Focus on developing lower-methane rice varieties
	3. Boost rice yields: Continue efforts in breeding and management to improve yields

[a](Ref. Zou et al. 2005; Itoh et al. 2011; Joshi et al. 2013; Su et al. 2015; Jiang et al. 2017; Aware 2023)

India, in particular, is a critical player in this global challenge. As of the latest data, India ranks as the third largest emitter of carbon dioxide in the world, trailing only China and the USA. The energy sector in India contributes a significant 75% of the nation's overall carbon emissions, totalling approximately 2129 $MtCO_2e$. With a rapidly growing population and expanding energy demands, India faces the dual challenge of improving its energy access while reducing its carbon emissions. As a result, India's role in the global effort to mitigate climate change has become ever more pivotal.

Recognizing this, at the COP26 summit held in Glasgow in 2021, India set forth a bold and ambitious vision for the country's climate future. There India announced this country's commitment to achieving net zero carbon emissions by 2070, a target that underscores the country's determination to be part of the global solution to climate change. In addition to this long-term goal, India also introduced the

Table 6.4 Reducing emissions through increased agricultural energy efficiency and shifting to non-fossil energy sources[a]

Features	Salient points
Fossil energy emissions in agriculture (2050)	Emissions from fossil energy use in agriculture expected to remain around 1.6 Gt CO_2e/year in 2050 in baseline scenario
Assumption on energy efficiency	A 25% increase in energy efficiency is offset by a 25% increase in energy use, resulting in no net reduction in emissions
Mitigation options	Rely on increasing energy efficiency and transitioning to renewable energy sources
Energy efficiency potential	Some country-level studies show large potential for efficiency improvements (e.g., alternative water pumps in India, cassava-drying in Africa)
Sources of agricultural energy emissions	65% of agricultural energy emissions in 2050 will come from on-farm energy use
Renewable energy for on-farm use	Solar and wind energy can provide heating and power on farms; innovative solar heating systems needed for replacing coal
Diesel fuel mitigation	Difficult to mitigate use of diesel for tractors; solutions may include fuel cells with hydrogen power, battery-powered equipment, or synthetic carbon-based fuels
Hydrogen for nitrogen fertilizer emissions	Renewable hydrogen could mitigate 85% of emissions from nitrogen fertilizer synthesis, which is currently energy-intensive
Research on hydrogen production	Significant research is ongoing in hydrogen production using solar electricity, with declining costs for solar electricity
Enhanced efficiency goals	A 75% reduction in emissions per unit of energy use (compared to 25% baseline) could reduce the GHG mitigation gap by 8%
Policy and research recommendations	– Integrate low-carbon energy and efficiency in development efforts and supply chains – Fund research on nitrogen production from renewable electricity and solar-based hydrogen – Governments should regulate emissions from fertilizer production once low-carbon technologies are viable

[a](Ref. Goodrich et al. 2012; RTB 2016; Aware 2023)

"One-Word Movement"-LIFE, which stands for Lifestyle for Environment. This initiative emphasizes the significant role that individual and collective lifestyle choices play in addressing climate change. By promoting sustainable living practices, reducing waste, and embracing environmentally friendly technologies, the LIFE movement aims to inspire a global shift toward a more sustainable and eco-conscious way of living.

Table 6.5 Reducing emissions through focussing on realistic options to sequester carbon in soils[a]

Salient points	Details
Soil carbon sequestration	Achieving soil carbon sequestration is challenging. The two main strategies are adding more carbon or reducing losses
No-till farming	Initially thought to prevent carbon loss, no-till farming offers minimal or no benefits at deeper soil levels and can negatively affect crop yields. Many farmers still plough periodically, reducing carbon gain
Mulch and manure	Adding mulch or manure to soils can double-count the carbon, which would have contributed elsewhere
Crop residues for soil carbon	Using crop residues for soil carbon requires replacing animal feed, leading to additional carbon costs for land used to grow that feed
Nitrogen and soil carbon	Nitrogen is essential for soil microorganisms that convert organic matter into soil carbon. Nitrogen limitations, especially in regions like Africa, hinder soil carbon buildup
Uncertainty in soil carbon storage	There is a lack of understanding regarding why some soil carbon is retained longer than others. Some croplands may actually be losing carbon overall
Mitigation strategy focus	Given the challenges, the focus should be on stabilizing soil carbon, avoiding further losses, and pursuing no-regrets strategies that offer co-benefits
Recommended no-regrets strategies	– Avoid conversion of carbon-rich ecosystems (e.g., forests, permafrost soils) – Increase productivity of grasslands and croplands (adds carbon in roots and residues) – Increase agroforestry to build above-ground carbon – Focus on soil carbon efforts in areas critical for food security

[a](Ref. Kirkby et al. 2011; Powlson et al. 2014, 2016; Sanderman et al. 2017; van Groenigen et al. 2017; Aware 2023)

To achieve these lofty goals, India has outlined a comprehensive set of five strategies, known as *Panchamrit*, to guide the nation toward a low carbon future. The first strategy focuses on achieving net zero emissions by 2070. While this is a long term objective, it serves as a clear signal of India's commitment to the global climate agenda. The second strategy aims to reduce the carbon intensity of India's economy by 45% by 2030, relative to 2005 levels. This target is particularly important as it directly addresses the efficiency with which the country uses energy, aiming to decouple economic growth from increased carbon emissions.

The third strategy involves meeting 50% of India's energy needs through renewable sources by 2030. This shift toward cleaner energy is pivotal, as India's growing

Table 6.6 Reducing emissions through application of flexible technology forcing regulations[a]

Key points	Recommendations	Expected benefits
GHG mitigation in agriculture	Voluntary approaches will not be sufficient to close the GHG mitigation gap. Flexible regulations are needed to spur technological development	Cost-effective mitigation, technological advancement, and addressing agricultural emissions
Fertilizer regulation	Create regulatory systems similar to the U.S. fuel efficiency standards for auto manufacturers. Mandate increasing percentages of "enhanced efficiency" fertilizers (e.g., nitrification inhibitors)	Incentivizes fertilizer manufacturers to improve products and technologies, benefiting farmers and reducing nitrogen release
Manure management	Phase in pollution controls for manure management, starting with new and large facilities and gradually extending to medium and small farms	Improved pollution control, reduced environmental impact, and gradual adaptation for farms of different sizes
Enteric methane inhibitors	Commit to requiring the use of cost-effective enteric methane inhibitors (e.g., \$25 per ton CO_2e) if effective solutions are developed	Incentivizes development of effective mitigation technologies, providing market certainty for the private sector
Costs and co-benefits	Many mitigation options may have initial costs but are cost-effective in the long run, with additional co-benefits like reducing water/ air pollution and controlling diseases from livestock waste	Long-term cost-effectiveness, environmental and health benefits, and potential to cover costs through technology evolution
Technological evolution	Technologies such as nitrification inhibitors and methane inhibitors may require regulation to evolve and be deployed effectively	Technological advancement, cost reduction, and broader deployment of effective mitigation solutions

[a](Ref. Padhee 2018; Aware 2023)

energy demands can no longer be met by fossil fuels alone without exacerbating the carbon crisis. By investing in solar, wind, hydro, and other renewable energy sources, India can reduce its reliance on coal and oil, which are major contributors to carbon emissions.

The fourth strategy is to generate 500 gigawatts (GW) of energy from non-fossil fuel sources by 2030. This ambitious target not only supports the transition to a cleaner energy grid but also positions India as a leader in renewable energy capacity, creating opportunities for green jobs and technological innovation.

Finally, the fifth strategy involves reducing carbon emissions by one billion tons by 2030. This short-term target plays a critical role in setting the stage for the longer term objectives and demonstrates India's commitment to making substantial reductions in emissions in the near future.

Among these targets, two key short-term strategies, *viz.* reducing carbon emissions by one billion tons and achieving a 45% reduction in carbon intensity by 2030, are particularly important in the journey toward net zero emissions by 2070. Achieving these targets will require a multifaceted approach that combines policy changes, technological innovation, shifts in industrial practices, and changes in consumer behavior. The success of these short-term goals will lay the foundation for the more ambitious targets in the long term.

One of the strategies that has gained increasing attention as an effective tool for reducing greenhouse gas emissions is agroforestry. Agroforestry involves integrating trees and shrubs into agricultural landscapes, creating a system that not only promotes biodiversity but also acts as a carbon sink. By sequestering carbon in the form of biomass and soil organic carbon, agroforestry systems have the potential to significantly reduce GHG emissions. Moreover, agroforestry can provide numerous co-benefits, including improved soil health, increased water retention, and enhanced resilience to climate change. In the context of India's carbon management strategy, agroforestry offers a cost effective and sustainable approach to both mitigate climate change and improve the livelihoods of rural communities (MoEFCC 2021, Verma et al. 2021).

The role of lifestyle choices in tackling climate change cannot be overstated. Through the LIFE movement, India is encouraging individuals, communities, and businesses to rethink their environmental impact and adopt more sustainable practices. By promoting energy efficiency, waste reduction, sustainable consumption, and eco-friendly technologies, the LIFE movement is aimed at fostering a culture of environmental responsibility. If embraced widely, this movement can lead to significant reductions in carbon emissions across multiple sectors, ranging from transportation and housing to food production and waste management.

India's commitment to addressing climate change is a significant step in the global effort to mitigate the impacts of rising temperatures. However, achieving the ambitious targets outlined in the *Panchamrit* will require concerted effort and collaboration from all sectors of society, *viz.* government, industry, civil society, and individuals. India's policies must be supported by robust infrastructure investments, technological innovation, and widespread public engagement. Additionally, international cooperation will be essential to ensure that developing nations like India receive the necessary financial and technological support to transition to low carbon economies while meeting their development goals.

In conclusion, carbon management and climate change mitigation are at the heart of India's development agenda for the coming decades. By setting ambitious targets for emission reductions, increasing renewable energy capacity, and promoting sustainable lifestyles through initiatives like LIFE, India is positioning itself as a key actor in the global effort to combat climate change. However, achieving net zero emissions by 2070 and meeting the intermediate targets set for 2030 will require continued commitment, innovation, and cooperation at all levels of society. Only through a holistic approach that integrates policies, technology, and behavior change can India hope to meet its climate goals and secure a sustainable future for its people and the planet.

6.4 Estimation of Carbon Sequestration in AFS

The process of carbon sequestration within agroforestry systems (AFS) begins by calculating the total biomass in AFS and then applying a factor of 0.55 (Carvalho et al. 1998; Malhi et al. 2002) for wood, and 0.5 for all other plant biomass (Ma et al. 2018) to estimate the carbon stock, as carbon typically constitutes about 55% of the dry biomass (Fig. 6.2).

The biomass in AFS is divided into distinct categories: tree biomass, crop biomass, and root biomass, with each category further classified into above ground and below ground components.

6.4.1 Tree Biomass (ton/m^3)

The tree trunk, provided as a trimmed cylindrical body or a wood log, biomass is estimated using an allometric method, where the volume of the tree trunk (V, m^3) is multiplied by the wood density (WD, kg/m^3). For estimating V (m^3), it is required to estimate the basal area of the tree trunk (Ab, sq. cm) from the measured value of girth (cm) at DBH, i.e., diameter at breast height, then multiplied with H, i.e. breast height (BH) considering the standard BH of the tree at 1.3 m (Chapman 1921; Condit 1998; Ponce-Hernandez et al. 2004; Dahdouh-Guedes and Codedam 2006) or 1.37 m (Chapman 1921; Brokawa and Thompson 2000) above the base of the

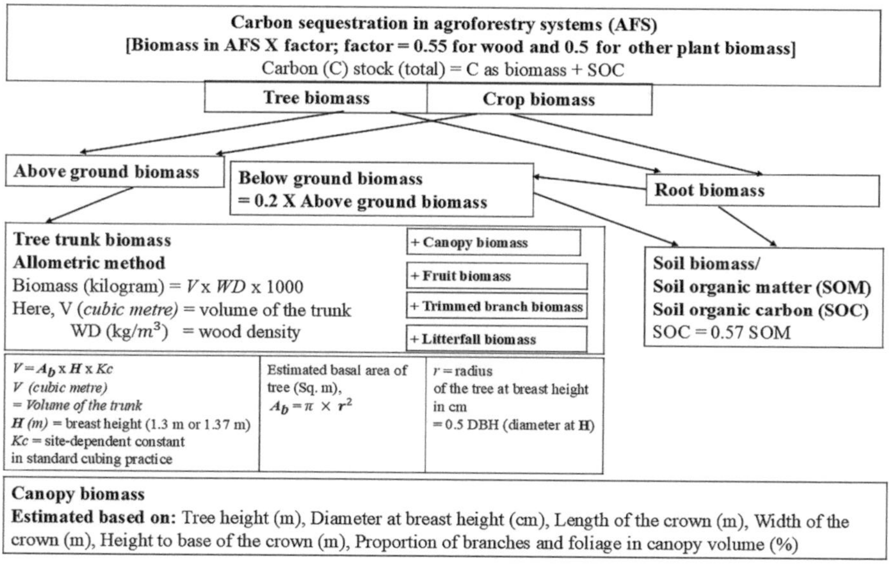

Fig. 6.2 Estimation of carbon sequestration in agroforestry systems—Flow chart. (Ref. Chapman 1921; MacDicken 1997; Carvalho et al. 1998; Condit 1998; Brokawa and Thompson 2000; Malhi et al. 2002; Ponce-Hernandez et al. 2004; Dahdouh-Guedes and Codedam 2006; Ma et al. 2018; Rai et al. 2021)

tree; and multiplied by a site dependent constant, kc in standard cubing practice. Other components of tree biomass, such as the canopy, fruit biomass, trimmed biomass, and litterfall biomass are also considered. These contribute to the overall carbon sequestration potential of the agroforestry system (Fig. 6.2).

"Quarter Girth Formula": It is in practice to estimate *V* by multiplying cross-sectional area of the log with breast height.

Here,

$$\text{Girth or}, g = 2\pi r, \tag{6.5}$$

or

$$r = \text{radius of the tree at } DBH = g/2\pi \tag{6.6}$$

Here, DBH is diameter at breast height, *H* considered at 1.3 m or 1.3 7 m above base.

It is conventionally followed that

$$V = \pi r^2 H = (g/4)^2 \times H \tag{6.7}$$

now, considering Eq. (6.6), V can be deduced as

$$V = \pi (g/2\pi)^2 \times H \tag{6.8}$$

or,

$$V = (g/4)^2 \times H \tag{6.9}$$

It is this regularly followed made easy, "Quarter Girth Formula," a simple practical formula, here, the Eq. (6.9), that approximates π as 4 instead of 3.14 results in a derived volume that is only 78.5% of the true volume. However, adjustments for factors like bark and loss during squaring compensate for this discrepancy, making the method accurate in the end. Despite this inaccuracy, its simplicity has made it widely adopted for practical use.

The point of measurement of diameter of the log in volume, i.e., DBH.

For most species, the point 1.37 m above ground has been accepted by foresters for measurement of diameter, as it falls above the swelling and at a convenient height for the use of callipers. This height is also used in England and India. In Continental Europe, 1.3 m is the standard height.

This measurement at 1.37 m is termed **diameter at breast height** and is abbreviated both in speech and record to **DBH** measurement outside bark is always indicated by this abbreviation.

In the Philippines and other tropical countries, it will be impossible to use a similar height for many species due to the development of buttresses on the trunks. Such species will probably have to be measured either above the flare, or at a height of 4.88 – 6.1 m, by eye, using the 1.37 m standard point only for species and types which permit it.

Where DBH is adhered to for species like Western larch, red cedar, or Douglas fir on the Pacific Coast, butt swelling greatly interferes with the uniformity of the volumes for these species for trees of given diameters when compared with other species like western yellow pine, as its swelling seldom reaches this height. This apparent difference in volume may be from 20 to 40% in favor of the pine (Chapman 1921).

6.4.2 Crop Biomass

Biomass from crops, whether above or below ground, contributes to the carbon stock in agroforestry systems but is calculated separately from tree biomass.

6.4.3 Root Biomass

Biomass associated with root systems is crucial, with below ground biomass often accounting for a significant portion of total biomass and contributing to soil organic matter (SOM).

6.4.4 Soil Biomass

The soil biomass component is tied to SOM, which is crucial in carbon sequestration. Soil organic carbon (SOC) is then derived as 0.57 of SOM. This means that approximately 57% of soil organic matter is considered as carbon, which plays a significant role in long term carbon storage in agroforestry systems (Ponce-Hernandez et al. 2004).

This detailed classification of biomass in agroforestry is very much helpful for an accurate and comprehensive estimation of carbon sequestration in AFS, emphasizing both above and below ground (roots, soil organic matter) contributions, as well as the critical relationship between soil organic matter and carbon storage. While soil organic matter significantly influences the carbon dynamics in AFS. Thus, such a comprehensive evaluation requires considering all biomass components to ensure precise carbon sequestration estimates (Fig. 6.2).

6.5 Applying AF Technology of Carbon Sequestration for Increasing Energy Efficiency Toward Managing Climate Change

The agricultural sector, facing escalating energy demands and environmental pressures, must evolve to address the twin challenges of growing food demand and increasing energy consumption. As energy use in agriculture is projected to rise sharply in the coming decades, driven by both the needs of food production and rising energy costs, the sector is confronted with the task of improving energy efficiency while maintaining or enhancing production outputs. This challenge is compounded by the expectation that global energy production will not keep pace with rising food demand, making the need for efficient energy use in agricultural systems more urgent than ever.

Agricultural energy consumption is diverse, encompassing direct energy use, such as fuel and electricity to power farm equipment, as well as indirect energy use for the production of fertilizers, pesticides, and other off-farm inputs. In India, agriculture accounts for roughly 18.5% of total energy consumption, highlighting the sector's substantial energy footprint. In high income countries, energy use is often more focused on the processing, transportation, and storage of agricultural products, while in low income countries, cooking energy consumption is dominant. Regardless of the income level, energy needs in agriculture is rising and creating an unsustainable strain on available resources.

At the heart of addressing this energy imbalance lies the need for increased energy efficiency. Energy efficiency refers to producing the same amount of agricultural output using less energy, a principle that can be applied across all agricultural practices. Techniques such as conservation agriculture, organic farming, integrated pest management, and improved soil and water management practices are fundamental to reducing energy consumption without compromising productivity. Furthermore, the adoption of innovative practices and technologies in crop cultivation and food processing, including more mechanized farming and advanced food processing methods, is essential for sustainable energy use in agriculture (Table 6.4).

6.5.1 Role of Agroforestry in Improving Energy Use Efficiency (EUE) in Agriculture

The Green Revolution, which significantly boosted agricultural production in India through the intensive use of chemical inputs such as fertilizers and pesticides, as well as mechanization, led to a surge in energy demand. While it achieved the goal of increased food production, it also contributed to significant greenhouse gas (GHG) emissions, a major concern in today's climate crisis. As a result, there is an urgent need to identify traditional agricultural practices that can reduce energy consumption and improve energy use efficiency (EUE). This is where the role of agroforestry and the integration of forestry with agricultural systems becomes crucial. Agroforestry practices, which combine agriculture with tree planting, can reduce the dependency on nonrenewable resources and help mitigate the negative environmental impacts of high energy agricultural systems.

6.5.2 Role of Agroforestry in Offsetting GHG Emissions Generated by Intensive Agriculture

Carbon sequestration in agroforestry systems offers a promising solution to both reducing energy consumption and mitigating climate change. Agroforestry practices contribute to carbon sequestration by increasing soil organic carbon (SOC) storage, enhancing soil fertility, and reducing the need for chemical fertilizers.

These systems also enhance biodiversity, improve water retention, and support sustainable livelihoods for rural communities. By capturing carbon dioxide in both plants and soil, agroforestry systems can help offset some of the GHG emissions generated by intensive agricultural practices.

In addition, the carbon footprint (CF) of agricultural systems, which measures the GHG emissions associated with different farming practices, plays a critical role in assessing the environmental impact of agriculture. By adopting low energy, resource efficient practices, including those found in agroforestry, farmers can lower their carbon footprint and contribute to global efforts to reduce GHG emissions. Practices such as legume and residue retention, crop rotation, and the careful management of soil and water resources are essential components of an agroforestry system that can reduce the overall carbon footprint of agricultural systems. These approaches are not only beneficial for the environment but also improve energy efficiency by reducing the need for high energy inputs like synthetic fertilizers and irrigation (Table 6.4).

Efforts to improve carbon footprint (CF) estimates are also essential in identifying specific actions that can mitigate GHG emissions at the farm level. For instance, using site-specific emission factors when calculating CF allows for more accurate assessments of GHG emissions and can guide farmers in adopting the most appropriate mitigation strategies for their local context. Additionally, incorporating multiple functional units, such as changes in SOC storage and land use practices, provides a more comprehensive understanding of the environmental impacts of different agricultural systems (Box 6.3, Tables 6.3, 6.5, and 6.6).

Ultimately, achieving a sustainable and energy efficient agricultural system requires a holistic approach that integrates innovative technologies, efficient resource management, and the incorporation of carbon sequestration practices such as agroforestry. It is imperative that agricultural policies and practices are adapted to meet the dual challenges of increasing energy demand and environmental sustainability. By improving energy use efficiency and reducing GHG emissions through best management practices, we can move toward a greener, more resilient food production system. The integration of agroforestry offers a pathway to not only reduce energy consumption but also enhance carbon sequestration, ultimately helping the agricultural sector play a pivotal role in mitigating climate change while ensuring food security for the growing global population (Kumar et al. 2021; Girijaveni et al. 2023; Vijayakumar et al. 2023; Box 6.3, Tables 6.1, 6.2, and 6.6).

6.6 Concluding Remarks

The key aspects of the role of agroforestry in mitigating climate change, enhancing soil health, and improving energy efficiency in agriculture can be stated in a classified way as follows.

1. **Soil carbon sequestration and climate change**: Soil carbon sequestration is a critical process for combating climate change by storing CO_2 in soil, particularly through soil organic matter (SOM). Soils rich in organic matter can store carbon for decades to centuries, acting as significant carbon sinks.

2. **Agroforestry and soil carbon sequestration**: Agroforestry systems, which integrate trees into farming landscapes, enhance soil carbon sequestration by increasing both aboveground and belowground carbon stocks. Trees in agroforestry systems capture more carbon than conventional farming due to their deep roots and high biomass production, improving ability of soil to store carbon.
3. **Synergies between agroforestry and soil health**: Agroforestry promotes the growth of soil organisms that help decompose organic material, enriching soil and creating a positive feedback loop for carbon storage. This results in both improved soil health and increased carbon sequestration, making agroforestry an effective strategy of nbs against climate change.
4. **Additional benefits of agroforestry**: Agroforestry reduces soil erosion, especially in areas prone to desertification and land degradation, providing soil conservation benefits. It also contributes to soil fertility, making it a promising land use strategy for enhancing productivity while mitigating climate impacts.
5. **Agroforestry's role in reducing greenhouse gas emissions**: Agroforestry reduces greenhouse gas emissions by sequestering carbon in both the biomass and soil, making it a key strategy to address rising CO_2, methane (CH_4), and nitrous oxide (N_2O) levels. Agroforestry helps to manage the loss of terrestrial carbon stocks and control greenhouse gas emissions.
6. **Carbon sequestration in specific agroforestry systems**: Coffee-rubber-based agroforestry systems in southern Brazil enhance carbon stocks compared to traditional farming systems exposed to sunlight. Eucalyptus-based agroforestry systems and other tree-based systems store significant carbon, with some storing up to eight times the CO_2 emissions of degraded savannahs.
7. **Variations in carbon sequestration across agroforestry systems**: Carbon sequestration rates vary with elevation, with higher biomass and soil organic carbon (SOC) at upper elevations. In the Eastern Amazon, oil palm and cacao-based agroforestry systems store more carbon than oil palm monocultures.
8. **Soil carbon and eucalyptus-based agroforestry systems**: Eucalyptus-based integrated crop-livestock systems (ICLS) store more carbon in deeper soil layers, while grass-based systems store carbon more in the upper soil layers.
9. **Economic considerations of agroforestry**: Despite its potential for large-scale carbon sequestration, economic incentives for adopting agroforestry practices, such as coffee agroforestry systems, are limited, which hinders widespread adoption.
10. **Agroforestry's contribution to sustainable agriculture**: Agroforestry practices can increase productivity through intercropping, mixed cropping, and livestock integration, enhancing both ecological and economic resilience. The integration of fruit and timber trees in farming systems provides higher carbon storage, contributing to climate change mitigation.
11. **Soil carbon sequestration management in agroforestry**: Carbon stock in agroforestry is estimated by calculating the biomass of trees, crops, and roots, applying conversion factors (e.g., 0.55 for wood) to estimate the carbon content. Soil organic carbon (SOC) is linked to soil organic matter (SOM), and carbon is typically about 57% of dry biomass.

12. **Energy consumption in agriculture**: Agriculture is a significant energy consumer, with rising energy demands for food production and increased energy costs posing a challenge. Energy efficiency in agriculture is crucial for sustaining food production while reducing environmental impacts.
13. **Agroforestry's role in reducing agricultural energy footprint**: Agroforestry can reduce energy consumption by improving soil fertility and decreasing reliance on chemical fertilizers. Practices such as agroforestry, organic farming, and conservation agriculture can lower the energy demand in agricultural systems.
14. **Green revolution and its energy impact**: The Green Revolution led to a surge in agricultural energy demand due to intensive use of fertilizers, pesticides, and mechanization, contributing to greenhouse gas emissions. A shift toward low energy practices like agroforestry is necessary to mitigate these impacts.
15. **Carbon footprint (CF) and agroforestry**: Agroforestry can lower the carbon footprint (CF) of farming by reducing GHG emissions and improving resource efficiency. Techniques such as legume and residue retention, crop rotation, and soil and water management can reduce the CF of agroforestry systems.
16. **Improving carbon footprint estimates**: Accurate carbon footprint assessments using site-specific emission factors can help identify appropriate mitigation strategies for reducing GHG emissions at the farm level.
17. **Holistic approach for sustainable agriculture**: Achieving a sustainable agricultural system requires integrating innovative technologies, energy efficient practices, and carbon sequestration strategies like agroforestry. Agricultural policies and practices should evolve to address the dual challenges of rising energy demand and environmental sustainability.
18. **Agroforestry as a pathway to climate change mitigation**: Agroforestry systems contribute to mitigating climate change by enhancing carbon sequestration, reducing energy consumption, and supporting biodiversity. The integration of agroforestry offers a pathway to reducing GHG emissions and improving food security, playing a pivotal role in addressing the climate crisis.

References

Albrecht A, Kandji ST (2003) Carbon sequestration in tropical agroforestry systems. Agric Ecosyst Environ 99:15–27. https://doi.org/10.1016/s0167-8809(03)00138-5

Arora G, Chaturvedi S, Kaushal R, Nain A, Tewari S, Alam NM, Chaturvedi OP (2014) Growth, biomass, carbon stocks, and sequestration in an age series of Populus deltoides plantations in Tarai region of central Himalaya. Turk J Agric For 2014(38):550–560. https://doi.org/10.3906/tar-1307-94

Aware M (2023) Methane emission mitigation strategies in livestock. https://baif.org.in/livestock-strategies

Barneze AS, Mazzetto AM, Zani CF, Misselbrook T, Cerri CC (2014) Nitrous oxide emissions from soil due to urine deposition by grazing cattle in Brazil. Atmos Environ 92:394–397. https://doi.org/10.1016/j.atmosenv.2014.04.046

Birhane E, Ahmed S, Hailemariam M, Negash M, Rannestad MM, Norgrove L (2020) Carbon stock and woody species diversity in homegarden agroforestry along an elevation gradient in Southern Ethiopia. Agrofor Syst 94:1099–1110. https://doi.org/10.1007/s10457-019-00475-4
Borges O, Raimundo F, Coutinho J, Gonçalves B, Oliveira I, Martins A, Madeira M (2017) Carbon fractions as indicators of organic matter dynamics in chestnut orchards under different soil management practices. Agrofor Syst 93:301–310. https://doi.org/10.1007/s10457-017-0088-3
Brokawa N, Thompson J (2000) The H for DBH. For Ecol Manage 129(2000):89–91. https://doi.org/10.1016/S0378-1127(99)00141-3
Byrnes RC, Nùñez J, Arenas L, Rao I, Trujillo C, Alvarez C, Arango J, Rasche F, Chirinda N (2017) Biological nitrification inhibition by Brachiaria grasses mitigates soil nitrous oxide emissions from bovine urine patches. Soil Biol Biochem 107:156–163. https://doi.org/10.1016/j.soilbio.2016.12.029
Carvalho JA Jr, Higuchi N, Araujo TM, Santos JC (1998) Combustion completeness in a rainforest clearing experiment in Manaus, Brazil. J Geophys Res Atmos 103:13195–13199. https://doi.org/10.1029/98JD00172
Chapman HH (1921) Forest mensuration. Braunworth & Co. Book Manufacturers, Brooklyn, NY, p 150
Condit R (1998) Taxonomy. In: Tropical Forest census plots. Environmental intelligence unit. Springer, Berlin, Heidelberg. https://doi.org/10.1007/978-3-662-03664-8_5
Dahdouh-Guedes F, Codedam N (2006) Empirical estimate of the reliability of the use of the point-centred quarter method (PCQM): solutions to ambiguous field situations and description of the PCQM + protocol. For Ecol Manage 228(1–3):1–18. https://doi.org/10.1016/j.foreco.2005.10.076
Dixon RK, Winjum JK, Andrasko KJ, Lee JJ, Schroeder PE (1994) Integrated land use systems: assessment of promising agroforest and alternative land use practices to enhance carbon conservation and sequestration. Clim Change 27:71–92. https://doi.org/10.1007/bf01098474
Doddabasawa BMC, Murthy MM (2019) Comparison of carbon footprint of traditional agroforestry systems under rainfed and irrigated ecosystems. Agrofor Syst 94:465–475. https://doi.org/10.1007/s10457-019-00413-4
Doole GJ, Paragahawewa UH (2011) Profitability of nitrification inhibitors for abatement of nitrate leaching on a representative dairy farm in the Waikato region of New Zealand. Water 3(4):1031–1049. https://doi.org/10.3390/w3041031
FAO (2022) I-GLEAM_3.0_Model_description. https://www.fao.org/gleam/resources/en/; https://www.fao.org/fileadmin/user_upload/gleam/docs/GLEAM_3.0_Model_description.pdf
Galbally IA, Meyer MCP, Wang Y-P, Smith CJ, Weeks IA (2010) Nitrous oxide emissions from a legume pasture and the influences of liming and urine addition. Agr Ecosyst Environ 136(3–4):262–272. https://doi.org/10.1016/j.agee.2009.10.013
Gerber P, Vellinga TV, Opio C, Steinfeld H (2011) Productivity gains and greenhouse gas emissions intensity in dairy systems. Livest Sci 139(1–2):100–108. https://doi.org/10.1016/j.livsci.2011.03.012
Gerber PJ, Steinfeld H, Henderson B, Mottet A, Opio C, Dijkman J, Falcucci A, Tempio G (2013) Tackling climate change through livestock: a global assessment of emissions and mitigation opportunities. FAO (Food and Agriculture Organization of the United Nations), Rome
Girijaveni V, Sammi Reddy K, Prasad JVNS, Singh VK (2023) Carbon footprint and sustainability of different agricultural production systems in climate change scenario. In: Rakshit A, Biswas A, Sarkar D, Meena VS, Datta R (eds) Handbook of energy management in agriculture. Springer, Singapore. https://doi.org/10.1007/978-981-19-7736-7_6-1
Goncalves N, Andrade D, Batista A, Cullen L, Souza A, Gomes H, Uezu A (2021) Potential economic impact of carbon sequestration in coffee agroforestry systems. Agrofor Syst 95:419–430. https://doi.org/10.1007/s10457-020-00569-4
Goodrich A, James T, Woodhouse M (2012) Residential, Commercial, and Utility-Scale Photovoltaic (PV) System Prices in the United States: Current Drivers and Cost-Reduction

Opportunities. In: Technical Report, NREL/TP-6A20–53347, February 2012. NREL: A National Laboratory of the U.S. Department of Energy, Office of Energy Efficiency & Renewable Energy, Operated by the Alliance for Sustainable Energy, LLC. https://www.nrel.gov/docs/fy12osti/53347.pdf. https://www.linkedin.com/in/al-goodrich/details/publications/

Goswami S, Verma KS, Kaushal R (2013) Biomass and carbon sequestration in different agroforestry systems of a Western Himalayan watershed. Biol Agric Hortic 30(2):88–96. https://doi.org/10.1080/01448765.2013.855990

Hristov AN, Oh J, Firkins JL, Dijkstra J et al (2013) Special topics—mitigation of methane and nitrous oxide emissions from animal operations: I. A review of enteric methane mitigation options. J Anim Sci 91(11):5045–5069. https://doi.org/10.2527/jas.2013-6583

Hristov AN, Oh J, Giallongo F, Frederick TW (2015) An inhibitor persistently decreased enteric methane emission from dairy cows with no negative effect on milk production. PNAS 112(34):10663–10668. https://doi.org/10.1073/pnas.15041241

IPCC (2006) 2006 IPCC guidelines for National Greenhouse gas Inventories. In: Agriculture, forestry and other land use, vol 4. IPCC, Geneva. https://www.ipcc-nggip.iges.or.jp/public/2006gl/vol4.html

Itoh M, Sudo S, Mori S et al (2011) Mitigation of methane emissions from paddy fields by prolonging midseason drainage. Agr Ecosyst Environ 141(3–4):359–372. https://doi.org/10.1016/j.agee.2011.03.019.WOS:000292534800012

Jiang Y, van Groenigen KJ, Huang S et al (2017) Higher yields and lower methane emissions with new rice cultivars. Glob Chang Biol 23(11):4728–4738. https://doi.org/10.1111/gcb.13737

Joshi E, Kumar D, Lal B et al (2013) Management of direct seeded rice for enhanced resource—use efficiency. Plant Knowl J 2(3):119–134. https://www.researchgate.net/publication/265378620_Management_of_direct_seeded_rice_for_enhanced_resource_-_use_efficiency

Kanime N, Kaushal R, Tewari SK, Raverkar KP, Chaturvedi S, Chaturvedi OP (2013) Biomass production and carbon sequestration in different tree-based systems of central Himalayan Tarai region. For Trees Livelihoods 22(1):38–50. https://doi.org/10.1080/14728028.2013.764073

Kirkby CA, Kirkegaard JA, Richardson AE, Wade LJ, Blanchard C, Batten G (2011) Stable soil organic matter: a comparison of C:N:P:S ratios in Australian and other world soils. Geoderma 163(3–4):197–208. https://doi.org/10.1016/j.geoderma.2011.04.010

Kumar S, Singh SP, Meena RS et al (2021) Reduction of energy consumption in agriculture for sustainable green future. In: Bhatt R, Meena RS, Hossain A (eds) Input use efficiency for food and environmental security. Springer, Singapore. https://doi.org/10.1007/978-981-16-5199-1_7

Lai R (2004) Soil carbon sequestration in natural and managed tropical forest ecosystems. J Sustain For 21:1–30. https://doi.org/10.1300/j091v21n01_01

Ma S, He F, Tian D et al (2018) Variations and determinants of carbon content in plants: global synthesis. Biogeosciences 15:693–702. https://doi.org/10.5194/bg-15-693-2018

MacDicken KG (1997) A guide to monitoring carbon storage in forestry and agroforestry projects. Forest carbon monitoring program, Winrock International Institute for Agricultural Development, Winrock, USA, Forest Carbon Monitoring and Verification Services. https://www.researchgate.net/publication/237434580_A_Guide_to_Monitoring_Carbon_Storage_in_Forestry_and_Agroforestry_Projects

Malhi Y, Baldocchi DD, Jarvis PG (2002) The carbon balance of tropical, temperate and boreal forests. Plant Cell Environ 22:715. https://doi.org/10.1046/j.1365-3040.1999.00453.x

Martínez-Fernández G, Abecia L, Arco A, Cantalapiedra-Hijar G, Martín-García AI, Molina-Alcaide E, Kindermann M, Duval S, Yáñez-Ruiz DR (2014) Effects of ethyl-3-nitrooxy propionate and 3-nitrooxypropanol on ruminal fermentation, microbial abundance, and methane emissions in sheep. J Dairy Sci 97(6):3790–3799. https://doi.org/10.3168/jds.2013-7398

Mathews K, Haley M (2015) Livestock, dairy, and poultry outlook. Economic Research Service. Situation and Outlook. LDP-M-248. https://www.ers.usda.gov/webdocs/outlooks/37598/51784_ldpm248.pdf?v=7684.1

Mazzetto AM, Barneze AS, Feigl BJ et al (2014) Use of the nitrification inhibitor dicyandiamide (DCD) does not mitigate N2O emission from bovine urine patches under Oxisol in Northwest Brazil. Nutr Cycl Agroecosyst 101(1):83–92. https://doi.org/10.1007/s10705-014-9663-4

MoEFCC (2021) India: third biennial update report to the United Nations framework convention on climate change. Ministry of Environment, Forest and Climate Change, Government of India. https://unfccc.int/sites/default/files/resource/INDIA_%20BUR-3_20.02.2021_High.pdf

Montagnini F, Nair PKR (2004) Carbon sequestration: an underexploited environmental benefit of agroforestry systems. Agrofor Syst 61:281–295. https://doi.org/10.1023/b:agfo.0000029005.92691.79

Noumi VN, Zapfack L, Hamadou MR, Djongmo VA, Witanou N, Nyeck B, Ngossomo JD, Mbobda RBT, Mapongmetsem PM (2018) Floristic diversity, carbon storage and ecological services of eucalyptus agrosystems in Cameroon. Agrofor Syst 92:239–250. https://doi.org/10.1007/s10457-017-0130-5

Padhee A (2018) DBT in fertilizers: game changing reforms. Blog Times of India, 4 May 2018. https://timesofindia.indiatimes.com/blogs/voices/dbt-in-fertilizers-game-changing-reforms/

Palm CA (1995) Contribution of agroforestry trees to nutrient requirements of intercropped plants. Agrofor Syst 30:105–124. https://doi.org/10.1007/BF00708916

Pandey DN (2002) Carbon sequestration in agroforestry systems. Clim Pol 2(4):367–377. https://doi.org/10.1016/S1469-3062(02)00025-6

Pelster DE, Gisore B, Goopy J, Korir D, Koske JK, Rufino MC, Butterbach-Bahl K (2016) Methane and nitrous oxide emissions from cattle excreta on an east African grassland. J Environ Qual 45(5):1531–1539. https://doi.org/10.2134/jeq2016.02.0050

Ponce-Hernandez R, Koohafkan P, Antoine J (2004) Assessing carbon stocks and modelling win-win scenarios of carbon sequestration through land-use changes. Food and Agriculture Organization of the United Nations (FAO), Rome, p 30. https://openknowledge.fao.org/server/api/core/bitstreams/fc0cd77f-93a4-4134-829a-9421591597c1/content

Powlson DS, Stirling C, Jat M et al (2014) Limited potential of no-till agriculture for climate change mitigation. Nat Clim Change 4:678–683. https://doi.org/10.1038/nclimate2292

Powlson DS, Stirling CM, Thierfelder C, White RP, Jat ML (2016) Does conservation agriculture deliver climate change mitigation through soil carbon sequestration in tropical agro-ecosystems? Agr Ecosyst Environ 220:164–174. https://doi.org/10.1016/j.agee.2016.01.005

Rai P, Vineeta, Shukla G, Manohar KA, Bhat JA, Kumar A, Kumar M, Cabral-Pinto M, Chakravarty S (2021) Carbon Storage of single tree and mixed tree dominant species stands in a reserve forest-case study of the Eastern Sub-Himalayan region of India. Land 10:435. https://doi.org/10.3390/land10040435

Rallings AM, Smukler SM, Mullinix K (2020) A novel methodology to characterize and quantify regional farmscape non-production perennial vegetation carbon storage and potential for loss in Southwest British Columbia. Agrofor Syst 94:1947–1958. https://doi.org/10.1007/s10457-020-00514-5

Ramos HMN, Vasconcelos SS, Kato OR, Castellani DC (2018) Above-and belowground carbon stocks of two organic, agroforestry-based oil palm production systems in eastern Amazonia. Agrofor Syst 92:221–237. https://doi.org/10.1007/s10457-017-0131-4.221

Reynolds CK, Humphries DJ, Kirton P, Kindermann M, Duval S, Steinberg W (2014) Effects of 3-nitrooxypropanol on methane emission, digestion, and energy and nitrogen balance of lactating dairy cows. J Dairy Sci 97(6):3777–3789. https://doi.org/10.3168/jds.2013-7397

Romero-Perez A, Okine EK, McGinn SM, Guan LL, Oba M, Duval SM, Kindermann M, Beauchemin KA (2015) Sustained reduction in methane production from long-term addition of 3-nitrooxypropanol to a beef cattle diet. J Anim Sci 93(4):1780–1791. https://doi.org/10.2527/jas.2014-8726

RTB (2016) Research for innovation and impact. Annual Report 2016. CGIAR Research Program on Roots, Tubers and Bananas, Lima, Peru. https://doi.org/10.4160/23085932/2016

Sanderman J, Hengl T, Fiske GJ (2017) Soil carbon debt of 12,000 years of human land use. PNAS 114(36):9575–9580. https://doi.org/10.1073/pnas.1706103114

Sarto MVM, Borges WLB, Sarto JRW, Rice CW, Rosolem CA (2020) Deep soil carbon stock, origin, and root interaction in a tropical integrated crop–livestock system. Agrofor Syst 94:1865–1877. https://doi.org/10.1007/s10457-020-00505-6

Sathaye JA, Ravindranath NH (1998) Climate change mitigation in the energy and forestry sectors of developing countries. Annu Rev Env Resour 23:387–437. https://doi.org/10.1146/annurev.energy.23.1.387

Schnepf R (2015) Farm-to-food Price dynamics. Congressional Research Service Report prepared for Members and Committees of Congress. 7-5700. www.crs.gov R40621. https://sgp.fas.org/crs/misc/R40621.pdf

Sordi A, Dieckow J, Bayer C et al (2014) Nitrous oxide emission factors for urine and dung patches in a subtropical Brazilian pastureland. Agr Ecosyst Environ 190:94–103. https://doi.org/10.1016/j.agee.2013.09.004

Su J, Hu C, Yan X et al (2015) Expression of barley SUSIBA2 transcription factor yields high-starch low-methane rice. Nature 523:602–606. https://doi.org/10.1038/nature14673

van Groenigen JW, van Kessel C, Hungate BA et al (2017) Sequestering soil organic carbon: a nitrogen dilemma. Environ Sci Technol 51(9):4738–4739. https://doi.org/10.1021/acs.est.7b01427

Verchot LV, Van Noordwijk M, Kandji S et al (2007) Climate change: linking adaptation and mitigation through agroforestry. Mitig Adapt Strat Glob Chang 12:901–918. https://doi.org/10.1007/s11027-007-9105-6

Verma A, Kaushal R, Alam NM et al (2014) Predictive models for biomass and carbon stocks estimation in Grewia optiva on degraded lands in western Himalaya. Agrofor Syst 88:895–905. https://doi.org/10.1007/s10457-014-9734-1

Verma K, Sharma P, Kumar D, Vishwakarma SP, Meena NK (2021) Strategies sustainable management of agroforestry in climate change mitigation and adaptation. Int J Curr Microbiol App Sci 10(01):2439–2449. https://doi.org/10.20546/ijcmas.2021.1001.282

Vijayakumar S, Chatterjee D, Subramanian E, Ramesh K, Saravanane P (2023) Efficient management of energy in agriculture. In: Rakshit A, Biswas A, Sarkar D, Meena VS, Datta R (eds) Handbook of energy management in agriculture. Springer, Singapore. https://doi.org/10.1007/978-981-19-7736-7_18-2

Ward GN, Kelly KB, Hollier JW (2016) Greenhouse gas emissions from dung, urine and dairy pond sludge applied to pasture: 1. Nitrous oxide emissions. Anim Prod Sci 58(6):1087–1093. https://doi.org/10.1071/AN15595

World Bank (2024) DataBank, Metadata Glossary. https://databank.worldbank.org/metadata-glossary/world-development-indicators/series/EN.ATM.GHGT.KT.CE

Zaro GC, Caramori PH, Junior GMY, Sanquetta CR, Filho AA, Nunes AL, Prete CE, Voroney P (2019) Carbon sequestration in an ecosystem management of degraded red and lateritic soil through agroforestry system of coffee with rubber trees compared to open-grown coffee in southern Brazil. Agrofor Syst 94:799–809. https://doi.org/10.1007/s10457-019-00450-z.807

Zhang S, Gao P, Tong Y, Norse D, Lu Y, Powlson D (2015a) Overcoming nitrogen fertilizer over-use through technical and advisory approaches: a case study from Shaanxi Province, Northwest China. Agr Ecosyst Environ 209:89–99. https://doi.org/10.1016/j.agee.2015.03.002

Zhang X, Davidson EA, Mauzerall DL, Searchinger TD, Dumas P, Shen Y (2015b) Managing nitrogen for sustainable development. Nature 528(7580):51–59. https://doi.org/10.1038/nature15743

Zou J, Huang Y, Jiang J (2005) A 3-year field measurement of methane and nitrous oxide emissions from rice paddies in China: effects of water regime, crop residue, and fertilizer application. Global Biogeochem Cycles 19:(GB2021)1–(GB2021)9. https://doi.org/10.1029/2004GB002401

Websites

https://www.inspirecleanenergy.com/blog/clean-energy-101/what-is-co2e#fn3
https://ec.europa.eu/eurostat/statistics-explained/index.php?title=Glossary:Carbon_dioxide_equivalent
https://klima.com/blog/CO2-vs-CO2e-what-is-the-difference/
https://www.lawinsider.com/dictionary/mtco2e
https://climatekids.nasa.gov/greenhouse-effect/
https://www.epa.gov/ghgemissions/understanding-global-warming-potentials
https://databank.worldbank.org/metadataglossary/world-development-indicators/series/EN.ATM.GHGT.KT.CE
https://research.wri.org/wrr-food/course/reduce-greenhouse-gas-emissions-agricultural-production-synthesis

Agroforestry for Sustainable and Improved Livelihoods Through Value Addition and Animal Husbandry

7

Abstract

Value addition to agroforestry products and services is a strategy for sustainable development, rural empowerment, and environmental conservation. Agroforestry, which integrates trees, shrubs, and agricultural practices, offers significant ecological, social, and economic benefits, such as soil conservation, biodiversity enhancement, and carbon sequestration. However, its full potential remains untapped in many rural communities dominated by subsistence farming. By transforming raw agroforestry products into higher-value goods, value addition can enhance marketability, diversify income streams, and provide rural populations with economic advancement opportunities. A proposed method for estimating carbon credit, requiring approval of the government, is also included for awareness among agroforesters for their extra benefits from agroforestry farming. This process not only increases profitability but also reduces reliance on environmentally degrading agricultural practices, creates employment, and strengthens local value chains. This chapter examines the ecological, economic, and social benefits of value-added agroforestry systems, focusing on its role in improving soil fertility, water regulation, and carbon sequestration with numerical problems and solutions from among the published literatures. This is no doubt an urgent need among the students, interns, and agroforestry start-ups for formulating and examining financial pros and cons about taking up agroforestry projects. It also addresses the challenges of implementing value addition strategies and presents potential solutions. Case studies of successful agroforestry value addition projects are analyzed to highlight key lessons and demonstrate the transformative potential of this approach in fostering resilient and sustainable economics of rural communities.

S. Panda, *Agroforestry*, https://doi.org/10.1007/978-981-96-6855-7_7

Keywords

Value addition · Agroforestry · Boundary plantation · Sacred grove · Sustainable development · Rural empowerment · Environmental preservation · Ecological benefits · Social benefits · Economic benefits · Economic resilience · Soil conservation · Biodiversity enhancement · Carbon sequestration · Carbon credit · Subsistence farming · Marketability · Income diversification · Livestock · Poultry · Aquaculture · Economic advancement · Fishery · Profitability · Environmental degradation · Employment creation · Local value chains · Forest bathing (*Shinrin-yoku*) · Soil fertility · Water regulation · Case studies · Resilient rural economies · Sustainable economics · Numerical problem · Problems and solutions

Abbreviations

AFS	Agroforestry system
B:C	Benefit-to-cost ratio
CO_2	Carbon dioxide
Nbs	Nature-based solution
NTFP	Non-timber forest product
SOC	Soil organic carbon
SOM	Soil organic matter

7.1 Introduction

In the pursuit of sustainable development and economic growth, the focus must shift toward innovative and practical solutions that not only enhance livelihoods but also ensure the preservation of the environment. One of the most promising strategies in this regard is the value addition to agroforestry products, a concept that lies at the intersection of environmental sustainability, rural empowerment, and economic resilience. This approach holds immense potential to foster long-term benefits for rural communities, contributing to both their economic well-being and on the health of the ecosystems they depend upon (Manna et al. 2003; Mekonnen et al. 2023).

Agroforestry, defined as the integration of trees and shrubs into agricultural systems, is a holistic land-use practice that combines the best of both forestry and agriculture. By combining crops, trees, and livestock in an integrated and mutually beneficial manner, agroforestry systems (AFS) provide a variety of ecological, social, and economic benefits. These include soil conservation, biodiversity enhancement (Julius et al. 2011), improved water retention, carbon sequestration, and the provision of diverse non-timber forest products (NTFPs). However, while agroforestry systems offer numerous environmental benefits, their full potential remains untapped in many rural communities where subsistence farming predominates (Honnayya et al. 2020; Mukangango et al. 2020; Mekonnen et al. 2023).

One of the key ways to unlock the full potential of agroforestry is through the concept of value addition. This involves transforming raw agroforestry products into higher-value goods that are more marketable, thereby increasing income streams and providing rural populations with opportunities for economic advancement. Value addition can take many forms, from simple processing techniques like drying, packaging, and preserving to more complex transformations such as the development of eco-friendly products, essential oils, herbal medicines, and biofuels. By focussing on value addition, agroforestry products can gain greater market appeal, increase in demand, and provide rural communities with more diversified sources of income (Dhanushkodi et al. 2019; Parthiban 2019; Mekonnen et al. 2023).

The process of adding value to agroforestry products does more than simply increase profitability; it also catalyzes a broader range of positive outcomes. For instance, by incorporating value-added agroforestry products into rural livelihoods, communities can reduce their dependence on conventional agricultural practices, which often degrade the environment and lead to soil depletion and biodiversity loss (Julius et al. 2011). Moreover, value addition can create new employment opportunities, fosters skill development, and enhance local value chains, thus helping to strengthen the social fabric of rural areas.

This chapter aims to explore in detail how the strategy of adding value to agroforestry products and services can significantly enhance the livelihoods of rural communities while promoting sustainable agroforestry practices. It will delve into the multifaceted ecological, economic, and social benefits that arise from these integrated systems and how value addition can further amplify these benefits. Special attention is given to the ecosystem services provided by agroforestry systems, such as improved soil fertility, water regulation, and carbon sequestration, which make these systems particularly well-suited for a sustainable development approach. Additionally, it is here highlighted the challenges faced in value addition processes and offer potential solutions to overcome these barriers (Jhariya et al. 2019; Parthiban 2019; Raj et al. 2019; Honnayya et al. 2020; Mukangango et al. 2020; Mekonnen et al. 2023).

The exploration of value addition in agroforestry is not just about boosting income but about creating a more resilient and sustainable future for rural communities. As such, the objectives of this chapter are classified as follows.

1. To examine the role of agroforestry in enhancing environmental sustainability and promoting biodiversity conservation.
2. To analyze the socio-economic benefits of agroforestry systems for rural communities, with a focus on livelihood improvement.
3. To explore the concept of value addition in agroforestry products and services; and its potential for increasing income, diversifying income sources, and improving market access for rural farmers.
4. To assess the ecosystem services provided by agroforestry systems, and how value addition contributes to environmental sustainability.
5. To identify challenges and barriers to the implementation of value addition strategies in agroforestry and propose for solutions to overcome them.

6. To explore case studies of successful agroforestry value addition projects and the lessons learned from them with numerical problems and solutions.

These objectives are aimed at providing a holistic view of how value addition to agroforestry products not only supports economic growth but also contributes to the broader goals of sustainable development, environmental protection, and rural empowerment. By examining these diverse facets, this chapter aims to illustrate the transformative potential of agroforestry as a catalyst for change in the rural economies and ecosystems alike.

7.2 Nature-Based Solution for Agroforestry-Based Livelihoods

Nature-based solution for agroforestry-based livelihoods requires practicing such lifestyles adopted with the ecosystem services provided by agroforestry systems. This requires classification of various agroforestry-based livelihoods w.r.t. various ecosystem services provided by agroforestry systems.

Nature-based solutions for agroforestry-based livelihoods represent a holistic approach to sustainable development, recognizing the symbiotic relationship between human well-being and the ecosystems they depend on. This approach emphasizes the importance of integrating the natural environment into livelihood practices, particularly through agroforestry systems that combine agriculture and forestry. Agroforestry systems are diverse and multifaceted farming systems, providing a range of ecosystem services that not only support agricultural production but also contribute to broader environmental and socio-economic resilience (Meena et al. 2017; Mekonnen et al. 2023).

A key component of nature-based solutions is the careful management and use of ecosystem services, which are the benefits people derive from the functioning ecosystems. In the context of agroforestry, these services can include soil fertility enhancement, water regulation, biodiversity conservation, carbon sequestration, and climate change mitigation. By adopting agroforestry practices, communities can sustainably harness these services, ensuring long-term productivity while maintaining ecological balance (Julius et al. 2011; Mukangango et al. 2020).

To effectively implement nature-based solutions, it is essential to classify and understand the various agroforestry-based livelihoods and their specific relationship to the ecosystem services provided by these systems. This classification allows for a deeper understanding of how different agroforestry practices contribute to diverse livelihood strategies, whether through the production of timber, non-timber forest products, food, fodder, or medicinal plants or extraction of medicinally useful plant parts. It also helps to assess how these practices support environmental goals, such as biodiversity enhancement (Julius et al. 2011), water conservation, and climate adaptation.

Moreover, recognizing the interplay between agroforestry systems and ecosystem services is being critical for ensuring the sustainability and resilience of these livelihoods. For example, agroforestry can provide multiple layers of income for farmers, reduce their vulnerability to climate change by buffering against extreme weather events, and contribute to the restoration of degraded lands. This makes agroforestry a vital component of nature-based solutions, particularly in regions where livelihoods are closely tied to natural resources and the impacts of climate change are increasingly felt (Dhanushkodi et al. 2019).

Therefore, the practice of agroforestry-based livelihoods, through the lens of nature-based solutions, is not merely about combining trees, crops, livestock, and fishery; it is about understanding and utilizing the array of ecosystem services that agroforestry systems offer. This requires a multi-disciplinary approach that integrates ecological knowledge, community needs, and economic viability, ensuring that agroforestry systems are both environmentally sustainable and socially equitable. By classifying these systems according to their ecosystem services, stakeholders can better align development strategies with the realities of local ecosystems, fostering resilience, improving livelihoods, and promoting long-term sustainability (Mekonnen et al. 2023).

7.3 Ecosystem Service-Based Livelihoods Generated Through Agroforestry

Agroforestry systems provide a range of regulating, supporting, provisioning, and cultural ecosystem services, with important implications for both environmental sustainability (through soil conservation, water management, and carbon sequestration) and agricultural productivity (through improved soil health and fertility (Dollinger and Jose 2018; Mukangango et al. 2020). These benefits make agroforestry a climate-smart solution to address global challenges related to food security, climate change, and ecosystem health. These services provided by agroforestry are based on types of benefits AFS can provide. These are:

7.3.1 Regulating Services

These services regulate ecological processes that help maintain the balance of ecosystems and the environment:

(a) **Soil conservation (erosion control)**: The root systems of trees and shrubs stabilize the soil, reducing erosion. This helps prevent soil degradation, preserves agricultural productivity, and maintains soil structure.
(b) **Water management**: Agroforestry systems contribute to better water retention in the soil, preventing water runoff and promoting groundwater recharge. This is especially important for drought-prone areas and regions with irregular rainfall patterns.
(c) **Carbon sequestration**: Trees and shrubs absorb carbon dioxide (CO_2) from the atmosphere through photosynthesis and store it as organic carbon (OC) in both

the biomass and the soil. This contributes to mitigating climate change by reducing the concentration of greenhouse gases.

7.3.2 Supporting Services

These services are essential for the functioning of ecosystems and the production of other services.

(a) **Soil fertility and structure**: The organic matter provided by tree litter (leaves and branches) and the decay of roots enhances the soil fertility and its structure, making it more resilient to agricultural stress and supporting healthy crop production.
(b) **Soil organic carbon (SOC) enhancement**: The increased SOC content improves soil health, fertility, and water retention capacity. These factors are critical for the sustainability of agriculture in the long term.

7.3.3 Provisioning Services

These are the services that directly provide resources for human use.

(a) **Agricultural productivity**: Although not explicitly mentioned in the passage, the benefits of soil fertility, water retention, and erosion control indirectly contribute to increased agricultural productivity, supporting food security.

7.3.4 Cultural Services

Consequently, it is of utmost importance to be noted that agroforestry systems can also provide cultural and aesthetic value by creating landscapes that support local traditions, biodiversity, and community well-being.

7.4 Ecosystem Services-Based Livelihood Benefits from Agroforestry *vis-a-vis* nbs for Livelihood Generation

7.4.1 Regulating Ecosystem Services-Based Livelihoods Provided by Agroforestry

Agroforestry offers a range of livelihoods that benefit from the regulating services provided by trees and other vegetation in the system. Here are some examples based on the regulating services provided by agroforestry. In each of these examples, livelihoods are not only just based on the direct sale of products but also on the environmental services agroforestry provides, such as soil conservation, water management, and carbon sequestration. These benefits allow farmers to sustain and diversify their income streams while improving the health of their ecosystems (Naugraiya and Puri 2001; Panda et al. 2018; Dhanushkodi et al. 2019; Mukangango et al. 2020). These services are listed as follows:

7.4.1.1 Soil Conservation (Erosion Control)

(a) **Agroforestry livelihood example: Terracing and agroforestry integration**
  - **Farmers practicing agroforestry in hilly or sloped areas** can combine tree planting with traditional terracing methods. This approach stabilizes the soil and reduces erosion while maintaining productivity. The trees provide both economically valuable products like fruits (Kumar and Chaturvedi 2017), timber, fuelwood; and environmental benefits through provisioning of soil stabilization, reduced runoff, etc.
  - **Income Sources**: Farmers can generate income through the sale of timber, nuts, fruits (e.g., macadamia, almonds), or medicinal plants grown in these systems (Kumar and Chaturvedi 2017; Singh et al. 2017; Newaj and Ram 2019).

(b) **Agroforestry livelihood example: Silvopasture (agroforestry with livestock, poultry, aquaculture)**
  - **In silvopasture systems, livestock** are grazed in areas with tree cover. The tree root systems protect the soil from erosion, while the land remains productive for both grazing and tree-based income. Farmers can sell livestock, poultry, and aquaculture products (meat, milk, wool, egg, fish, and pearl) and tree-based products.
  - **Income Sources**: Livestock and poultry products (meat, milk, wool, and egg), aquaculture products (fish, pearl) timber, fodder, and fuelwood.

7.4.1.2 Water Management

(a) **Agroforestry livelihood example: Riparian buffers and agroforestry**
  - **In areas near rivers or water bodies**, farmers can establish riparian buffers (vegetation along waterways) using trees and shrubs. These systems help reduce water runoff, increase water retention, and improve water quality. This is particularly useful in areas suffering from water-scarcity or erratic rainfall.
  - **Income sources**: Farmers can sell non-timber forest products (NTFPs) such as medicinal herbs, fruits, and timber. Additionally, improving water availability can boost crop yields, leading to greater agricultural income (Naugraiya and Puri 2001; Sharma 2004; Banerjee and Dhara 2010; Kumar and Chaturvedi 2017; Singh et al. 2017; Panda et al. 2018; Newaj and Ram 2019).

(b) **Agroforestry livelihood example: Dryland agroforestry systems**
  - **In drought-prone areas**, integrating drought-resistant trees (e.g., moringa or baobab) or bamboos with crops helps retain moisture in the soil, reduce evaporation, and recharge groundwater. This method can stabilize crop production in areas with inconsistent rainfall, providing a sustainable livelihood

(Naugraiya and Puri 2001; Panda et al. 2018; Shukla et al. 2018; Dev et al. 2020).
- **Income Sources**: Sale of drought-tolerant fruits and tree products (e.g., moringa leaves, baobab fruit), along with improved crop yields from better water retention.

7.4.1.3 Carbon Sequestration

(a) **Agroforestry livelihood example: Carbon credit programs**
- **Farmers who adopt agroforestry practices**, such as planting trees for carbon sequestration, can participate in carbon credit programs. In these programs, they earn credits for storing carbon. That carbon credit can then be sold to businesses or governments aiming to offset their carbon emissions, or carbon footprint (Yadav et al. 2018).
- **Income Sources**: Income from the sale of carbon credits as well as income from tree products like timber, fruits, and non-timber forest products (e.g., honey, mushrooms).

(b) **Agroforestry livelihood example**: **Timber and tree crop production incarbon-rich systems**
- **Agroforestry systems that focus on fast-growing trees for carbon sequestration**(e.g., eucalyptus, poplar, or other species) provide timber and other tree products while also capturing carbon. Over time, these trees buildup carbon stores, making the system economically and ecologically valuable (Naugraiya and Puri 2001; Panda et al. 2018).
- **Income Sources**: Timber harvest, tree crops (e.g., nuts, fruits), and non-timber forest products.

In each of these examples on regulating ecosystem services of agroforestry, livelihoods are not just based on the direct sale of products but also on the environmental services agroforestry provides, such as soil conservation, water management, and carbon sequestration. These benefits allow farmers to sustain and diversify their income streams while improving the health of their ecosystems (Naugraiya and Puri 2001; Borges et al. 2018; Kumar and Chaturvedi 2017; Singh et al. 2017; Panda et al. 2018; Newaj and Ram 2019; Mukangango et al. 2020).

7.4.2 Supporting Ecosystem Services-Based Livelihoods Provided by Agroforestry

Agroforestry systems, which integrate trees with crops or livestock, can provide a variety of livelihoods that are directly or indirectly supported by the supporting services of ecosystems, such as soil fertility and structure and soil organic carbon (SOC) enhancement (Manna et al. 2003; Lal 2014; Mukangango et al. 2020). Some of these examples of agroforestry-based livelihoods are stated here.

7.4.2.1 Agroforestry Crop Farmers

Livelihood source: Growing crops such as maize, cassava, vegetables, or fruits in combination with trees like nitrogen-fixing legumes or fruit trees, e.g., mango, avocado (Bellow et al. 2008; Dhara 2016; Singh et al. 2017; Kumar and Chaturvedi 2017; Mohapatra et al. 2019; Newaj and Ram 2019).

Connection to supporting services: The organic matter from tree litter (fallen leaves and branches) enhances soil fertility, improving nutrient availability for crops. The increased SOC also improves soil structure and water retention, making crops more resilient to drought and improving yields in the long term.

Example: A farmer growing beans or com alongside nitrogen-fixing trees (like *Gliricidia sepium*)to improve soil fertility, reduce the need for synthetic fertilizers, and increase crop production.

7.4.2.2 Tree-Based Crop Producers (Agroforestry with Fruit Trees)

Livelihood source: Growing fruit trees such as cocoa, coffee, or citrus, alongside food crops like vegetables or legumes.

Connection to supporting services: The presence of deep-rooted trees can enhance soil structure and fertility by cycling nutrients from deeper soil layers to the surface, while the organic matter from tree litter increases the amount of organic material in the soil. This makes the soil healthier, promoting sustainable fruit and crop production (Dhara 2016; Kumar and Chaturvedi 2017; Singh et al. 2017; Kumar et al. 2019; Newaj and Ram 2019).

Example: Coffee farmers in Central America growing shade trees like *Inga* or *Erythrina* that improve soil health and provide natural mulch, enhancing long-term coffee production.

7.4.2.3 Livestock and Pastoralists (Silvopasture Systems)

Livelihood source: Grazing livestock (cattle, goats, and sheep) in silvopastoral systems where trees, shrubs, and grasses are integrated.

Connection to supporting services: Trees in silvopasture systems help prevent soil erosion, improve water retention, and increase organic carbon in the soil (Qaiser et al. 2016; Mukangango et al. 2020). The deep roots of trees also stabilize soil structure and provide shelter and fodder for livestock. As a result, livestock productivity improves due to better pasture quality and reduced vulnerability to soil degradation.

Example: Cattle ranchers in Brazil's "Cerrado" region practicing silvopastoral systems with *Acacia* and *Leucaena* trees that enhance pasture health and reduce the risk of soil erosion, increasing cattle grazing potential.

7.4.2.4 Forest Products and Non-Timber Forest Product (NTFP) Collectors

Livelihood source: Harvesting products such as honey, mushrooms, medicinal herbs, and edible fruits in agroforestry systems.

Connection to supporting services: The increase in soil organic carbon enhances soil quality and the health of plants (Mukangango et al. 2020), making agroforestry environments ideal for growing medicinal plants, herbs, and other NTFPs. Additionally, improved soil fertility contributes to higher yields of tree fruits or nuts, which can be harvested for sale (Naugraiya and Puri 2001; Manna et al. 2003; Sharma 2004; Kumar and Chaturvedi 2017; Singh et al. 2017; Panda et al. 2018; Kumar et al. 2019; Newaj and Ram 2019).

Example: In agroforestry systems in India, farmers harvest medicinal plants like *Aloe vera* or *Tulsi* under the canopy of nitrogen-fixing trees, benefiting from the improved soil health and fertility promoted by the trees.

7.4.2.5 Compost and Biofertilizer Production

Livelihood source: Production and sale of compost or organic fertilizers made from tree litter, animal manure, and crop residues.

Connection to supporting services: The decomposition of tree litter and organic material increases soil organic matter (SOM) and enriches soil organic carbon content (Borges et al. 2018; Mukangango et al. 2020). The resulting compost or biofertilizers are high in nutrients and can be sold to other farmers to improve agricultural productivity (Naugraiya and Puri 2001; Lal 2014; Panda et al. 2018).

Example: Farmers and local entrepreneurs in East Africa may collect tree litter (from *Acacia* or *Calliandra* species) and combine it with farm manure to produce organic compost, which is then sold to other farmers to improve crop yields (Figyantika et al. 2020).

7.4.2.6 Agroforestry for Erosion Control and Land Restoration

Livelihood source: Restoring degraded land using agroforestry practices (e.g., planting trees or bamboos on sloping land or reforesting bare soils) to prevent erosion and improve land productivity (Dev et al. 2020).

Connection to supporting services: Trees planted in erosion-prone areas enhance soil structure by increasing organic matter, promoting water infiltration, and stabilizing the soil with their root systems. These practices help restore degraded land, creating conditions for sustainable farming and grazing, which can improve livelihoods (Manna et al. 2003).

Example: In the Philippines, farmers plant *Flemingia* and *Leucaena* trees on sloped lands to prevent soil erosion and enhance soil fertility, which then allow for better crop production and sustainable income from farming.

7.4.2.7 Agroforestry Entrepreneurs in Soil Fertility Products

Livelihood source: Developing and marketing soil fertility products, such as tree-based biochar, mulching materials, or organic compost.

Connection to supporting services: As soil fertility improves due to the decomposition of tree matter and the addition of organic carbon (Mukangango et al. 2020), products derived from agroforestry systems (such as biochar or mulch) can be processed and sold to other farmers.

Example: In regions like Southeast Asia, entrepreneurs collect tree biomass from agroforestry systems (e.g., *Carya* or *Mangifera* trees) to produce biochar, which is sold as a soil amendment to enhance agricultural productivity.

7.4.2.8 Beekeeping (Apiculture) in Agroforestry Systems

Livelihood source: Beekeepers who operate in agroforestry systems where trees provide forage for pollinators, enhancing honey production.

Connection to supporting services: Tree species that support biodiversity, such as flowering trees and shrubs, provide essential pollen and nectar for bees. This symbiotic relationship enhances honey production and improves crop yields due to better pollination.

Example: Beekeepers in agroforestry systems that feature Citrus, Eucalyptus, or Tamarind trees can produce high-quality honey, benefiting from the improved soil and ecological health provided by these trees.

7.4.2.9 Agroforestry for Sustainable Fuelwood and Biomass Energy Production

Livelihood source: Harvesting sustainable fuelwood from fast-growing trees like *Leucaena*, *Gliricidia*, or *Moringa*, grown alongside crops.

Connection to supporting services: Trees in agroforestry systems not only help improve soil health but also provide a renewable source of biomass for fuel wood and energy production. The increased soil fertility and SOC content from the trees ensure that the land remains productive, supporting both energy and food needs.

Example: In parts of Africa, farmers raising fast-growing trees like *Leucaena* or *Moringa* for fuelwood and biomass energy benefit from improved soil fertility and can sell their wood to meet local energy needs.

In such various ways, agroforestry systems provide a wide range of livelihoods that are intimately linked to the supporting services of soil fertility and structure, and SOC enhancement. These services contribute to the long-term sustainability and productivity of agricultural systems, benefiting farmers, beekeepers, land restorers, and entrepreneurs in various sectors.

7.4.3 Provisioning Ecosystem Services-Based Livelihoods Provided by Agroforestry

Some of provisioning ecosystem services of agroforestry based livelihoods can be listed as follows.

7.4.3.1 Agroforestry Farmer (Diversified Crop and Livestock Production)

How it works: Farmers combine trees with crops and livestock. Trees help with soil fertility, water retention, and reducing erosion, directly benefiting crop yields.

Livelihoods: These farmers earn from both agricultural products (e.g., grains, vegetables, and fruits) and tree products (e.g., timber, nuts, and medicinal plants). The improved soil quality leads to more resilient and higher-yielding crops.

Example: A farmer in East Africa grows maize and beans while also managing an area of fruit trees like mangoes or avocados. The tree roots improve the soil and reduce the need for artificial fertilizers, which increases their agricultural productivity and food security (Mohapatra et al. 2019, Kumar et al. 2019, Newaj and Ram 2019).

7.4.3.2 Tree Crop Producer (Fruits, Nuts, and Timber)

How it works: Agroforestry systems include tree crops that are intercropped with food crops. These trees might include fruit-bearing trees (like citrus, apples, or bananas) or nut trees (e.g., cashews, almonds), which enhance the soil quality and provide food and income.

Livelihoods: These producers sell fruits and nuts to local or regional markets, contributing to the food system while benefiting from increased crop productivity.

Example: A farmer in Central America integrates cacao trees with maize or cassava. The cacao trees add organic matter to the soil, improving the overall soil health, which increases both the yield of the cacao and the companion crops (Naugraiya and Puri 2001; Dubiez et al. 2018; Panda et al. 2018; Kumar et al. 2019; Newaj and Ram 2019; Figyantika et al. 2020).

7.4.3.3 Smallholder Integrated Farming System (Mixed Crops and Livestock with Trees)

How it works: Smallholders integrate trees with food crops and livestock, creating a diversified and resilient farming system. The trees offer shade, improve the microclimate, and help in retaining soil moisture and fertility (Ahmed et al. 2014; Shukla et al. 2018).

Livelihoods: Livelihoods come from both crop sales (e.g., rice, vegetables) and livestock products (e.g., eggs, milk, meat). Trees may provide additional income from timber or fruit. The increased soil fertility helps improve overall agricultural productivity, benefiting their food security and market sales.

Example: In Southeast Asia, a farmer raises chickens and ducks while planting rice in the flooded fields. Tree lines planted around the field, i.e., boundary plantations provide shade and help maintain water retention, which increases rice yields. The trees also provide timber and fruit.

7.4.3.4 Agroforestry Specialist (Herbal and Medicinal Plant Cultivation)

How it works: Trees in agroforestry systems can include medicinal plants or herbs. These trees help improve the soil and provide valuable herbs for medicinal or nutritional purposes, contributing to the health and well-being of the community.

Livelihoods: Individuals who specialize in growing and selling medicinal plants, either for local use or for the pharmaceutical industry, can profit from this agroforestry approach.

Example: A farmer in India grows turmeric and ginger alongside forest trees like *neem* or *tulsi* (holy basil). The trees improve soil health and water retention, which boosts the growth of medicinal plants and farmer's income.

7.4.3.5 Silvopastoralist (Livestock and Forest Products Integration)

How it works: Silvopastoral systems combine livestock grazing with tree cultivation. The trees provide shade, shelter, and forage, which benefit animal health and productivity, while the livestock manure enhances soil fertility.

Livelihoods: The sale of meat, milk, and wool from livestock, as well as products from the trees (e.g., timber, fuelwood), creates multiple streams of income. The improved pasture quality from agroforestry practices also helps maintain a healthy livestock population.

Example: A livestock farmer in Brazil might raise cattle in pasture areas interspersed with trees such as acacia or eucalyptus. The trees protect the animals from the sun, improve soil fertility through nitrogen fixation, and provide timber for sale.

7.4.3.6 Beekeeping and Pollination Services Provider

How it works: Trees in agroforestry systems often provide crucial habitat for pollinators such as bees. The presence of a diversity of flowering trees supports honey production.

Livelihoods: Beekeepers in agroforestry systems can profit from honey production and the sale of bee products (e.g., beeswax, propolis), while farmers benefit from improved pollination of crops, leading to higher yields.

Example: A beekeeper working in a forested agroforestry area in Ethiopia places hives near trees like acacia or fruit-bearing trees. The trees support pollinators, which in tum enhances the production of honey, benefiting both the beekeepers and the surrounding farmers who rely on pollination for their crops.

In all of these examples, provisioning services, such as improved soil fertility, water retention, and erosion control help create sustainable agricultural systems that boost productivity, increase food security, and provide diversified income sources (Manna et al. 2003).

7.4.4 Cultural Ecosystem Services-Based Livelihoods Made Possible Through Agroforestry

Agroforestry systems, by integrating trees and shrubs with agricultural crops or livestock, often offer cultural services that go beyond direct economic or ecological benefits (Mekonnen et al. 2023). These systems can play an important role in local traditions, spirituality, and community identity, while also enhancing the beauty and esthetic value of the landscape (WAC (World Agroforestry Centre) 2010). Here are some examples of livelihoods based on the cultural services provided by agroforestry: (i) Traditional craft production, (ii) Sacred groves and spiritual practices, (iii) Cultural tourism and agroforestry landscapes, (iv) Traditional knowledge and education, (v) Ceremonial and festive uses, (vi) Biodiversity conservation and community well-being, (vii) Agroforestry and local festivals, (viii) Ethnobotanical products.

7.4.4.1 Traditional Craft Production

Livelihood example: Many communities use trees from agroforestry systems for crafting traditional goods. For instance, woodworkers may rely on timber from agroforestry systems to produce furniture, carvings, or handicrafts, which are integral to local culture.

Cultural connection: The specific trees used in these crafts can be tied to cultural symbols, heritage, or religious practices, enhancing the community's connection to the land.

Impact: This not only preserves cultural traditions but also creates an income stream for local artisans, promoting the continuity of traditional knowledge and craftsmanship.

7.4.4.2 Sacred Groves and Spiritual Practices

Livelihood example: In many cultures, certain agroforestry areas, particularly those with unique tree species, are sacred groves or are associated with spiritual practices. These sites may attract visitors for pilgrimages or community ceremonies.

Cultural connection: These sacred spaces are often preserved for cultural and spiritual reasons, and they can generate income through tourism, such as guided tours, offerings, or events centered around local rituals and festivities.

Impact: The spiritual significance of these agroforestry systems reinforces community well-being by offering a sense of identity and continuity.

7.4.4.3 Cultural Tourism and Agroforestry Landscapes

Livelihood example: Agroforestry systems create beautiful, scenic landscapes that attract tourists interested in cultural and natural heritage. A community might offer agro-tourism experiences, where visitors can explore agroforestry systems, learn about local farming traditions, and participate in cultural activities such as festivals or farm-to-table experiences. This is suitable for inclusion of forest bathing (*Shinrin-yoku*) in this program (Hansen et al. 2017; Wen et al. 2019).

Cultural connection: Agroforestry landscapes often hold historical or esthetic significance, becoming symbols of local identity or heritage. This can involve showcasing local biodiversity, traditional farming practices, and landscapes shaped by generations of cultivation.

Impact: This form of tourism not only provides economic opportunities but also strengthens cultural pride and supports the conservation of traditional landscapes.

7.4.4.4 Traditional Knowledge and Education

Livelihood example: Agroforestry systems are often used to pass on traditional agricultural knowledge. Communities may create educational programs or workshops for younger generations or visitors on how to manage trees and crops in a sustainable, culturally appropriate manner.

Cultural connection: This knowledge is often tied to specific local traditions, rituals, and seasonal cycles. Teaching these practices allows communities to preserve their cultural identity while simultaneously providing an income from education and training programs.

Impact: By educating others, these systems help maintain cultural values, practices, and skills while empowering the younger generation with livelihood opportunities.

7.4.4.5 Ceremonial and Festive Uses

Livelihood example: Trees in agroforestry systems may be used for ceremonial purposes, for example, leaves, fruits, or flowers may be harvested for use in religious ceremonies, festivals, or other community celebrations. Specific plants might have symbolic or ritual significance.

Cultural connection: In many regions, agroforestry landscapes support specific plants that are crucial for community events, whether for cooking, decoration, or symbolic acts.

Impact: This connection to ceremony enhances community cohesion and provides opportunities for trade, as the sale of ceremonial products (such as flowers, fruits, or other culturally significant plants/plant parts) can be a source of income.

7.4.4.6 Biodiversity Conservation and Community Well-Being

Livelihood example: Agroforestry can be designed to support biodiversity, providing habitats for native plants and animals that are central to local cultures and traditions. For instance, indigenous communities may rely on local plants for medicinal purposes, food, or ceremonial uses (Julius et al. 2011).

Cultural connection: The conservation of native species and the ecosystem health are often linked to cultural beliefs in environmental stewardship and the interdependence of humans and nature.

Impact: In addition to ecological benefits, such biodiversity can offer direct and indirect income opportunities through the collection and sale of wildcrafted plants, medicinal herbs, and/or honey produced by agroforestry systems.

7.4.4.7 Agroforestry and Local Festivals

Livelihood example: In some regions, agroforestry systems are used as daily private (i.e., opensky drawing room space0 or public (e.g., opensky restaurant) purposes or to organize local festivals, such as harvest festivals or tree planting events. These festivals often revolve around the celebration of the harvest and the planting of new crops or trees, creating a source of income through event management, local food sales, and crafts.

Cultural connection: These festivals are steeped in tradition and may involve cultural performances, traditional dances, music, and feasting, all of which provide opportunities for local artisans and performers to sell their products or services.

Impact: These events not only generate economic activity but also serve to reinforce cultural identity and community spirit.

7.4.4.8 Ethnobotanical Products

Livelihood example: Some agroforestry systems incorporate ethnobotanical plants that have value for local medicine, food, or cosmetics. These plants are often harvested and sold locally or internationally, providing a livelihood for the community.

Cultural connection: The knowledge of these plants, and the sustainable ways of harvesting them, are part of cultural traditions passed down through generations.

Impact: Through ethical harvesting and trade of ethnobotanical products, communities can earn income while also preserving their traditional knowledge about plant use and conservation.

In those ways, the cultural services provided by agroforestry can play a significant role in sustaining local livelihoods, fostering community identity, and preserving cultural heritage, while contributing to economic development in a way that is rooted in tradition and respect for the environment (Garrity 2004).

7.5 Agroforestry and Livelihood Security

Thorough scrutiny of any sorts of ecosystem services-based agroforestry-related livelihoods would reveal that agroforestry is certainly a guarantee toward maintaining sustainable and productive agroecosystem. This can be outlined as follows:

Agroforestry is a land management approach that combines elements of agriculture and forestry to create a more sustainable, resilient, and productive farming system. By integrating trees, shrubs, crops, and sometimes livestock, agroforestry helps address global challenges such as food insecurity, biodiversity loss (Julius et al. 2011), and climate change.

At its core, agroforestry improves both ecological and economic outcomes. It enhances food security by diversifying income streams, with farmers benefiting from crops, timber, fruit, fodder, and other forest products. This diversification reduces the risks associated with market volatility and climate change, making farmers more resilient to external shocks. Agroforestry systems also provide valuable ecosystem services, such as improved soil health, water retention, and carbon sequestration, which contribute to the overall health of the environment (Kumar et al. 2019; Mukangango et al. 2020; Mekonnen et al. 2023).

Agroforestry can be broadly classified into two systems: agrosilvopastoral, which integrates crops, animals, and trees, and agrosilvicultural, which focuses on combining crops and trees. Both systems foster synergistic relationships between different components, optimizing land use while conserving ecological balance. For example, trees can provide shade for crops, reduce soil erosion, and improve biodiversity (Julius et al. 2011), while animals can help with pest control and fertilization (Singh 1996; Kang and Akinnifesi 2000).

Beyond the environmental and economic benefits, agroforestry also enhances cultural and community connections to the land. Agroforestry systems can provide spaces for recreation, education, and cultural practices, fostering stronger ties between people and their environments. However, the adoption of agroforestry is not without challenges. Implementing these practices often requires significant investment in knowledge, resources, and changes to traditional farming practices. Farmers may face barriers such as lack of technical support, limited access to markets for agroforestry products, and insufficient policy frameworks that support the adoption of agroforestry. Despite these challenges, agroforestry represents a forward-thinking solution for creating more sustainable agricultural systems. As climate change, resource depletion, and population growth continue to pose significant threats to food security and ecological integrity, the widespread adoption of agroforestry practices is crucial. Supporting agroforestry through research, education, and policy can help ensure that agriculture remains both productive and environmentally responsible, benefiting both people and the planet in the long term.

7.5.1 Agroforestry for Strengthening Food Security and Revenue Generation

Agroforestry systems offer a powerful solution to the twin challenges of food security and rural economic development. By integrating trees with traditional farming practices, they not only diversify the range of products a farm can produce but also build resilience to the volatility of climate change linked market fluctuations. In regions dependent on monoculture crops, where pests, diseases, or adverse weather can lead to devastating losses, diversified approach of agroforestry reduces these risks by spreading production across multiple sources, such as fruits, nuts, timber, and fuelwood. This diversification is key in addressing food insecurity, as it increases the chances of consistent harvests, even when some crops might fail. With multiple outputs from the same land, farmers are more likely to find at least one successful harvest. This steady supply of food not only supports the farm household but can help stabilize local food systems and communities at large. From a financial standpoint, agroforestry is a game changer for rural economies. Beyond food, the additional revenue streams generated from timber, fruits, or other tree-based products add significant value to what may otherwise be a modest agricultural operation. Furthermore, agroforestry ecosystems provide essential services like improved soil fertility, water retention, and carbon sequestration, all of which can have direct economic benefits through the sale of environmental credits or increased land productivity (Kumar and Chaturvedi 2017; Singh et al. 2017; Kumar et al. 2019; Newaj and Ram 2019; Honnayya et al. 2020; Mukangango et al. 2020; Mekonnen et al. 2023).

Overall, agroforestry is a win-win for food security and revenue generation. It empowers farmers to maximize their land potential, creating sustainable livelihoods and contributing to the broader local economy. The integration of trees into agricultural systems is not just a practice of diversification—it's a pathway to long-term resilience and prosperity (Mekonnen et al. 2023).

7.5.2 Safeguarding Biodiversity Through Practicing Ecosystem-Based Agroforestry-Related Livelihoods

Agroforestry represents a transformative approach to land management, bridging the gap between agricultural productivity and environmental sustainability. A key advantage of agroforestry is its capacity to safeguard biodiversity, an aspect increasingly threatened by conventional farming methods. In traditional agriculture, large-scale monoculture crops and deforestation lead to the destruction of natural habitats, resulting in a dramatic loss of wildlife. Agroforestry, however, integrates trees into farming systems, creating multi-layered ecosystems that support a wide variety of plant and animal species (Julius et al. 2011).

This diverse environment nurtures not only wildlife but also beneficial organisms like pollinators and natural pest predators. The presence of trees offers food and shelter, creating a haven for insects, birds, and other wildlife that would otherwise be displaced by monoculture farming. In tum, this biodiversity boosts crop yields by providing natural services such as pollination and pest control, reducing the need for synthetic inputs like pesticides and fertilizers. This is not just a win for the environment but for farmers as well, improving the resilience and profitability of their agricultural systems (Julius et al. 2011).

Beyond wildlife, agroforestry also plays a vital role in preserving genetic diversity, particularly in crops. By incorporating native tree species alongside traditional agricultural crops, farmers can maintain local varieties that are naturally adapted to the local environment. These varieties tend to be more resilient to climatic changes and pests, enhancing food security in the face of global challenges like climate change and biodiversity loss. Agroforestry, therefore, helps to create a more stable and resilient agricultural landscape, ensuring that both ecosystems and human populations benefit from a diversified and sustainable approach to food production (Naugraiya and Puri 2001; Julius et al. 2011; Panda et al. 2018).

Those deliberations reveal that agroforestry presents a promising model for conserving biodiversity while meeting the growing demand for food. By harmonizing agricultural practices with nature, agroforestry can provide ecological benefits, economic advantages, and greater food security, making it an essential strategy for the future of agriculture (Julius et al. 2011; Mekonnen et al. 2023).

7.6 Farm Accountant's Book-Keeping Sheet on Analyzing Cost-Benefit from Agroforestry-Based Various Livelihoods

A truthful cost-benefit accounting is a must to establish the economically viability of ecosystem services-based agroforestry related livelihoods. For this reason, an example of a farm accountant's book keeping sheet and some other livelihood-related numerical problems and solutions are given here. This would be very much helpful for beginners on practicing agroforestry as well as extracting ecosystem-based livelihood scopes out of those agroforestry practices.

7.6.1 Preparation and Filling-Up of Farm Accountant's Book-Keeping Sheet (Table 7.1)

7.6.1.1 Instructions for Filling Up the Sheet

1. Cost columns:
   (a) All capital costs (one-time investments, non recurring) should be filled in the "Cost" column for each item.
   (b) Operating costs (recurring) such as labor, inputs, and maintenance should also be filled in the "Cost" column for each relevant item.

Table 7.1 Farm accountant's bookkeeping sheet[a]

Category	Subcategory	Details	Cost (₹)	Benefit (₹)	Net benefit (₹)
Capital costs	Land preparation	Soil clearing, ploughing, tilling	[Amount]		
	Planting materials	Seeds, sapling, nursery setup	[Amount]		
	Infrastructure development	Irrigation systems, fencing. sheds for animals	[Amount]		
	Livestock purchase	Cattle, poultry, goats, pigs, etc.	[Amount]		
Operating costs	Labor costs	Wages for field workers, animal caretakers	[amount]		
	Inputs	Fertilizers, pesticides, feed for animals	[Amount]		
	Maintenance	Tools, equipment, upkeep of farm infrastructure	[Amount]		
Management costs	Marketing and evaluation	Data collection, farm management, consultancy	[Amount]		
	Training and capacity building	Farmer training, technical knowledge sharing	[Amount]		
Revenues from agroforestry	Crop sales	Timber, grains, pulses, nuts, fruits, vegetables, etc.		[Amount]	
	Non-timber forest product sales	Medicinal plant parts, honey, mushrooms		[Amount]	
Revenues from aquaculture	Aquaculture product sales	Fishes, edible frogs (*Pelophylax kl. esculentus*), oysters, water chestnut, lotus, water lily (esculent stalk)		[Amount]	
Revenues from animal husbandry	Animal products	Meat, eggs, milk, wood, leather, horn		[Amount]	
	Sale of livestock	Sale of livestock (cattle, goats, etc.)		[Amount]	
Other income	Eco-tourism	Revenue from agro-tourism, workshops		[Amount]	

(continued)

Table 7.1 (continued)

Category	Subcategory	Details	Cost (₹)	Benefit (₹)	Net benefit (₹)
Subsidies and grants	Government or NGO support	Financial support for sustainable practices		[Amount]	
Intangible benefits	Biodiversity enhancement	Improvement in local biodiversity, ecosystem services			Value
	Carbon sequestration	Value of carbon credits or environmental benefits			Value
	Community welfare	Improved quality of life, educational benefits, etc.			Value
Risk Factors	Uncertainty and contingencies	Risk of crop failure, disease outbreaks	[Amount]		

[a] Depending on farm activities, other categories may be included or some of which may be deleted

   (c) Management costs (recurring) should include any ongoing expenses related to training, monitoring, and consultancy.
2. Benefit columns:
   (a) Revenues from agroforestry (e.g., crop sales, timber, non-timber products), animal husbandry (e.g., milk, eggs), aquaculture (e.g., fishery, duckery, oyster, water chestnut, and many more) should be filled in the "Benefit" column.
   (b) Other income such as eco-tourism or government subsidies should be added to the "Benefit" column.
   (c) Intangible benefits such as improved biodiversity or carbon sequestration can be estimated qualitatively or monetarily (based on local/governmental approved valuation methods) and filled under the "Net Benefit" column.
3. Net Benefit Calculation:
   (a) Subtract total costs from total benefits to determine the net benefit.
   (b) The net benefit is the key indicator of the success of the agroforestry-animal husbandry-aquaculture integration (Table 7.1a).

7.6.2 Numerical Problems and Solutions Related to Ecosystem-Based Agroforestry Practices

(Based mostly on Design and Diagnostic survey among farmers in various places in India and published data: Newaj and Rai 2005; Singh et al. 2014; Sridhar et al. 2015; Handa et al. 2019; Raj et al. 2019; Dev et al. 2020; ICFRE 2020; Das and Panda 2022; Das et al. 2022).

Table 7.1a Example of breakdown (partial) of farm accountant's book keeping sheet

Category	Subcategory	Details	Cost (₹)	Benefit (₹)	Net benefit (₹)
Capital costs	Land preparation	Soil clearing, ploughing, tilling	₹ 50,000		
	Planting materials	Seeds, sapling, nursery setup	₹ 30,000		
	Livestock purchase	Cattle, poultry, goats, pigs, etc	₹ 20,000		
Operating costs	Labor costs	Wages for field workers, animal caretakers	₹ 40,000		
	Feed	Animal feed for 6 months	₹ 15,000		
Revenue from agroforestry	Production sales	Timber, grains, pulses, nuts, fruits, vegetables, etc		₹ 60,000	
Revenues from animal husbandry	Milk sales	Milk from cows, goats		₹ 25,000	
	Eco-Tourism	Revenue from farm tours & workshops		₹ 10,000	
Net benefit					₹[Final Net Benefit]

Problem 7.1 Calculation of Income from Lac Production

Problem: A farmer in central India plants 5 *Palash* trees under an agroforestry system (AFS). Each tree yields 2 kg of lac per year. The annual income per tree is 750 rupees. How much total income will the farmer generate from these 5 trees?

Solution:

1. Income per tree:
   The income generated by one *Palash* tree is 750 rupees per year.
2. Income from 5 trees
   Multiply the income per tree by the number of trees.
   Total income = 750 rupees/tree × 5 trees = 3750 rupees.

Answer:

The farmer will generate a total of 3750 rupees per year from 5 *Palash* trees.

Problem 7.2 Benefit-to-Cost Ratio Calculation for Aonla-Based Agroforestry

A farmer adopts an aonla-based agroforestry system on marginal land. The benefit-to-cost ratio (B:C) for the 13-year-old aonla-based system is reported to be 3.28. If the total cost of implementing the system is 50,000 rupees, what is the total benefit or income the farmer earns from the system?

Solution:

1. Benefit-to-Cost Ratio (B:C)
   The B:C ratio is 3.28, meaning for every 1 rupee spent, the farmer earns 3.28 rupees.
2. Total income (benefit)
   Multiply the total cost by the B:C ratio to find the total income.
   Total income = Cost × B:C ratio.
   Or, Total income = 50,000 rupees × 3.28 = 1,64,000 rupees.

Answer:
The farmer will earn a total income of 1,64,000 rupees from the aonla-based agroforestry system.

Problem 7.3 Comparing Profitability of Agroforestry Systems
A farmer in the upper Gangetic plain adopts a poplar-paddy agroforestry system, which has a B:C ratio of 3.31. If the farmer's total investment in the system is 80,000 rupees, what is the total benefit (income) generated from this system?
Solution:

1. Benefit-to-Cost Ratio (B:C)
   The B:C ratio is 3.31, meaning for every 1 rupee spent, the fanner earns 3.31 rupees.
2. Total income (benefit)
   Multiply the total cost by the B:C ratio to find the total income.
   Total income = Cost × B:C ratio.
   Or, Total income = 80,000 rupees × 3.31 = 264,800 rupees.

Answer:
The farmer will generate a total income of 264,800 rupees from the poplar-paddy agroforestry system. This can be compared with the total income of 1,64,000 rupees from the aonla-based agroforestry system as in Problem 7.2.

Problem 7.4 Calculation of Profit from Eucalyptus-Based Agroforestry System
An eucalyptus-hybrid paddy agroforestry system in the lower Gangetic plains has a B:C ratio of 2.33. If a farmer invests 60,000 rupees in this system, what will be the total income (benefit) the farmer earns?
Solution:

1. Benefit-to-Cost Ratio (B:C)
   The B:C ratio is 2.33, meaning for every 1 rupee spent, the farmer earns 2.33 rupees.
2. Total income (benefit)
   Multiply the total cost by the B:C ratio to find the total income.
   Total income = Cost x B:C ratio = 60,000 rupees x 2.33 = 1,39,800 rupees.

Answer:

The farmer will generate a total income of 1,39,800 rupees from the Eucalyptus-hybrid paddy agroforestry system.

Problem 7.5 Profit Comparison Between Aonla-Based and Poplar-Paddy Agroforestry Systems

A farmer is considering two agroforestry systems: the aonla-based system with a B:C ratio of 1.09 and the poplar-paddy system with a B:C ratio of 3.3 l. If the farmer has the same initial investment of 70,000 rupees for both systems, calculate the total income for each system and determine which one is more profitable.

Solution:

1. Aonla-based AF system income
   B:C ratio = 1.09.
   Investment = 70,000 rupees.
   Total income = Investment × B:C ratio = 70,000 rupees × 1.09 = 76,300 rupees.
2. Poplar + paddy system income
   B:C ratio = 3.31 Investment = 70,000 rupees.
   Total income = 70,000 rupees × 3.31 = 2,31,700 rupees.

Answer:

The poplar-paddy agroforestry system is more profitable with a total income of 2,31,700 rupees, compared to 76,300 rupees from the aonla-based system.

7.6.2.1 Problems Related to Various Ecosystem Services from Agroforestry

7.6.2.1.1 Agroforestry's Contribution to Biodiversity

Problem 7.6

A study shows that agroforestry systems can increase biodiversity by 40% compared to conventional agricultural systems (Udawatta et al. 2019; Boinot et al. 2022). If a conventional system supports 50 species per hectare, how many species would be supported per hectare in an agroforestry system?

Solution:

Increase in biodiversity = 40%.

Species supported in conventional system = 50 species.

Increase in species = 50 species × 40% = 20 increased species supported in agroforestry system.

= 50 species +20 species = 70 species per hectare.

Answer:

70 species per hectare.

7.6.2.1.2 Potential Benefits of Pollinators in Agroforestry Systems

Problem 7.7

A farm with 50 hectares of agroforestry has seen an increase in pollinators (bees and butterflies) as a result of the diverse flowering plants present. Research suggests that the increase in pollinator populations leads to a 15% increase in crop yields for certain crops. The farm grows three crops on 60% of the land area, with the remaining 40% dedicated to trees and natural habitat.

If the average yield of these crops is 2000 kilograms per hectare, calculate the total increase in crop yield due to the enhanced pollination in the agroforestry system.

Solution:

1. Calculate the total land area dedicated to crops
   Total land area = 50 hectares
   Area for crops = 60% of 50 hectares = 0.60 × 50 = 30 hectares.
2. Calculate the total yield of the crops before the increase
   Yield per hectare = 2000 kg
   Total crop yield = 30 hectares × 2000 kg/hectare = 60,000 kg.
3. Calculate the increase in crop yield due to pollination
   Increase in crop yield = 15% of total yield.

Answer:
Increased yield = 0.15 × 60,000 kg = 9000 kg.
The total increase in crop yield due to enhanced pollination = 9000 kg.

7.6.2.1.3 Genetic Diversity Preservation in Agroforestry

Problem 7.8

In an agroforestry system, farmers integrate 5 native tree species into their fields to preserve genetic diversity among crops. These trees also support 3 different local varieties of maize, each planted across 10 hectares of land.

If each variety of maize has a yield of 1500 kg per hectare and the integration of the trees reduces the susceptibility of maize crops to pests by 25%, how much additional maize yield is generated by the reduction in pest damage?

Solution:

1. Calculate the total maize yield before the 'reduction in pest damage
   Yield per hectare = 1500 kg
   Area per variety = 10 hectares.
   Total maize yield = 3 varieties × 10 hectares × 1500 kg/hectare = 45,000 kg.
2. Calculate the additional maize yield due to reduction in pest damage
   Reduction in pest damage = 25% of the total yield.
   Additional yield = 0.25 × 45,000 kg = 11,250 kg.

Answer:
The additional maize yield due to reduced pest damage = 11,250 kg.

Problem 7.9

A farm invests \$500 per hectare to establish an agroforestry system, which includes planting trees, maintaining natural habitats, and enhancing biodiversity. In return, the farm receives a 10% increase in crop yield and an additional income from the sale of timber and non-timber forest products, totaling \$800 per hectare per year. If the farm has 100 hectares, calculate the total return on investment (ROI) after 5 years from the biodiversity benefits alone (excluding other benefits such as timber income).

Solution:

1. Calculate the total investment
   Investment per hectare = \$500
   Total investment for 100 hectares = 100 × \$500 = \$50,000.
2. Calculate the total increased crop yield over 5 years
   Total yield increase per hectare
   = 10% of 2000 kg/hectare (assuming original yield is 2000 kg/ hectare).
   Increased yield per hectare = 0.10 × 2000 kg = 200 kg/ hectare.
   Total increased yield for 100 hectares = 100 × 200 kg = 20,000 kg.
   Value of increased yield (assuming market price of \$1 per kg) = 20,000 kg × \$1 = \$20,000.
3. Calculate the total return on investment
   Total return over 5 years = \$20,000
   ROI = [(Total return – Total investment)/ Total investment] × 100.
   = [(\$20,000–\$50,000)/ \$50,000] × 100 = (–) 60%.

Answer:

The ROI from biodiversity benefits alone after 5 years is (–) 60%.

7.6.2.1.4 Impact of Agroforestry Tree Root System on Soil Erosion

Problem 7.10

In an agroforestry system, a plot of land is protected by trees and shrubs whose root systems help stabilize the soil. A certain plot of land has an area of 10 hectares and a soil erosion rate of 5 tons per hectare per year without trees. With trees in place, the erosion rate is reduced by 70%.

How much soil erosion is reduced in total (in tons per year) for the entire 10 hectare plot due to the presence of trees?

Solution:

Soil erosion without trees = 5 tons/hectare/year.
Erosion reduction rate = 70% = 0.70.
Erosion reduction per hectare = 5 tons × 0.70 = 3.5 tons/hectare/year.
Total reduction for 10 hectares = 3:5 tons/hectare/year × 10 hectares = 35tons/year.

Answer:

35 tons of soil erosion are reduced per year.

Problem 7.11

In an agroforestry system, the root systems of trees reduce soil erosion by 30% compared to conventional farming methods. If a field normally loses 100 tons of soil per hectare per year through erosion, how much soil is saved per hectare per year by using agroforestry?

Solution:

Erosion reduction = 30% of 100 tons = $100 \times 0.30 = 30$ tons.

Therefore, the agroforestry system saves 30 tons of soil per hectare per year.

Answer:

This agroforestry system saves 30 tons of soil per hectare per year.

7.6.2.1.5 Impact of Agroforestry Tree Coverage on Water Retention in Soil

Problem 7.12

In an agroforestry system, trees are planted to improve water retention in the soil. In a study plot, it was observed that the water retention capacity of the soil increased by 40% after the establishment of trees. The plot of land has a total water retention capacity of 200,000 L per hectare without trees.

What is the total increase in water retention (in L) for the entire 15 hectare plot after trees are planted?

Solution:

Water retention without trees = 200,000 L/hectare.

Water retention increase = 40% = 0.40.

Increase in water retention per hectare = 200,000 L × 0.40 = 80,000 L/hectare.

Total increase for 15 hectares = 80,000 L × 15 hectares = 1,200,000 L.

Answer:

The total increase in water retention is 1,200,000 L.

7.6.2.1.6 Impact of Agroforestry on Soil Organic Carbon (SOC) and Crop Yield Improvement

Problem 7.13

An agroforestry system improves soil organic carbon (SOC) by 0.25% per year. In a specific agricultural area, the SOC content is currently 2.5% in the soil. Over the next 5 years, the SOC content is expected to increase due to agroforestry practices.

What will the SOC content be after 5 years, and by how much will it increase in total over the 5-year period?

Solution:

Such problem is needed to be solved through compound growth formula:

$= (1 + 0.0025)^5 \approx 1.01\%$.

Answer:

The 'SOC content will increase to 1.01% after 5 years, with a total increase of 1.25%.

7.6.2.1.7 Impact of Agroforestry on Carbon Sequestration *vis-a-vis* Climate Mitigation

Problem 7.14

An agroforestry system in a 50 hectare plot is capable of sequestering 2.5 tons of carbon per hectare per year. Over the course of 10 years, how much total carbon will be sequestered by this agroforestry system in terms of tons?

Calculate the total amount of carbon sequestered by the agroforestry system over 10 years.

Solution:

Carbon sequestration per hectare per year = 2.5 tons.

Total area = 50 hectares.

Number of years = 10 years.

Total carbon sequestration per year = 2.5 tons/hectare/year × 50 hectares = 125 tons/ year.

Total carbon sequestration over 10 years = 125 tons/year × 10 years = 1250 tons.

Answer:

The agroforestry system will sequester a total of 1250 tons of carbon over 10 years.

7.6.2.1.8 Impact of Agroforestry on Organic Matter Contribution *vis-a-vis* Soil Fertility

Problem 7.15

In a specific agroforestry plot, the organic matter contributed by the trees (in the form of leaf litter and decaying roots) is 3 tons per hectare per year. A study shows that the organic matter contribution improves soil fertility, which leads to a 15% increase in crop yield. The original crop yield without agroforestry is 50 tons per hectare per year.

What is the total increase in crop yield per hectare per year due to the presence of trees?

Solution:

Original crop yield = 50 tons/hectare/year.

Increase in crop yield = 15% = 0.15.

Increase in crop yield per hectare = 50 tons × 0.15 = 7.5 tons/ hectare/year.

Answer:

The total increase in crop yield per hectare per year is 7.5 tons.

7.6.2.1.9 Impact of Agroforestry on Soil Fertility Improvement through Improvement of Soil Organic Matter

Problem 7.16

In an agroforestry system, the amount of organic matter added to the soil from leaf litter and root decay increases by 2 tons per hectare per year. If this results in a 15% increase in soil fertility (in terms of soil available nitrogen), how much more fertile is the soil per hectare after 5 years?

Solution:

1. Increase in organic matter per year
   The organic matter added to the soil increases by 2 tons per hectare per year.
2. Increase in soil fertility per year
   This organic matter increases the soil fertility by 15% per year.
3. Calculate the total increase in soil fertility after 5 years

Since the soil fertility increases by 15% each year due to the added organic matter, we can use this percentage to determine how much more fertile the soil is after 5 years.

After the first year: 15% increase

After the second year: another 15% increase on the already increased fertility.

This process repeats for 5 years.

We can calculate the cumulative increase in fertility using the formula for compound growth:

Total fertility after 5 years = $(1 + 0.15)^5 \approx 20\%$.

7.6.2.2 Problems Related to Project Preparation and Maintenance of Agroforestry Projects

Agroforestry projects require ensuring economic benefits of integrating various tree species into agricultural systems, highlighting both the profitability and sustainability of such systems. By using B:C ratios, income calculations, and profitability analysis, farmers can make informed decisions about which agroforestry practices best suit their land and economic goals.

Problem 7.17 Economic Viability of Agroforestry Systems: Benefit-to-Cost Analysis

Problem Statement:

Analyze and compare the economic profitability of various agroforestry systems based on the benefit-to-cost (B:C) ratio provided in the data. Specifically, calculate and interpret the B:C ratios for different agroforestry models mentioned in the project description.

Steps:

1. B:C Ratio: The B:C ratio is a measure of profitability that compares the benefits (revenues) from the agroforestry system to the costs incurred for its establishment and maintenance

   B:C ratio = Total Benefits (Revenue)/ Total Costs (Investment).

2. Given data for B:C Ratios
   Poplar + Paddy + Wheat (Upper Gangetic plain, 7 years): 3.31
   Babool + Bajra + Guar (Trans-Gangetic plains, 20 years): 2.61
   Khejri + Bajra (Western dry regions, 20 years): 2.53
   Eucalyptus + Hybrid Paddy/ Wheat (Lower Gangetic plains, 10 years): 2.33
   Aonla + Groundnut + Gram (Central plateau and hill regions, 8 years): 1.09
3. Calculate expected annual benefits: For each system, let's assume the following
   Total Benefits (Revenue) = Total income generated from the sale of crops and products (e.g., timber, fruits, honey, and many more)
   Total Costs (Investment) = Expenses for inputs, labor, maintenance, and other costs.
   Assume hypothetical values for the total benefits and costs for each agroforestry system.
   For example: Poplar + Paddy + Wheat:
      Total Benefits = 1,00,000 per year
      Total Costs = 30,000 per year.
      B:C ratio = 3.31.
      Using the formula above, confirm the B:C ratio:
      B:C = 1,00,000/ 30,000 = 3.33 (approx. 3.31 as per data).
4. Interpretation of B:C Ratios
   High B:C Ratio indicates a highly profitable agroforestry system.
   B:C ratio greater than 1 means the system is profitable.

Project validation

Poplar + Paddy + Wheat system (3.31) is the most profitable, whereas Aonla + Groundnut + Gram (1.09) is the least profitable among the listed systems.

Higher B:C ratios correlate with agroforestry systems that produce more marketable products, such as timber and fruits, which are sold at higher prices, or have longer-term yields that improve profitability.

Problem 7.18 Income Generation from Specific Agroforestry Species (Palash Tree for Lac Production)

Statement:

Calculate the total income generated from a palash tree (*Butea monosperma*) based on lac production.

Analyze the sustainability and income potential of planting palash trees in agroforestry systems.

Steps:

1. Data on lac production
   A single palash tree produces 1.5–2.5 kg of lac per year.
   Each kg of lac generates an income of 500 INR.
2. Calculation of annual income per tree
   For the lower production estimate of 1.5 kg:
   Income = 1.5 kg × 500 INR = ₹750 per year

For the higher production estimate of 2.5 kg:
Income = 2.5 kg × 500 INR = ₹1250 per year.

3. Potential for multiple trees
   Let us assume a farmer plants 100 palash trees.
   For the lower estimate (1.5 kg per tree):
   Total Income = 100 trees × ₹750 = ₹75,000 per year.
   For the higher estimate (2.5 kg per tree):
   Total Income = 100 trees × ₹1250 = ₹1,25,000 per year.

Project validation:

Planting palash trees in an agroforestry system offers a viable source of income through lac production. The income can range from ₹75,000 to ₹125,000 annually from just 100 trees, demonstrating the economic potential of integrating this species into agroforestry systems.

Problem 7.19 Profitability of Boundary-Based Agroforestry with Babul Tree
Problem Statement:

Evaluate the profitability of boundary-based agroforestry using babul trees for gum exudation.

Steps:

1. Gum-exudation process: The babul tree is highly effective for gum exudation when scientifically induced (using ethephon-induced gummosis). The gum is then harvested and sold as a valuable product Box 7.1.
2. Calculate expected revenue from gum sales:
   Assume each babul tree produces 5 kg of gum per year.
   The market price for gum is ₹300 per kg.
   For one babul tree: Revenue from gum = 5 kg × ₹300 = ₹1500 per year.
3. Timber and Other Products
   Assume the babul tree also produces timber worth ₹2000 per year.
   Total revenue per tree per year:
   Total Revenue = ₹1500 (gum) + ₹2000 (timber) = ₹3500 per year.
4. Cost Considerations
   Assume the annual maintenance cost for one babul tree (labor, water, ethephon application, and many more) = ₹1000.
5. Profit Calculation
   Profit = ₹3500 (revenue) – ₹1000 (cost) = ₹2500 per tree per year

Project validation:

Babul-based bund agroforestry is a highly profitable venture with a net income of ₹2500 per tree per year, considering both gum and timber production.

Box 7.1 Ethephon: Chemistry, Ethylene Release, and Toxicity[a]

Chemical identity

- IUPAC name: 2-Chloroethylphosphonic acid
- Molecular formula: $C_2H_6ClO_3P$
- CAS number: 16672-87-0
- Appearance: White crystalline solid
- Solubility: Highly soluble in water
- Melting point: ~74 °C

Mechanism of action

Ethephon decomposes in plant tissues to release ethylene gas, a natural plant hormone that regulates growth, ripening, and senescence. This conversion occurs non-enzymatically under acidic conditions (pH < 3.5) and is catalyzed by bases, producing phosphoric acid and chloride ions as byproducts

$$\underset{\text{(Ethephon)}}{\mathrm{ClCH_2}CH_2PO_3^{2-}} + \mathrm{H_2O} \rightarrow Cl^- + \underset{\text{(Ethylene)}}{\mathrm{C_2H_4}} + \mathrm{H_2}PO_4^-$$

Reaction conditions

- Medium: Water (aqueous)
- pH: Typically <3.5 for efficient decomposition
- Catalysis: The reaction is non-enzymatic and can be enhanced by acids or bases
- Mechanism highlights (at pH > 3.5):

The reaction is slower because protonation of the ethephon molecule is less efficient

At neutral or slightly basic pH, the phosphonic acid group resists protonation, which delays the intramolecular nucleophilic substitution needed to release ethylene

In plant tissues, natural cellular acidity or enzymatic activity can still trigger decomposition even at pH > 3.5

pH of aqueous extract of *Babool* (*Acacia arabica* Willd.) bark powder: 8.93

Agricultural uses

Ethephon is applied to various crops to:

- Accelerate fruit ripening (e.g., tomatoes, apples, and pineapples)
- Induce flowering and fruit set in pineapples
- Promote defoliation and boll opening in cotton
- Synchronize flowering and fruiting in crops like cucumbers and cherries
- Enhance female flower formation in cannabis

Toxicity profile

- Acute toxicity: Low oral LD_{50} in rats (3400–4229 mg/kg), classified as EPA Toxicity Category III
- Cholinesterase inhibition: Inhibits plasma and brain cholinesterase activity, indicating potential neurotoxic effects
- Corrosive effects: Can cause burns and irritation upon contact with skin, eyes, or mucous membranes
- Chronic exposure risks: May lead to hepatotoxicity, nephrotoxicity, and reproductive toxicity in animals
- Human health concerns: Improper use, such as excessive application for fruit ripening, can result in pesticide residues on produce, posing health risks like respiratory and gastrointestinal issues

Regulatory notes

The Food Safety and Standards Authority of India (FSSAI) has set guidelines for ethephon use in fruit ripening, limiting the permissible quantity to 0.5 g per 10 kg of fruit. Excessive use beyond these limits is prohibited due to health concerns

[a] Ref:
https://chem.hbcse.tifr.res.in/wp-content/uploads/2024/08/Ethephon-English.pdf
Sharma and Johari (2023)
https://www.fssai.gov.in/upload/uploadfiles/files/Guidance_Note_Ver2_Artificial_Ripening_Fruits_03_01_2019_Revised_10_02_2020.pdf
https://fssai.gov.in/upload/uploadfiles/files/artificial_fruit_ripening_english.pdf

7.6.2.3 Problems Related to Financial Data from Agroforestry Research Project for Establishment of Best Agroforestry Practices

Problem 7.20 Comparison of the Economic Performances of Various Agroforestry Systems for Recommendation

To compare the economic performances of various agroforestry systems between the years 2017-18 and 2018–19, we can analyze the data for each treatment and focus on the key parameters, e.g., Cost of Cultivation (ha^{-1}), Gross Return (ha^{-1}), Net Return (ha^{-1}), and Benefit-Cost (B:C) ratio.

Let us go step-by-step through the comparison of data from nineteen agroforestry research treatments, as given in Table 7.2 (Table 5 in Das et al. 2022).

Step 1: Overview of the parameters.

1. Cost of cultivation (ha^{-1}): This is the total expenditure required to cultivate the agroforestry system.
2. Gross return (ha^{-1}): This represents the total income generated from the agroforestry system, including all harvested products (timber, crops, etc.).
3. Net return (ha^{-1}): This is calculated by subtracting the cost of cultivation from the gross return showing the actual profit made from the system.
4. B:C ratio: The Benefit-Cost ratio is calculated as the ratio of the gross return to the cost of cultivation. A higher B:C ratio indicates better economic performance.

Step 2: Treatment-wise comparison for 2017–18 *vs.* 2018–19.

Now, let us look at the performance of each treatment in both years:

T1: Gamhar + Sweet Orange + Pigeon Pea

2017–18: Cost of cultivation = ₹66,651, Gross return = ₹95,517, Net return = ₹28,866

B:C = Cost of cultivation/Gross return = 95,517/66,651 = 1.43.

2018–19: Cost of cultivation = ₹66,651, Gross return = ₹1,03,849, Net return = ₹37,198

B:C = Cost of cultivation/Gross return = 1,03,849/ 66,651 = 1.56.

Analysis: There is an increase in both gross return (8332) and net return (8332) from 2017–18 to 2018–19. The B:C ratio improves from 1.43 to 1.56, indicating better economic efficiency in 2018– 19.

Recommendation: Gamhar tree on attaining tenth year interspaced with sweet orange after third year are beneficial for production in the Gamhar + sweet orange + pigeon pea agroforestry system. Other treatments also need such peer B:C ratio analysis before recommendation for further cultivation as follows.

T2: Gamhar + Sweet Orange + Green Gram-Toria

2017–18: Cost of cultivation = ₹66,537, Gross return = ₹99,462, Net return = ₹32,925, B:C = 1.49.

2018–19: Cost of Cultivation = ₹66,537, Gross Return = ₹1,10,458,Net Return = ₹43,921, B:C = 1.66.

Analysis: Gross return and net return increased significantly. The B:C ratio improved from 1.49 to 1.66, indicating a more profitable system in 2018–19.

Table 7.2 Benefit-Cost (B:C) ratio analysis from nineteen agroforestry treatments (10-years-old mango-gamhar, 10-years-old eucalyptus, 3-years-old sweet orange, Jhargram, West Bengal, India, 2017–18 to 2018–19[a])

	Cost of cultivation (₹ ha^{-1})		Gross return (₹ ha^{-1})		Net return (₹ ha^{-1})		B:C	
	Cropping years		Cropping years		Cropping years		Cropping years	
Treatments	2017–18	2018–19	2017–18	2018–19	2017–18	2018–19	2017–18	2018–19
T1 Gamhar + sweet orange + pigeon pea	66,651	66,651	95,517	103,849	28,866	37,198	1.43	1.55
T2: Gamhar + sweet orange + green gram-toria	66,537	66,537	99,462	110,458	32,925	43,921	1.49	1.66
T3: Gamhar + sweet orange + cowpea-toria	66,889	66,889	107,263	118,601	40,374	51,712	1.60	1.77
T4: Gamhar + mango + pigeon pea	66,651	66,651	201,342	283,383	134,691	216,732	3.02	4.25
T5: Gamhar + mango + green gram-toria	66,537	66,537	211,168	299,891	144,631	233,354	3.17	4.50
T6: Gamhar + mango + cowpea-toria	66,889	66,889	221,171	313,947	154,282	247,058	3.30	4.69
T7: Eucalyptus + mango + pigeon pea	66,651	66,651	259,884	342,369	193,233	275,718	3.89	5.13
T8: Eucalyptus + mango + green gram-toria	66,537	66,537	264,066	344,363	197,529	277,826	3.96	5.17
T9: Eucalyptus + mango + cowpea-toria	66,889	66,889	272,638	355,610	205,749	288,721	4.07	5.31
T10: Gamhar + sweet orange	58,620	58,620	81,308	88,536	22,688	29,916	1.38	1.51
T11: Gamhar + mango	58,620	58,620	161,734	221,980	103,114	163,360	2.75	3.78
T12: Eucalyptus + mango	58,620	58,620	221,556	284,736	162,936	226,116	3.77	4.85
T13: Gamhar	25,970	25,970	19,716	20,034	−6254	−5936	0.75	0.77
T14: Eucalyptus	25,970	25,970	78,154	81,984	52,184	56,014	3.00	3.15
T15: Mango	32,650	32,650	121,000	156,300	88,350	123,650	3.70	4.78
T16: Sweet orange	32,650	32,650	44,165	49,980	11,515	17,330	1.35	1.53
T17: Pigeon pea	8031	8031	7740	8113	−291	82	0.96	1.01
T18: Green gram-toria	7912	7912	8768	9488	856	1576	1.10	1.19
T19: Cowpea-toria	8269	8269	14,595	16,021	6326	7752	1.76	1.93

[a](Ref. Das et al. 2022)

T3: Gamhar + Sweet Orange + Cowpea-Toria

2017–18: Cost of cultivation = ₹66,889, Gross return = ₹1,07,263, Net return = ₹40,374, B:C = 1.60.

2018–19: Cost of cultivation = ₹66,889, Gross return = ₹1,18,601, Net return = ₹51,712, B:C = 1.77.

Analysis: There is an increase in both gross return (₹11,338) and net return (₹11,338). The B:C ratio improves from 1.60 to 1.77, showing better profitability.

T4: Gamhar + Mango + Pigeon Pea

2017–18: Cost of cultivation = ₹66,651, Gross return = ₹2,01,342, Net Return = ₹1,34,691, B:C = 3.02.

2018–19: Cost of cultivation = ₹66,651, Gross return = ₹2,83,383, Net return = ₹2,16,732, B:C = 4.25.

Analysis: Significant improvement in gross return (₹82,041) and net return (₹82,041). The B:C ratio increases' from 3.02 to 4.25, reflecting a much more profitable system in 2018–19.

T5: Gamhar + Mango + Green Gram-Toria

2017–18: Cost of cultivation = ₹66,537, Gross return = ₹2,11,168, Net return = ₹1,44,631, B:C = 3.17.

2018–19: Cost of Cultivation = ₹ 66,537, Gross Return = ₹2,99,891, Net Return = ₹2,33,354, B:C = 4.50.

Analysis: There is a substantial increase in gross return (₹88,723) and net return (₹88,723). The B:C ratio increases from 3.17 to 4.50, indicating a strong improvement in profitability.

T6: Gamhar + Mango + Cowpea-Toria

2017–18: Cost of cultivation = ₹66,889, Gross return = ₹2,21,171, Net return = ₹1,54,282, B:C = 3.30.

2018–19: Cost of cultivation = ₹66,889, Gross return =3₹,13,947, Net return = ₹2,47,058, B:C = 4.69.

Analysis: The increase in gross return (₹92,776) and net return (₹92,776) is significant. The B:C ratio rises from 3.30 to 4.69, showing a substantial economic gain.

T7: Eucalyptus + Mango + Pigeon Pea

2017–18: Cost of cultivation = ₹66,651, Gross return = ₹2,59,884, Net return = ₹1,93,233, B:C = 3.89.

2018–19: Cost of cultivation = ₹66,651, Gross return = ₹3,42,369, Net return = ₹2,75,718, B:C = 5. 13.

Analysis: There is a significant increase in both gross return (₹82,485) and net return (₹82,485). The B:C ratio improves from 3.89 to 5.13, indicating a marked increase in economic performance.

T8: Eucalyptus + Mango + Green Gram-Toria

2017–18: Cost of cultivation = ₹66,537, Gross return = ₹2,64,066, Net return = ₹1,97,529, B:C = 3.96.

2018–19: Cost of cultivation = ₹66,537, Gross return = ₹3,44,363, Net return = ₹2,77,826, B:C = 5.17.

Analysis: The gross return increases by ₹80,297 and the net return increases by ₹80,297. The B:C ratio increases from 3.96 to 5.17, reflecting a notable improvement.

T9: Eucalyptus + Mango + Cowpea-Toria

2017–18: Cost of cultivation = ₹66,889, Gross return = ₹2,72,638, Net return = ₹2,05,749, B:C = 4.07, 2018–19: Cost of cultivation = ₹66,889, Gross return = ₹3,55,610, Net return = ₹2,88,721, B:C = 5.31.

Analysis: There is an increase in gross return (₹82,972) and net return (₹82,972). The B:C ratio rises from 4.07 to 5.31, signifying improved economic performance.

Treatments from T10 to T19 were either with trees or with crops. As in those treatments agroforestry practices were not followed, economic analysis was not performed for treatments from T10 to T19.

Step 3: General Trends and Observations

1. Increased profitability: For almost all the agroforestry systems, both the gross return and net return increased from 2017–18 to 2018–19. This resulted in a higher B:C ratio in the majority of the AF treatments, indicating better economic efficiency.
2. Higher returns in agroforestry systems with mango and eucalyptus: The AF treatments that include mango and eucalyptus consistently showed higher gross returns, net returns, and B:C ratios in both years, suggesting that these species may have a strong economic advantage.
3. Best performing AF systems: AF Treatments involving Eucalyptus + Mango + Cowpea-Toria (T9), Eucalyptus + Mango + Green Gram-Toria (T8), and Gamhar + Mango + Cowpea-Toria (T6) showed the most significant improvements, with high B:C ratios in both years (over 5 in 2018–19).

Step 4: Conclusion

The economic performance of the agroforestry systems improved in 2018–19 compared to 2017–18, with most AF systems showing better returns and higher B: C ratios. This suggests that certain agroforestry systems, especially those involving high-value species like mango and eucalyptus, are becoming more economically viable over time. Additionally, the inclusion of leguminous crops (pigeon pea, green gram, and cowpea) appears to enhance the overall profitability of these systems.

7.6.2.4 Problems Related to Financial Data from Agroforestry Research Project for Ranking of Best Performing Agroforestry Practices

Based on the analysis of the economic performance of the various agroforestry systems for the years 2017–18 and 2018–19, the systems can be ranked based on their Benefit-Cost (B:C) ratio for 2018–19, as it best reflects the economic performance (Table 7.2). Here's the ranking from highest to lowest B:C ratio:

Top-ranked agroforestry systems

1. **Eucalyptus + Mango + Cowpea-Toria (T9)**

2018–19: B:C ratio: 5.31.

Analysis: The highest B:C ratio, indicating the most profitable system.

2. **Eucalyptus + Mango + Green Gram-Toria (T8)**

2018–19: B:C ratio: 5.17.
Analysis: Strong performance, just behind T9 in terms of profitability.

3. **Gamhar + Mango + Cowpea-Toria (T6)**

2018–19: B:C ratio: 4.69.
Analysis: A substantial increase in net returns and a strong B:C ratio.

4. **Eucalyptus + Mango + Pigeon Pea (T7)**

2018–19: B:C ratio: 5.13.
Analysis: Shows strong improvement with significant returns.

5. **Gamhar + Mango + Green Gram-Toria (T8)**

2018–19: B:C ratio: 4.50.
Analysis: A very profitable system, with a marked improvement over 2017–18.

6. **Gamhar + Mango + Pigeon Pea (T4)**

2018–19: B:C ratio: 4.25.
Analysis: A well-performing system with a solid B:C ratio.
Mid-ranked agroforestry systems

7. **Gamhar + Sweet Orange + Cowpea-Toria (T3)**

2018–19: B:C ratio: 1.77.
Analysis: A decent system with strong improvement in profitability, though lower than the top-.
performing systems.

8. **Gamhar + Sweet Orange + Green Gram-Toria (T2)**

2018–19: B:C ratio: 1.66.
Analysis: A moderately profitable system, with notable improvement in both gross and net returns.

9. **Gamhar + Sweet Orange + Pigeon Pea (T1)**

2018–19: B:C ratio: 1.55.
Analysis: The lowest B:C ratio among the systems, although still showing an improvement over 2017–18.
Summary observations
Top performers:

Agroforestry systems involving Eucalyptus + Mango consistently outperformed the other systems, especially in terms of B:C ratio.

Strong profitability: Systems that combine Mango with legumes like Cowpea or Green Gram-Toria performed exceptionally well.

Improvement across the observation years: Nearly all agroforestry systems showed improved profitability from 2017–18 to 2018–19, with substantial increases in both gross and net returns.

Thus, Eucalyptus + Mango + Cowpea-Toria (T9) is thc most economically valuable agroforestry system, followed closely by Eucalyptus + Mango + Green Gram-Toria (T8), and Gamhar + Mango + Cowpea- Toria (T6).

7.6.2.5 Problems Related to Financial Data from Agroforestry Research Project for Screening of Economically Unsuitable Agroforestry Practices, Based on Their B:C Ratio, Gross Return and Net Return

Based on data given in Table 7.2, here is a ranking of the treatments from least to most economically suitable (w.r.t. B:C ratio and overall profitability trends):

7.6.2.5.1 Economically Unsuitable Agroforestry Systems (From Less to More Suitable)

1. **Gamhar + Sweet Orange + Pigeon Pea (T1)**

2017–18: B:C = 1.43, Net Return = ₹28,866

2018–19: B:C = 1.55, Net Return = ₹37,198

Analysis: While there is improvement from 2017–18 to 2018–19, the B:C ratio remains relatively low compared to other systems, making it less economically attractive.

2. **Gamhar + Sweet Orange + Green Gram-Toria (T2)**

2017–18: B:C = 1.49, Net Return = ₹32,925

2018–19: BC = 1.66, Net Return = ₹43,921.

Analysis: Similar to T1, although there is a positive trend in returns, the B:C ratio is still on the lower side when compared to other treatments.

3. **Gamhar + Sweet Orange + Cowpea-Toria (T3)**

2017–18: B:C = 1.60, Net Return = ₹40,374

2018–19: B:C = 1.77, Net Return = ₹51,712.

Analysis: There is a notable improvement in profitability, but it still lags behind other treatments in terms of B:C ratio and overall returns.

7.6.2.5.2 More Economically Suitable Agroforestry Systems (Improving Profitability)

4. **Gamhar + Mango + Pigeon Pea (T4)**

2017–18: B:C = 3.02, Net Return = ₹1,34,691.

2018–19: B:C = 4.25, Net Return = ₹2,16,732.

Analysis: This system shows a significant improvement in economic performance, though its B:C ratio is still lower than those of the top-performing systems.

5. **Gamhar + Mango + Green Gram-Toria (T5)**

2017–18: B:C = 3.17, Net Return = ₹1,44,631
2018–19: B:C = 4.50, Net Return = ₹2,33,354.

Analysis: This system shows a strong improvement in profitability and a relatively good B:C ratio, making it more economically suitable than T4.

6. **Gamhar + Mango + Cowpea-Toria (T6)**

2017–18: B:C = 3.30, Net Return = ₹1,54,282
2018–19: B:C = 4.69, Net Return = ₹2,47,058.

Analysis: A substantial increase in returns and a good B:C ratio. It shows significant improvement, making it more economically suitable.

7.6.2.5.3 Best Performing Agroforestry Systems (Most Suitable Economically)

7. **Eucalyptus + Mango + Pigeon Pea (T7)**

2017–18: B:C = 3.89, Net Return = ₹1,93,233
2018–19: B:C = 5.13, Net Return = ₹2,75,718.

Analysis: One of the top performers with a high B:C ratio and a strong increase in net returns. This system is economically very efficient.

8. **Eucalyptus + Mango + Green Gram-Toria (T8)**

2017–18: B:C = 3.96, Net Return = ₹1,97,529
2018–19: B:C = S.17, Net Return = ₹2,77,826.

Analysis: Another top performer with a remarkable increase in profitability and a high B:C ratio.

9. **Eucalyptus + Mango + Cowpea-Toria (T9)**

2017–18: B:C = 4.07, Net Return = ₹2,05,749
2018–19: B:C = 5.31, Net Return = ₹2,88,721.

Analysis: The best performing system with the highest B:C ratio, substantial increases in returns, and overall superior economic performance.

Conclusion: The AF systems with T1, T2, and T3 are the least economically suitable based on the B:C ratio and overall returns. AF Systems that include Eucalyptus and Mango tend to show the best economic performance, with T7, T8, and T9 being the most economically viable.

7.6.2.6 Problems Related to Examining Suitability of AF Project Proposal, Based on Economic Value of Agroforestry Practices for Health Purposes Other Than Agroforestry Products

Project: Economic Valuation of Forest Tourism—Forest Bathing (Shinrin-yoku)

Objectives

The goal of this project is to estimate the economic value of forest tourism through the practice of *Shinrin-yoku* (forest bathing as followed in Japan). This is achieved by determining the benefits of forest bathing in terms of visitor spending, improved health outcomes, and the impact on local economies.

The economic valuation of forest tourism through forest bathing yields an estimated total value of EUR 8700 annually in a forest travel zone of Italian Alps in northern Italy (Paletto et al. 2024). This valuation includes direct visitor spending, health benefits, and the value of the forest environment. Policymakers and forest managers can use this information to justify investments in forest conservation, promote sustainable tourism practices, and advocate for the therapeutic benefits of forest environments in promoting public health.

Steps for the project proposal:

1. **Introduction to forest bathing (*Shinrin-yoku*)**: *Shinrin-yoku* (forest bathing) as a therapeutic practice that involves immersing oneself in a forest environment to engage the senses and promote well-being. The global rise in interest in nature-based therapies, gave importance on *Shinrin-yoku*, particularly in Japan, where it originated.

   This project requires identifying the key components of forest bathing, such as sensory engagement with the environment (sight, sound, touch, and smell).
2. **Identifying benefits of forest bathing:** Forest bathing has various benefits, including:
   (a) Physical health: It reduces stress levels, lowers blood pressure, improves immune system function, and enhances overall physical well-being.
   (b) Mental health: It improves mood, reduces anxiety and depression, and promotes mental clarity and relaxation.
   (c) Social Benefits: Forest bathing encourages social interaction in nature and promotes community wellness. The economic valuation aims to quantify these intangible benefits in monetary terms.
3. **Methodology for economic valuation of forest bathing**:

   There are various approaches to valuing non-market services such as forest bathing. We will focus on the following methods:
   (a) Contingent valuation method (CVM): A survey-based approach asking individuals how much they would be willing to pay for the benefits derived from forest bathing.
   (b) Travel cost method (TCM): Uses the cost of traveling to the forest site as a proxy for the value of the forest experience.
   (c) Health benefits valuation: The economic value of improved health outcomes through reduced healthcare costs due to stress reduction and other health benefits.

7.6.2.7 Numerical Problem on Economic Valuation of Forest Tourism Through *Shinrin-yoku* (Forest Bathing) in an Agroforestry Project Site Based on Assumed Data

Visitor data: 1000 tourists annually visit an agroforestry site specifically for forest bathing.

Average spending: Each visitor spends $100 on travel, accommodation, food, and other expenses during their stay.

Health benefits: Forest bathing can reduce healthcare costs by $500 per person per year due to stress reduction and overall health improvements (Hansen et al. 2017; Wen et al. 2019).

Environmental amenities: The value of the agroforest environment (scenic beauty, biodiversity) is estimated at $50 per visitor (willingness to pay).

Step-by-Step Solution:

Step 1: Estimate total visitor spending.

First, calculate the total economic contribution of visitors spending money at the agroforest site.

Visitor spending per person = $100.

Number of visitors = 1000.

Total spending = $100 × 1000 = $1,00,000.

This amount represents the direct economic impact on the local economy from tourists.

Step 2: Estimate the health benefits of forest bathing.

Health benefit per person = $500 (reduced healthcare costs).

Number of visitors = 1000.

Total health benefit = $500 × 1000 = $5,00,000.

This amount represents the indirect benefits of forest bathing in terms of healthcare savings.

Step 3: Estimate the value of environmental amenities.

The forest environment itself holds intrinsic value that people are willing to pay for.

Value per person (i.e., willingness to pay) = $50.

Number of visitors = 1000.

Total value of environmental amenities = $50 × 1000 = $50,000.

This represents the non-market value of the forest's scenic and ecological qualities.

Step 4: Calculate the total economic value (TEV).

The total economic value is the sum of the direct, indirect, and non-market benefits associated with forest bathing.

Total Economic value = Total visitor spending + Health benefits + Environmental amenities.

= $1,00,000 + $5,00,000 + $50,000.

= $6,50,000.

Thus, the total economic value of the agroforest tourism through forest bathing is $6,50,000 per year.

Step 5: Cost-benefit analysis and conclusion.

Based on the calculated economic value, we can now analyze the cost-benefit ratio of implementing and maintaining a forest bathing program in the agroforestry project site. Here are some considerations:

1. Costs:
   (a) Establishment and maintenance of forest bathing facilities.
      - Marketing and promotional costs to attract visitors.
      - Operational costs (e.g., staffing, upkeep of trails and amenities, health and safety measures). Let us assume the total annual operational cost for the forest site is $1,50,000.
2. Benefits:
   Total Economic Value of Forest Bathing: $6,50,000 annually (from Step 4).
3. Cost-Benefit Ratio (CBR):
   CBR = Total Benefits/ Total Costs = $6,50,000/ $1,50,000 = 4.33.

The cost-benefit ratio of 4.33 indicates that for every dollar spent on the forest bathing program, there is a return of approximately $4.33 in economic value. This represents a highly favorable outcome, suggesting that the project is financially viable and delivers significant social and environmental benefits.

Conclusion: Forest bathing offers considerable economic value, both directly through visitor spending and indirectly through health benefits and environmental amenities. The project demonstrates a strong positive return on investment, making it a viable option for sustainable tourism and nature-based health interventions. By quantifying the benefits of forest bathing, this project underscores the importance of nature conservation and highlights the growing potential for nature-based therapies in the wellness industry.

7.6.2.8 Problem Related to Diversifying Income Sources in Agroforestry

A rural community cultivates 3000 square meters of land using agroforestry practices. The community produces a variety of products, including fruits, nuts, and herbs. The revenue breakdown is as follows:

1. Fruits: $1000 per 1000 square meters.
2. Nuts: $500 per 1000 square meters.
3. Herbal products: $300 per 1000 square meters.

Calculate the total annual income from each product and the overall income.

Solution:

1. Income from fruits = (5000/1000) × 1000 = $5000.
2. Income from nuts = (5000/1000) × 500 = $2500.
3. Income from herbal products = (5000/1000) × 300 = $1500.

Answer:

Total annual income = $5000 + $2500 + $1500 = $9000.

7.6.2.9 Problem Related to the Economic Benefits of Value Addition in Agroforestry

In a rural community, farmers harvest 1000 kg of non-timber forest products (NTFPs) such as medicinal herbs, fruits, and nuts per year. By processing and packaging these products into value-added goods, they can sell the processed goods at 50% higher prices than raw products.

1. If the raw product is sold for \$2 per kilogram, calculate the total annual income from selling the raw product.
2. After adding value, if the price increases by 50%, what will be the new income for the same quantity of value-added products?

Solution:

1. Income from raw product =1000 kg × \$2 = \$2000.
2. Income from Value-added product:

If the price increases by 50%, the new price is:
New price = \$2 + (\$2 × 0.50) = \$3/kg.
Now, income from value-added products = 1000 kg × \$3 /kg = \$3000.

Answer:
1. Income from raw product = \$2000.
2. Income from value-added products = \$3000.

Thus, the value addition increases income by \$1000 annually.

7.6.2.10 Problem Related to Ecological Benefits of Agroforestry (Carbon Sequestration)

An agroforestry system with 200 trees per hectare sequesters 5 tons of CO_2 per year per hectare. If a country implements agroforestry on 50,000 hectares, how much CO_2 will be sequestered annually?

Solution:
CO_2 sequestered per hectare per year = 5 tons.
Total hectares of agroforestry = 50,000 hectares.
Total CO_2 sequestration
= 5 tons/hectare CO_2 per year × 50,000 hectares.
= 2,50,000 tons of CO_2 per year in the country.

7.6.2.10.1 Proposed Method of Raising Carbon Credit Through Agroforestry-Based Carbon Sequestration

Table 7.3 is prepared based on the following step-by-step methods as elaborated in the following problems 7.6.2.10.1a–c.

Problem 7.6.2.10.1 (a): Method of Calculating Money Value of C-Sequestered by Gamhar Wood

Biomass equivalent of 1 m^3 of gamhar wood = 1 × 0.98 = 0.98 t ≈ 1 t of gamhar wood. [Note that actual wood density of Gamhar is 0.45 t m^(-3). For a simple calculation, this value is taken, here, as 0.98 ≈ 1; Ref: https://www.fao.org/4/w4095e/w4095e0c.htm]

= (0.98/ 2) = 0.49 t of carbon sequestered as gamhar wood.

Sale price of 1m³ gamhar wood = ₹10,600.

or, money value of 0.98 t of gamhar wood = ₹10,600.

So, money value of 0.49 t of carbon sequestered as 1 m³ of gamhar wood = ₹10,600.

Problem 7.6.2.10.1 (b): Method of Calculating Money Value of C-Sequestered by Eucalyptus Wood

Biomass equivalent of 1 m³ of eucalyptus wood = 1 × 0.79 = 0.79 t of eucalyptus wood = (0.79/2) ≈ 0.40 t of carbon sequestered as eucalyptus wood. [Note that wood density of eucalyptus is approx. 0.64 t m^(-3), For a simple calculation, this value is taken, here, as 0.79; https://www.fao.org/4/w4095e/w4095e0c.htm]

Sale price of 1 m³ eucalyptus wood = ₹4000 or, money value of 0.98 t of eucalyptus wood = ₹4000.

So, money value of 0.49 t of carbon sequestered as 1 m³ of eucalyptus wood = ₹4000.

Problem 7.6.2.10.1 (c): Proposed Method of Calculating Money Value of C-Sequestered by One ha AF System, Silviculture, and Crop Production Followed by step-by-step process as elaborated in Problems 7.6.2.8.1a–c and following Tables 7.2, 7.3, Table 7.4 is prepared for assessing year-to-year market price dependent money value of carbon sequestration-based carbon credit (i.e., equivalent to year-wise market price-based gross return, ₹ ha^{-1}).

7.6.2.10.2 Analysis for Comparing Agroforestry System, Silviculture, and Crop Production for Best Carbon Credit

Carbon credits of 19 numbers of various agricultural production systems including AF (Table 7.4) are compared for evaluating best affordable carbon credits. From Table 7.4, it is evident that 19 production systems involving agroforestry system, silviculture and crop production based on carbon sequestration can be classified in four groups with the performing orders as follows:

1. **Top performers (High carbon sequestration):**

   T9: Eucalyptus + Mango + Cowpea-Toria > T8: Eucalyptus + Mango + Green Gram-Toria >

   T7: Eucalyptus + Mango + Pigeon Pea> T6: Gamhar + Mango + Cowpea-Toria > T 5: Gamhar + Mango + Green Gram-Toria
2. **Mid-range performers:**

   T4: Gamhar + Mango + Pigeon Pea> T12: Eucalyptus + Mango> T11: Gamhar + Mango
3. **Low performers:**

   T3: Gamhar + Sweet Orange + Cowpea-Toria > T2: Gamhar + Sweet Orange + Green Gram-Toria >.

   T1: Gamhar +-Sweet Orange + Pigeon Pea> T10: Gamhar + Sweet Orange

Table 7.3 Carbon sequestration of different fruit-based AF systems during the years of 2017–18 and 2018–19 (after attaining the ages of 10th, and 11th year of gamhar, eucalyptus, mango; and 3rd, and 4th year of sweet orange in the corresponding cropping years of 2017–18 and 2018–19)[a]

	Carbon stock as trees and fruit trees biomass (t ha^{-1} yr^{-1})		Crop carbon stock (t ha^{-1} yr^{-1})		SOC (t ha^{-1})		Carbon stock total (t ha^{-1} yr^{-1})	
	Cropping years		Cropping years		Cropping years		Cropping years	
Treatments	2017–18	2018–19	2017–18	2018–19	2017–18	2018–19	2017–18	2018–19
T1 Gamhar + sweet orange + pigeon pea	2.11	2.19	13.77	14.23	7.98	8.54	23.86	24.96
T2: Gamhar + sweet orange + green gram-toria	2.16	2.28	38.73	41.37	7.84	8.12	48.73	51.77
T3: Gamhar + sweet orange + cowpea-toria	2.18	2.30	11.24	12.37	8.12	8.68	21.54	23.35
T4: Gamhar + mango + pigeon pea	3.07	3.32	14.51	15.05	8.54	8.96	26.12	27.33
T5: Gamhar + mango + green gram-toria	3.12	3.45	41.19	44.23	8.26	8.68	52.57	56.36
T6: Gamhar + mango + cowpea-toria	3.14	3.50	11.64	12.99	8.82	9.24	23.60	25.73
T7: Eucalyptus + mango + pigeon pea	15.84	16.94	12.59	12.90	7.70	8.12	36.13	37.96
T8: Eucalyptus + mango + green gram-toria	15.60	16.69	36.63	37.80	7.56	7.84	59.79	62.33
T9: Eucalyptus + mango + cowpea-toria	15.51	16.34	10.69	11.61	7.84	8.12	34.04	36.07
T10: Gamhar + sweet orange	2.00	2.08	-	-	7.14	7.42	9.14	9.50
T11: Gamhar +mango	2.77	2.96	-	-	7.42	7.70	10.19	10.66
T12: Eucalyptus + mango	14.34	14.74	-	-	7.01	7.28	21.35	22.02
T13: Gamhar	0.98	1.01	-	-	6.72	7.00	7.70	8.01
T14: Eucalyptus	10.20	10.25	-	-	6.58	6.86	16.78	17.11

[a](Ref. Das et al. 2022)

Table 7.4 Assessing year-to-year market price-dependent money value of carbon sequestration-based proposed carbon credit (i.e., equivalent to year-wise market price-based gross return, ₹ ha^{-1})[a]

	Gross return (₹ ha^{-1}) (from Table 7.2)		Carbon stock total up to the year (t ha^{-1}) (from Table7.3)		Carbon sequestration-based carbon credit (equivalent to year-wise market price-based gross return) (₹ ha^{-1})	
	Cropping years		Cropping years		Cropping years	
Treatments	2017–18	2018–19	2017–18	2018–19	2017–18	2018–19
T1 Gamhar + sweet orange + pigeon pea	95,517	1,03,849	23.86	24.96	95,517	1,03,849
T2: Gamhar + sweet orange + green gram-toria	99,462	1,10,458	48.73	51.77	99,462	1,10,458
T3: Gamhar + sweet orange + cowpea-toria	1,07,263	1,18,601	21.54	23.35	1,07,263	1,18,601
T4: Gamhar + mango + pigeon pea	2,01,342	2,83,383	26.12	27.33	2,01,342	2,83,383
T5: Gamhar + mango + green gram-toria	2,11,168	2,99,891	52.57	56.36	2,11,168	2,99,891
T6: Gamhar + mango + cowpea-toria	2,21,171	3,13,947	23.60	25.73	2,21,171	3,13,947
T7: Eucalyptus + mango +pigeon pea	2,59,884	3,42,369	36.13	37.96	2,59,884	3,42,369
T8: Eucalyptus + mango + green gram-toria	2,64,066	3,44,363	59.79	62.33	2,64,066	3,44,363
T9: Eucalyptus + mango +cowpea-toria	2,72,638	3,55,610	34.04	36.07	2,72,638	3,55,610
T10: Gamhar + sweet orange	81,308	88,536	9.14	9.50	81,308	88,536
T11: Gamhar + mango	1,61,734	2,21,980	10.19	10.66	1,61,734	2,21,980
T12: Eucalyptus + mango	2,21,556	2,84,736	21.35	22.02	2,21,556	2,84,736
T13: Gamhar	19,716	20,034	7.70	8.01	19,716	20,034
T14: Eucalyptus	78,154	81,984	16.78	17.11	78,154	81,984
T15: Mango	1,21,000	1,56,300	8.11	8.37	1,21,000	1,56,300
T16: Sweet orange	44,165	49,980	6.96	7.26	44,165	49,980
T17: Pigeon pea	7,740	8,113	16.64	17.02	7,740	8,113
T18: Green gram-toria	8,768	9,488	36.06	36.81	8,768	9,488
T19: Cowpea-toria	14,595	16,021	14.77	15.34	14,595	16,021

[a]As on 2017–18 financial year at Jhargram, West Bengal, India: (i) Sale Price (based on local market query): Gambar wood = Rs 10,600/cum, Eucalyptus wood = Rs 4000/cum, Mango fruit = Rs 25,000/t, Sweet orange fruit = Rs 60/ dozen, *Arhar* = Rs 1200/q, Cowpea = Rs 900/q, Toria = Rs 1400/q. (ii) Carbon equivalent of 1 ton of wood = 1 ton × 0.5 = 0.5 ton. (iii) Prepared on the basis of Tables 7.2 and 7.3.

4. **Standalone (i.e., sole) trees and crops (Lowest carbon Sequestration):**
T19: Cowpea-Toria > T18: Green Gram-Toria > T17: Pigeon Pea> Tl6: Sweet Orange> T15: Mango> T14: Eucalyptus> T13: Gamhar.

Thus, from the above analysis on best affordable carbon credits following salient findings can be documented:

1. **Highest carbon sequestration (Affordable):** If the goal is to maximize carbon sequestration at an affordable price, the treatments combining Eucalyptus, Mango, and legume-based crop rotation like Cowpea-Toria, Green Gram-Toria, and Pigeon Pea (T7, T8, T9) are the top performers, offering the highest returns in terms of carbon sequestration.
2. **Balanced systems:** For systems balancing affordability and efficiency, Gamhar + Mango systems (T11, T4, T5) perform well, with significant carbon sequestration values in both years, making them a cost-effective choice.
3. **Lower carbon sequestration systems:** Simple sole tree or crop systems like Mango, Sweet Orange, or standalone legume-based crop rotationlike Cowpea-Toria, Pigeon Pea, and Green Gram-Toria (T15, T16, T17, T18, T19) have the lowest C-sequestration and carbon credit returns. They might be affordable in terms of investment but are less effective in maximizing carbon credits.

   Therefore, from this analysis it can be conjectured that for maximizing carbon credits efficiently while considering the market price for carbon sequestration, systems involving Eucalyptus + Mango + Legumes (T7, T8, T9) are the most effective and profitable. However, for more affordable yet still reasonable carbon sequestration options, Gamhar + Mango systems (T5, T6, T4) strike a good balance in agroecosystem at Jhargram, West Bengal, India. Such analysis should be based on local agroecosystem situations.

7.7 Concluding Remarks: Enhancing Ecosystem Services Through Agroforestry and Aquaculture Integration

Agroforestry systems have proven to be powerful tools in the restoration and enhancement of various ecosystem services. By integrating trees, shrubs, and crops, these systems offer numerous ecological benefits, addressing some of the most critical environmental challenges faced by agriculture today. One such challenge is soil erosion, a major threat to agricultural productivity. Agroforestry systems, with their deep-rooted vegetation, act as a natural barrier against soil degradation. Tree and shrub root systems help anchor the soil, reducing erosion and promoting water retention. This improved soil stability not only prevents topsoil loss but also enhances water infiltration, ensuring that water is more readily available for crops, particularly in areas prone to drought.

Additionally, the organic matter provided by trees through leaf litter, root decay, and other forms of biomass greatly improves soil fertility and structure. This organic input nourishes the soil, increasing microbial activity and enriching the soil nutrient status, which benefits both crops and trees. The accumulation of soil organic carbon (SOC) is one of the most significant results of agroforestry, offering multifaceted advantages. SOC plays an essential role in soil health by improving soil texture, enhancing water retention, and boosting nutrient cycling. Furthermore, the increase in SOC is a critical tool in climate change mitigation, as it sequesters carbon dioxide from the atmosphere, reducing the greenhouse gases contributing to global warming.

Beyond the soil, agroforestry systems also contribute to water management. Trees can intercept rainwater and reduce surface runoff, allowing more water to percolate into the soil and recharge groundwater supplies. This makes agroforestry not only a climate-resilient solution but also a key player in sustainable water management, helping maintain ecosystem balance in regions facing water scarcity.

Incorporating aquaculture into agroforestry systems further enhances the sustainability of these integrated systems. Aquaculture, particularly through fish farming, provides valuable organic nutrients through fish waste, which can be used for manuring plants, trees, and crops in the surrounding areas. Fish metabolites, such as nitrogen and phosphorus, are rich in nutrients that can promote plant growth and improve soil fertility. Additionally, fish sludge can be utilized as an organic fertilizer, reducing the need for chemical fertilizers and enhancing nutrient cycling in agroecosystems. This integration not only creates a circular economy within the farming system but also contributes to sustainable nutrient management, and improves ecosystem services related to main focus of agroforestry-based livelihood opportunities.

The combination of agroforestry and aquaculture maximizes resource use efficiency, creating a symbiotic relationship between terrestrial and aquatic environments. This integrated approach supports biodiversity, enhances soil fertility, improves water retention, and promotes climate resilience. By embracing both agroforestry and aquaculture, we can address food security challenges while mitigating the effects of climate change, improving ecosystem services, and promoting long-term environmental sustainability (Madende and Hayes 2020; Gao et al. 2021; Cerozi et al. 2022; Brod et al. 2023; Fruscella et al. 2023; Hettige et al. 2023; Jaies et al. 2023).

In conclusion, agroforestry systems, enriched by aquaculture practices, offer a holistic solution to the pressing issues of soil erosion, water management, and carbon sequestration. This synergy between land and water-based land use practices provides a sustainable, climate-smart pathway forward for agriculture, ensuring healthier ecosystems and more resilient communities.

References

Ahmed A, Kumar S, Shukla AK (2014) Effect of different soil moisture conservation techniquesfor establishment of Aonla based Hortipasture systems in Bundelkhand region. In: Proceedings of National symposium on climate resilient forage production and its utilization. BCKV, West Bengal, pp 13–14

Banerjee H, Dhara PK (2010) Evaluation of fruit-based agroforestry model for rainfed upland under red & lateritic tract of West Bengal. Crop Res (Hisar) 39:200–206. https://www.cabdirect.org/cabdirect/abstract/20103243463

Bellow G, Hudson RF, Nair PKR (2008) Adoption potential of fruit-tree-based agroforestry onsmall farms in the subtropical highlands. Agrofor Syst 73:23–36

Boinot S, Barkaoui K, Mézière D et al (2022) Research on agroforestry systems and biodiversity conservation: what can we conclude so far and what should we improve? BMC Ecol Evol 22:24. https://doi.org/10.1186/s12862-022-01977-z

Borges O, Raimundo F, Coutinho J, Gonçalves B, Oliveira I, Martins A, Madeira M (2018) Carbon fractions as indicators of organic matter dynamics in chestnut orchards under different soil management practices. Agrofor Syst 92(2):301–310. https://doi.org/10.1007/s10457-017-0088-3

Brod E, Henriksen TM, Ørnsrud R, Eggen T (2023) Quality of fish sludge as fertiliser to spring cereals: nitrogen effects and environmental pollutants. Sci Total Environ 875:162541. https://doi.org/10.1016/j.scitotenv.2023.162541

Cerozi BS, Arlotta CG, Richardson ML (2022) Fish effluent as a source of water and nutrients for sustainable urban agriculture. Agriculture 2022(12):1975. https://doi.org/10.3390/agriculture12121975

Das P, Panda S (2022) Ecosystem management of degraded red and lateritic soil through agroforestry. In: Jatav HS, Rajput VD (eds) Ecosystem services: types, management and benefits. Agriculture issues and policies. Nova Science, Hauppauge, NY. https://doi.org/10.52305/PFZA6988. Edition: ISBN: 978-1-68507-747-1. Chapter: 18. https://www.researchgate.net/publication/359310869_Ecosystem_management_of_degraded_red_and_lateritic_soil_through_agroforestry

Das P, Dhara PK, Panda S (2022) Fruit based agroforestry systems—potential means for sustaining carbon sequestration, improving soil health and diet of community in red and lateritic zone of West Bengal, India. Int J Sci Eng Res 13(12):832–842. https://doi.org/10.14299/ijser.2022.12.03. https://www.researchgate.net/publication/367207106_Fruit_based_agroforestry_systems_-potential_means_for_sustaining_carbon_sequestration_improving_soil_health_and_diet_of_community_in_red_and_lateritic_zone_of_West_Bengal_India

Dev I, Ram A, Ahlawat SP, Palsaniya DR, Singh R, Dhyani SK, Kumar N, Tewari R, Singh M, Sridhar KB, Newaj R, Dwivedi R, Kumar R, Yadav R, Chand L, Kumar D, Prasad J (2020) Bamboo-based agroforestry system (*Dendrocalamus strictus* + sesame–chickpea) for enhancing productivity in semi-arid tropics of Central India. Agrofor Syst 94:1725–1739. https://doi.org/10.1007/s10457-020-00492-8

Dhanushkodi V, Easwaran S, Amuthaselvi G (2019) Agroforestry: a novel techniques to enhance farmers' income under changing climate. J Pharmacogn Phytochem SP2:204–206. https://www.phytojournal.com/archives/2019/vol8issue2S/PartF/Sp-8-2-55-476.pdf

Dhara PK (2016) Humid and sub humid region—BCKV, Jhargram (West Bengal). In: Chaturvedi OP, Sikka AK, Handa AK, Bajpai CK (eds) Agroforestry technologies for different agroclimatic zones of the country. All India Coordinated Research Projecton Agroforestry, ICAR-Central Agroforestry Research Institute, Jhansi, pp 27–28

Dollinger J, Jose S (2018) Agroforestry for soil health. Agrofor Syst 92:213–219. https://doi.org/10.1007/s10457-018-0223-9

Dubiez E, Freycon V, Marien J-N, Peltier R, Harmand J-M (2018) Long term impact of Acacia auriculiformis woodlots growing in rotation with cassava and maize on the carbon and nutrient contents of Savannah sandy soils in the humid tropics (Democratic Republic of Congo). Agrofor Syst 93:1167–1178. https://doi.org/10.1007/s10457-018-0222-x

Figyantika A, Daniel SM, Marcus AH, Eko BH, Mark AH (2020) Productivity Benefits from Integrating A cacia auriculiformis and Agricultural Cropping in Java, Indonesia. Agrofor Syst 94:2109–2123. https://doi.org/10.1007/s10457-020-00534-1

Fruscella L, Kotzen B, Paradelo M, Milliken S (2023) Investigating the effects of fish effluents as organic fertilisers on onion (*Allium cepa*) yield, soil nutrients, and soil microbiome. Sci Hortic 321:112297. https://doi.org/10.1016/j.scienta.2023.112297

Gao Y, Zhang H, Peng C, Lin Z, Li D, Lee CT, Wu W-M, Li C (2021) Enhancing nutrientrecovery fromfish sludge using a modified biological aerated filter with sponge media with extended filtration in aquaponics. J Clean Prod 320:128804. https://doi.org/10.1016/j.jclepro.2021.128804

Garrity DP (2004) Agroforestry and the achievement of the millennium development goals. Agrofor Syst 61:5–17. https://doi.org/10.1023/b:agfo.0000028986.37502.7c

Handa AK, Dev I, Rizvi RH, Kumar N, Ram A, Kumar D, Kumar A, Bhaskar S, Dhyani SK, Rizvi J (eds) (2019) Successful agroforestry models for different agroecological regions in India. Jointly published by the Central Agroforestry Research Institute (CAFRI), Jhansi, and the South Asia Regional Programme (based in New Delhi) of World Agroforestry (ICRAF).

Hansen MM, Jones R, Tocchini K (2017) *Shinrin-Yoku* (Forest Bathing) and nature therapy: a state-of-the-art review. Int J Environ Res Public Health 14(8):851. https://doi.org/10.3390/ijerph14080851. PMID: 28788101; PMCID: PMC5580555. https://pmc.ncbi.nlm.nih.gov/articles/PMC5580555/

Hettige FS, Madhusanki S, Kankanamge Epa UP (2023) Use of aquaculture pond sludge and fish waste to produce manure worm, Eisenia fetida (Lumbricidae), vermiwash. J Fish Environ 47(3):126–137

Honnayya, Chittapur BM, Doddabasawa (2020) Productivity of Pigeonpea (*Cajanus Cajan* L. Millsp.) in neem (*Azadirachta indica* A. Juss.) based agroforestry system on Alfisols in semi-arid tropics. Agrofor Syst 94:1879–1889. https://doi.org/10.1007/s10457-020-00507-4

ICFRE (2020) Agroforestry models developed by Indian Council of Forestry Research and Education (a compilation). Dehradun, India. www.icfre.org

Jaies I, Qayoom I, Saba F, Khan S (2023) Fish wastes as a source of fertilizers and manures. In: Maqsood S et al (eds) Fish waste to valuable products, sustainable materials and technology, pp 329–338. https://doi.org/10.1007/978-981-99-8593-7_15

Jhariya MK, Banerjee A, Meena RS, Yadav DK (eds) (2019) Sustainable agriculture, forest and environmental management. Springer, Singapore. https://doi.org/10.1007/978-981-13-6830-1_4

Julius AJ, Iluyomade ON, Akinyemi GO, Oniroko NS, Onilude QA, Aderemi AM (2011) Assessment of the impact of agroforestry education on biodiversity conservation and community development. Int J Cross Discip Subj Educ 1:577–582. https://doi.org/10.20533/ijcdse.2042.6364.2011.0080

Kang BT, Akinnifesi FK (2000) Agroforestry as alternative land-use production systems for theTropics. Nat Res Forum 24:137–151. https://doi.org/10.1111/j.1477-8947.2000.tb00938.x

Kumar S, Chaturvedi OP (2017) Agroforestry interventions in fruit orchards for improving productivity, profitability and sustainability. In: Sing SK, Kalia P, Dhilon WS, Bahera TK, Prakash J (eds) Doubling farmers income through horticulture. Daya Publishing House, a Division of Astral International Pvt. Ltd, New Delhi, pp 569–576

Kumar S, Prasad R, Kumar A, Dhyani SK (2019) Integration of fruit trees in agroforestry for sustainability and profitability of farming systems in arid and semi-arid regions. Indian J Agrofor 21:95–99. https://www.researchgate.net/publication/335220173_Integration_of_fruit_trees_in_agroforestry_for_sustainability_and_profitability_of_farming_systems_in_arid_and_semi-arid_regions

Lal R (2014) Societal value of soil carbon. J Soil Water Conserv 69(6):186A–192A. https://doi.org/10.2489/jswc.69.6.186A. www.swcs.org

Madende M, Hayes M (2020) Fish by-product use as biostimulants: an overview of the current state of the art, including relevant legislation and regulations within the EU and USA. Molecules 25(5):1122. https://doi.org/10.3390/molecules25051122

Manna MC, Ghosh PK, Acharya CL (2003) Sustainable crop production through management ofSoil organic carbon in semiarid and tropical India. J Sustain Agric 21:85–114. https://doi.org/10.1300/j064v21n03_07

Meena RS, Gogaoi N, Kumar S (2017) Alarming issues on agricultural crop production andenvironmental stresses. J Clean Prod 142:3357–3359

Mekonnen M, Worku T, Yitaferu B, Cerdà A, Keesstra S (2023) Economics of agroforestry land use system, upper Blue Nile Basin, Northwest Ethiopia. Agrofor Syst 97(3):305–317. https://doi.org/10.1007/s10457-021-00612-y

Mohapatra SC, Behera S, Mishra P, Behera BB (2019) Mango based agrihorticultural system. In: Handa AK, Dev I, Rizvi RH, Kumar N, Ram A, Kumar D, Kumar A, Bhaskar S, Dhyani SK, Rizvi J (eds) Successful agroforestry models for different agro-ecological regions in India. Central Agroforestry Research Institute (CAFRI), Jhansi, pp 113–116

Mukangango M, Nduwamungu J, Naramabuye FX, Nyberg G, Sigrun Dahlin AS (2020) BiomassProduction and nutrient content of three agroforestry tree species growing on an acid anthropic Ferralsol under recurrent harvesting at different cutting heights. Agrofor Syst 94:857–867. https://doi.org/10.1007/s10457-019-00455-8

Naugraiya MN, Puri S (2001) Performance of multipurpose tree species under agroforestry systems on Entisols of Chattisgarh plains. Range Manag Agrofor 22:164–172. https://www.cabdirect.org/cabdirect/abstract/20023061834

Newaj R, Rai P (2005) Aonla-based agroforestry system: a source of higher income under rainfed conditions. Ind Farming 55:24–27

Newaj R, Ram A (2019) Aonla based agri-horticultural System. In: Handa AK, Dev I, Rizvi RH, Kumar N, Ram A, Kumar D, Kumar A, Bhaskar S, Dhyani SK, Rizvi J (eds) Successful agroforestry models for different agro-ecological regions in India. Central Agroforestry Research Institute (CAFRI), Jhansi, pp 31–36

Paletto A, Notaro S, Sergiacomi C, Di Mascio F (2024) The economic value of forest bathing: an example case of the Italian Alps. Forests 15(3):543. https://doi.org/10.3390/f15030543

Panda S, Dhara PK, Sarkar S, Das NC (2018) Agroforestry: a multipurpose multi—StoreyedRenewable plant treasure with medicinal values giving multi—returns for livelihood and benefits over climate change. In: Pandey CB, Gaur MK, Goyal RK (eds) Climate change and agroforestry: adaptation, mitigation and livelihood security. New India Publishing Agency, New Delhi, pp 627–637

Parthiban KT (2019) Melia dubia based agroforestry systems: mini clonal technology for mass multiplication. In: Handa AK, Dev I, Rizvi RH, Kumar N, Ram A, Kumar D, Kumar A, Bhaskar S, Dhyani SK, Rizvi J (eds) Successful agroforestry models for different agro-ecological regions in India. Central Agroforestry Research Institute (CAFRI), Jhansi, pp 71–84

Qaiser KN, Dutt V, Mughal AH, Mugloo JA (2016) Apple based Horti-Agri-pasture system under temperate conditions of Kashmir valley. In: Chaturvedi OP, Sikka AK, Handa AK, Bajpai CK (eds) Agroforestry technologies for different agro-climatic zones of the country. AllIndia Coordinated Research Project on Agroforestry, ICAR-Central Agroforestry Research Institute, Jhansi, pp 1–177

Raj A, Jhariya MK, Banerjee A, Meena RS (2019) Agroforestry: a holistic approach for agricultural sustainability. In: Jhariya MK et al (eds) Sustainable agriculture, forest and environmental management, pp 101–131. https://doi.org/10.1007/978-981-13-6830-1_4

Sharma SK (2004) Hortipasture based land use systems for enhancing productivity of degraded lands under rain fed and partially irrigated conditions. Uganda J Agric Sci 9:320–325

Sharma S, Johari R (2023) Critical study on physiochemical and preliminary phytochemical parameters of *Bakul* (*Mimusops elengi* Linn.) stem bark and *Babool* (*Acacia arabica* Willd.) stem bark for their gum disorders. Int J Pharm Res Appl 8(4):229–232. www.ijprajournal.com

Shukla AK, Kumar S, Ram SN (2018) Yield, soil health and moisture status in Bael (*Aeglemarmelos*)-based Hortipasture system in semi-arid ecosystem of India. Agrofor Syst 93:1275–1283. https://doi.org/10.1007/s10457-018-0241-7

Singh RP (1996) Alternate land use system for sustaining development. Range Manag Agrofor 17(2):155–178

Singh NR, Jhariya MK, Loushambam RS (2014) Performance of soybean and soil properties under poplar based agroforestry system in Tarai Belt of Uttarakhand. Ecol Environ Conserv 20(4):1569–1573

Singh NR, Arunachalam A, Bhusara JB, Dobriyal MJ, Gunaga RP (2017) Diversification of agroforestry systems in Navsari district of South Gujarat. Indian J Hill Farming 30(1):70–72

Sridhar KB, Dhyani SK, Kumar S, Dwivedi RP, Singh M, Venkatesh A, Monobrullah Goshal S, Dev I, Tewari RK, Singh R, Chavan S, Uthappa AR, Singh R, Tripati VD (2015) A tree with a purpose: Butea monosperma (Lam.) (Improving livelihood of disadvantaged rural people of central India) In: XIV World Forestry Congress, Durban, South Africa, 7–11 September 2015

Udawatta RP, Rankoth L, Jose S (2019) Agroforestry and biodiversity. Sustain For 11(10):2879. https://doi.org/10.3390/su11102879

WAC (World Agroforestry Centre) (2010) Transforming lives and landscapes, pp 1–5

Wen Y, Yan Q, Pan Y et al (2019) Medical empirical research on forest bathing (Shinrin-yoku): a systematic review. Environ Health Prev Med 24:70. https://doi.org/10.1186/s12199-019-0822-8. https://environhealthprevmed.biomedcentral.com/articles/10.1186/s12199-019-0822-8#citeas

Yadav GS, Das A, Lal R, Babu S, Meena RS, Saha P, Singh R, Datta M (2018) Energy budget and carbon footprint in a no-till and mulch based rice–mustard cropping system. J Clean Prod 191:144–157

Concluding Remarks: Agroforestry—Nature-Based Solution for Food Security

8

Abstract

Agroforestry is gaining recognition as a powerful strategy for addressing global environmental and socio-economic challenges. By integrating trees with crops and livestock, agroforestry improves soil health, boosts biodiversity, and enhances water management, while also contributing to carbon sequestration, a key tool in mitigating climate change. This approach offers significant benefits such as improved food security, poverty reduction, and sustainable agricultural productivity, making it a key player in advancing the UN 2030 Sustainable Development Goals. Despite its growing potential, agroforestry adoption remains limited in all classes of income levels of countries, due to economic barriers and insufficient policy support. To fully harness capacity of agroforestry, greater investment in research, education, and policy frameworks is essential. With its ability to diversify farming systems and build resilience to climate impacts, agroforestry represents a crucial step toward a more sustainable, productive, and climate-resilient agricultural future.

Keywords

Agroforestry · Environmental challenges · Socio-economic challenges · Trees · Crops · Livestock · Soil health · Biodiversity · Water management · Carbon sequestration · Climate change mitigation · Food security · Food forests · Poverty reduction · Agricultural landscape · Agricultural and aquatic productivity · Sustainable agricultural productivity · Air pollution · Harmful pollutants · Aesthetic benefits · Tree selection criteria · Livestock farming · Aquaculture · Mud and effluent · Fish-friendly trees · Water-loving plants · Biodiversity · Aquatic plants · Ecosystem · Crop production · UN SDG 2030 Sustainable Development Goals · Adoption barriers · High-income countries · Economic barriers · Policy support · Investment in research and education · Policy frameworks · Farming systems · Resilience to climate impacts · Rural areas · Smallholder farmers · Diversifying incomes · Climate change · Droughts · Planting water · Floods · Sustainable agricultural future

S. Panda, *Agroforestry*, https://doi.org/10.1007/978-981-96-6855-7_8

Abbreviations

EU	European Union
GOI	Government of India
HICs	High-income countries
ICRAF	International Centre for Research in Agroforestry
L&MICs	Low—medium—income countries
SDG	Sustainable development goal
UN	United Nations
USDA	United States Department of Agriculture

8.1 Concluding Remarks

8.1.1 Rise of Agroforestry as a Key Solution for Climate Change, Food Security, and Sustainable Development

Agroforestry, once a niche concept in land-use strategies, has firmly taken center stage in the global conversation on sustainable agriculture and environmental conservation. With the pressing need to combat climate change, restore biodiversity, and secure food for future generations, agroforestry presents a dynamic solution that is rapidly gaining traction worldwide. This integrated approach to land management combines trees with crops and/or livestock on the same piece of land, offering numerous environmental, economic, and social benefits. At its core, agroforestry holds immense promise for carbon sequestration, making it a critical tool in the fight against global warming. The practice naturally captures and stores carbon dioxide in trees, which helps mitigate the rising levels of greenhouse gases in the atmosphere. Beyond its climate-saving capabilities, agroforestry plays a key role in reducing soil erosion and controlling runoff—two major environmental challenges exacerbated by conventional farming practices. In regions prone to flooding and soil degradation, agroforestry helps stabilize landscapes and prevent the loss of fertile soil. Furthermore, it enhances nutrient cycling and water management, ensuring that agricultural lands remain productive and resilient over time. The appeal of agroforestry goes beyond the environment; its socio-economic advantages are equally compelling. By diversifying agricultural production, it offers farmers in both high-income and low- to middle-income countries (HICs and L&MICs) a stable income stream, providing a buffer against market volatility and climate-related shocks. It promotes food security by enhancing the availability and diversity of crops, particularly in areas with fragile ecosystems or limited access to resources. For communities living in poverty or at risk of economic hardship, agroforestry can serve as a nature-based solution to building resilience, improving livelihoods, and reducing vulnerability. On the global stage, agroforestry is increasingly being recognized as a crucial strategy for achieving the United Nations' 2030 Sustainable Development Goals

(SDGs). Its potential to foster sustainable agricultural productivity, improve water quality, and combat climate change aligns directly with several SDGs, including those related to climate action, life on land, and ending poverty. As a result, both policy support and scholarly attention to agroforestry have surged, particularly in high-income countries where the focus is on advancing scientific research and scaling the practice. However, the potential of agroforestry is not limited to developed nations. It is particularly transformative for low- and middle-income countries, where it is seen as an essential mechanism for sustainable development. For farmers in these regions, the practice is not only just about environmental benefits, but also about a tangible means of securing long-term income and promoting community development. With the growing recognition of far-reaching impacts of agroforestry, both local and international stakeholders are beginning to prioritize it in policy frameworks and funding allocations. Despite the growing enthusiasm, challenges remain in fully implementing agroforestry across diverse landscapes and agricultural systems. Knowledge gaps exist regarding the socio-economic impacts of different agroforestry practices, and there is a need for more research on the biophysical outcomes to better inform policies. Nevertheless, the expanding evidence base across disciplines and regions provides a strong foundation for knowledge sharing and informed decision-making. As we look toward the future, the increasing adoption of agroforestry promises to reshape our relationship with the land. By embracing this sustainable, multi-benefit approach, we are not only ensuring a more resilient environment but also laying the groundwork for a more equitable, prosperous global society. Agroforestry, it seems, is not just a land-use strategy–it is a cornerstone of our shared future (Garrity 2004; Jose 2009; Lassoie et al. 2009; Jose and Bardhan 2012; Jose et al. 2012; Smith et al. 2012; Tsonkova et al. 2012; Udawatta and Jose 2012; UN SDG 2030 2015; Fagerholm et al. 2016; Wilson and Lovell 2016; Waldron et al. 2017; Agroforestry Network 2018; Lovell et al. 2018; Razafindratsima et al. 2021; Grow Billion Trees 2025).

Agroforestry, the practice of integrating trees and shrubs with crops and livestock on a single land unit, is gaining recognition for its potential to diversify production systems and provide numerous environmental, economic, and social benefits. By fostering complementary interactions between these components, agroforestry systems, such as agrisilviculture, silvopasture, and agrosilvopasture, not only enhance biodiversity but also improve soil health, water retention, or the idea of 'Planting water' upheld Ernst Götsch (2018) in Bahia, Brazil and, carbon sequestration. These systems can vary from simple tree retention in pastures to complex multistrata food forests with a diverse range of species. Agroforestry can be applied across scales—from individual plots to entire landscapes and offers significant ecosystem services. The impact of these services, like climate regulation or enhanced biodiversity, often becomes more evident when considered at a larger landscape scale. With the flexibility to integrate into various agricultural operations, including urban as well as rural agroforestry farming, this approach is poised to reshape sustainable land management for the future (USDA 2011; Atangana et al. 2014; Van Noordwijk et al. 2019; Castle et al. 2022).

8.1.2 Agroforestry and Aquaculture Integration

Similarly, like integration of livestock farming with agroforestry, integrating aquaculture with agroforestry farming systems is potentially a promising and sustainable approach to enhancing both biodiversity and productivity. By planting fish-friendly trees along pond banks, such as the gum Arabic tree (*Vachellia nilotica*), Indian almond tree (*Terminalia catappa*), jujube tree (*Ziziphus mauritiana*), acacia coral tree (*Adenanthera pavonine*), drumstick tree (*Moringa oleifera*), mango tree (*Mangifera indica*), and coconut palm (*Cocos nucifera*), and many more, farmers can create a synergistic environment that benefits both aquatic life and soil health. These trees, combined with local water-loving grasses, sedges, and herbaceous plants like the canna lily (*Canna indica* L.) and garden commelina (*Commelina benghalensis* L.), purple bladderwort (*Utricularia purpurascens*), net veined bladderwort (*Utricularia reticulata* Sm.), mat sedge (*Cyperus pangorei* Rottb.), etc. provide crucial support for erosion control while offering habitats for fish. Additionally, incorporating aquatic plants like the Indian lotus (*Nelumbo nucifera*), water lily (e.g., *Nymphaea rubra*), and water chestnut (*Trapa natans* L.), feathered waterfern (*Azolla pinnata*), etc. further enriches the ecosystem, improving water quality and fostering a diverse, resilient aquaculture system. Mud and effluent from naturally reared aquaculture ponds are also fertile which, in tum, is beneficial for crop production. This method, when tailored to local conditions, presents a sustainable, multi-functional model for enhancing both agricultural and aquatic productivity (BBG 1985; Cook 1996; Roy et al. 2006; Sileshi et al. 2009; Illera-Vives et al. 2015; Moyo et al. 2015; Kambale et al. 2016; Zhou et al. 2016; Lynch et al. 2018; Arisdason et al. 2019; Yuan et al. 2019; Madende and Hayes 2020; Gao et al. 2021; Zhang et al. 2020; Lall and Kaushik 2021; Cerozi et al. 2022; Brod et al. 2023; Fruscella et al. 2023; Sanduni Madhusanki et al. 2023; Jaies et al. 2023; Miller 2025).

8.1.3 Potentiality of Agroforestry in All Countries of Various Income Groups As Well As Addressing Ecological Balance

Agroforestry research has evolved from studying traditional practices used by local populations globally, providing the foundation for more structured and experimental investigations. As this research progressed, scientists uncovered the significant potential of agroforestry to tackle pressing environmental and social issues, including climate change and food security. Recognizing these benefits, enthusiasts for agroforestry began advocating for policies and programs that support the integration of trees into agricultural landscapes, aiming to create more sustainable and resilient farming systems. This shift highlights agroforestry as a vital tool in addressing contemporary global challenges (Nair and Garrity 2012).

Agroforestry practices, which have long been used in low-and middle-income countries (L&MICs) as a solution for environmental sustainability and food security, are only relatively recent phenomena in high-income countries (HICs). While international organizations like World Agroforestry (ICRAF) have championed agroforestry since the 1960s and 1970s, formal agroforestry policies in countries like the USA and the European Union emerged much later. In the USA, agroforestry policy began to take shape in the mid-1980s, with more substantial efforts only occurring in the 1990s, such as the establishment of the National Agroforestry Center. Likewise, the EUs approach to agroforestry shifted in the early 1990s, following the reform of the Common Agricultural Policy, which had historically discouraged the integration of trees in farming systems. It is only in the past decade that agroforestry has gained more traction in HICs, with growing institutional support for these practices as part of broader efforts to promote conservation and sustainable agricultural development. This delayed adoption highlights the evolving recognition of agroforestry as a viable and essential component of modern agricultural and environmental policy (Garrity et al. 2010; Jose et al. 2012; Mosquera-Losada et al. 2012; Waldron et al. 2017; USDA 2019; Santiago-Freijanes et al. 2021).

Agroforestry has garnered significant attention for its potential to offer a wide range of ecological, environmental, economic, and social benefits, encouraging landowners to adopt this practice. As interest in agroforestry grows, there is increasing recognition of its role in creating a diversified, resilient farming system. However, despite its promising benefits, the widespread adoption of agroforestry remains limited, with the USDA estimating that it is implemented on less than 1% of agricultural land eligible for agroforestry programs. This highlights a crucial opportunity for countries to enhance efforts to promote agroforestry through education and support initiatives. Notably, India took a step forward by adopting the National Agroforestry Policy in 2014. Recent research has underscored the potential of agroforestry in areas such as climate change adaptation, carbon sequestration, soil health, and food production. Studies also focus on the integration of agroforestry with livestock, like silvopasture, to improve farm productivity, water and air quality, and overall sustainability, as well as the positive effects of vegetated strips like windbreaks and hedgerows on soil retention, biodiversity, and socioeconomics in temperate systems (Garrett and Buck 1997; Montambault and Alavalapati 2005; Jose 2009; Jose and Bardhan 2012; Jose et al. 2012; GOI 2014; Lovell et al. 2018; Tsonkova et al. 2012; Holzmueller and Jose 2012; Jordan and Warner 2013; USDA 2013; Dollinger and Jose 2018; Wolz et al. 2018; Jordon et al. 2020).

8.1.4 Trees with Potentiality for Reducing Air Pollution: An Obvious One of the Criteria for Agroforestry Tree Selection

To neutralize ill effects of climate change as well as reducing air pollution in the locality, one of the agroforestry tree selection criteria should be inclusion of trees with potentiality for reducing air pollution and trees suitable for bringing rain like *arjuna* tree (*Terminalia arjuna* (Roxb.) Wight & Arn., arjun), though *arjuna* tree not to be included in the homestead agroforestry, as mentioned in the *slokas* 30 and 323 of *Surapala's Vrikshyurveda* (Sadhale 1996). As urbanization rapidly expands, air pollution has become an urgent environmental challenge, but trees present a natural solution. Indian tree species such as the Indian redwood (*Caesalpinia sappan*), shisham (*Dalbergia sissoo*), and neem (*Azadirachta indica*) are emerging as vital allies in combating air pollution while enhancing urban spaces. These trees absorb harmful pollutants like sulfur, lead, and carbon dioxide, improving air quality and providing valuable aesthetic benefits. The banyan tree (*Ficus benghalensis*), with its extensive roots and canopy, efficiently filters pollutants, while the *gulmohar* tree (*Delonix regia*) offers much-needed shade and helps reduce temperatures. Other trees, like *ashoka* (*Saraca asoca*), *deodar* or cedar (*Polyalthia longifolia*), guava (*Psidium guajava*), and mango (*Mangifera indica*), further contribute to absorbing harmful chemicals and promoting long-term ecological health. Integrating these species into urban landscapes is an essential step toward fostering greener, healthier environments for future generations. (Sadhale 1996; Chaturvedi et al. 2017; Agroforestry Network 2019; Traverso 2020; Dubey et al. 2023; Grover 2023; Nature WorldWide 2024; Neem Foundation 2025).

8.1.5 Overall Views on the Previous Seven Chapters

From Chap. 1, one can agree that agroforestry systems are emerging as a critical solution to two of the most pressing global challenges: food security and climate change. By integrating trees into agricultural landscapes, agroforestry offers a unique approach that supports the diversification of crops, enhances nutritional value, and strengthens the resilience of food systems against climate impacts like droughts and floods. This practice not only boosts farmers' income through the sale of both crops and forest products but also contributes significantly to environmental conservation by improving soil health, increasing biodiversity, and sequestering carbon. Furthermore, agroforestry plays a key role in public health by providing fresh produce free from the pesticides, often used in monoculture farming. To unlock the full potential of agroforestry, it is essential to invest in education, research, and policies that support smallholder farmers. Strengthening access to resources, providing technical training, and fostering market development for agroforestry products will enable these systems to thrive and address the interconnected issues of food security, public health, and environmental sustainability. As the world faces unprecedented environmental challenges, agroforestry presents a pathway to a more sustainable, resilient, and healthy future.

From Chap. 2, it is visible that agroforestry is rapidly emerging as a sustainable solution for cereal-based cropping systems, offering numerous ecological and economic benefits. Agroforestry systems offer a promising solution to the intertwined challenges of climate change, food security, and public health by integrating trees into agricultural landscapes. This sustainable land-use practice not only enhances biodiversity, improves soil health, and sequesters carbon, but also boosts food security and income generation for farmers. By diversifying crops and incorporating perennial trees, agroforestry systems ensure a more resilient multipurpose food supply, protect ecosystems, and help mitigate the effects of climate change. Moreover, these systems offer access to nutritious foods like fruits and nuts while reducing reliance on harmful chemicals. However, to fully realize these benefits, it is crucial for governments and organizations to support research, invest in farmer education, and foster policy frameworks that promote agroforestry. With concerted effort, agroforestry can become a cornerstone of sustainable development, benefiting both the environment and public health while supporting rural economies.

From Chap. 3, it is clear that the integration of agroforestry systems into agricultural landscapes is gaining recognition as a vital strategy for fostering sustainability. By integrating trees with agricultural crops or livestock, agroforestry promotes ecological balance, reduces dependency on synthetic fertilizers and pesticides, and enhances farm resilience. This diversified approach improves soil conservation, increases biodiversity, mitigates climate change, and boosts crop yields. In regions like the Indo-Gangetic Plains, poplar trees (*Populus* sp.), in particular, are being cultivated alongside wheat, showing promising results with increased wood yields and profitability. Additionally, agroforestry offers solutions for challenging environments like the Trans-Himalayan region, where trees can improve crop production by providing shade and increasing humidity. Despite its advantages, designing effective agroforestry systems requires careful planning, including tree species selection, soil management, and irrigation strategies. As India continues to evaluate and refine agroforestry models, it holds great potential to transform farming across diverse ecological regions, offering a pathway to more sustainable, resilient agricultural practices.

From Chap. 4, it is manifested that the interconnectedness between soil health, agroforestry, and climate resilience is crucial for sustainable land management, particularly as climate change intensifies. By promoting soil health and microbial activity, agroforestry enhances soil organic carbon (SOC) levels, reduces the need for chemical fertilizers, and mitigates soil erosion, ultimately boosting productivity while preserving the environment. The presence of trees and shrubs improves nutrient cycling, water retention, and soil structure, creating a more resilient agricultural system. Moreover, the potential of agroforestry for carbon sequestration makes it an essential tool in combating climate change. However, challenges such as market access, economic feasibility, and insufficient policy support hinder its widespread adoption. To overcome these barriers, targeted interventions like improved infrastructure and farmer education are crucial. With the backing of research and community engagement, agroforestry has the potential to transform global agriculture, aligning productivity with ecological health and advancing sustainable development goals.

From Chap. 5, it is undoubtedly established that agroforestry plays a pivotal role in mitigating climate change, improving soil health, and enhancing energy efficiency in agriculture. Agroforestry, which combines trees with crops and livestock, enhances soil health by improving water retention, nutrient cycling, and erosion resistance, offering a natural buffer against extreme weather events like droughts and floods. This approach not only stabilizes agricultural yields but also boosts biodiversity and carbon sequestration, addressing both ecological and climatic challenges. The adaptability of agroforestry systems to diverse climates makes it a versatile solution for farmers and conservationists, enhancing food security and reducing dependency on external inputs. However, despite its proven benefits, research gaps remain, especially regarding role of agroforestry in climate adaptation and mitigation in high-income countries. To fully unlock the potential of agroforestry, it is essential for governments, policymakers, and farmers to prioritize research, expand adoption, and implement supportive policies, ensuring agroforestry becomes a cornerstone of resilient and sustainable agricultural systems worldwide.

From Chap. 6, it is resolved that agroforestry systems, with their innovative integration of trees, shrubs, and crops, are proving to be essential tools in addressing key environmental challenges in agriculture today. These systems combat soil erosion by using deep-rooted vegetation to stabilize the soil, promoting water retention, and ensuring that crops thrive, even in drought-prone areas. By integrating trees into farming landscapes, agroforestry systems increase both aboveground and belowground carbon stocks, making them highly effective for carbon sequestration and soil conservation. These systems not only capture more carbon compared to conventional farming, but they also foster soil organisms that promote decomposition, enriching soil and creating a positive feedback loop for carbon storage. Moreover, agroforestry reduces greenhouse gas emissions by sequestering carbon and decreasing reliance on chemical fertilizers, thus reducing the energy footprint of agricultural practices. Despite its potential, economic barriers hinder widespread adoption, limiting its full impact. However, as a sustainable agricultural practice, agroforestry contributes to biodiversity, improves productivity, and offers a valuable strategy to address both climate change and food security challenges.

From Chap. 7, it is explored that value addition to agroforestry products presents a promising strategy for sustainable development, rural empowerment, and environmental preservation. Agroforestry, which blends trees, shrubs, and agricultural practices, offers a range of ecological benefits, including soil conservation, biodiversity enhancement, and carbon sequestration. Despite these advantages, its potential remains underutilized in many rural communities, where subsistence farming predominates. By processing raw agroforestry products into higher-value goods, value addition can significantly improve marketability, diversify income sources, and provide new economic opportunities for rural populations. This approach not only boosts profitability but also reduces dependence on harmful agricultural practices, fosters job creation, and strengthens local economies. Furthermore, value-added agroforestry systems contribute to enhanced soil fertility, better water regulation, and improved carbon capture. However, challenges in implementation exist, which can be overcome through strategic planning and community engagement, as evidenced by successful case studies demonstrating the transformative impact of this approach on rural livelihoods and environmental resilience. Some numerical problems with solutions will invigorate all farming communities, students, interns and researchers to systematically learn, to take up and harvest the benefits of agroforestry through hands-on practice on their farms.

Thus, lastly, it can be resolved that agroforestry is emerging as a powerful solution to the intertwined global issues of food security, climate change, and environmental degradation. By integrating trees into agricultural systems, agroforestry not only enhances biodiversity and soil health but also increases resilience against climate impacts like droughts and floods. This system provides a sustainable alternative to monoculture farming, boosting crop yields while reducing the reliance on harmful chemicals. Furthermore, the role of agroforestry in carbon sequestration makes it a key player in mitigating climate change. In addition to its environmental benefits, agroforestry offers significant economic opportunities, particularly in rural areas, by diversifying incomes through the sale of both crops and forest products. However, to fully realize its potential, there is a need for increased investment in research, education, and supportive policies, especially for smallholder farmers. With strategic planning, agroforestry could transform global agriculture, providing a sustainable and resilient pathway for the future.

References

Agroforestry Network (2018) Achieving the global goals through agroforestry. Agroforestry Network and VI-Skogen Stockholm 2018. https://agroforestrynetwork.org/database_post/achieving-the-global-goals-through-agroforestry/

Agroforestry Network (2019) In: Gustafsson M (ed) Agroforestry for adaptation and mitigation to climate change. Stockholm International Water Institute, SIWI and Vi Agroforestry/Vi-skogen, Stockholm. https://siwi.org/wp-content/uploads/2020/02/Agroforestry-for-adapation-and-mitigation-to-climate-change_web.pdf?utm_source

Arisdason W, Sampath V, Karthigeyan K, Maurya ON, Avinash (2019) Aquatic and wetland plants of India [pamphlet]. ENVIS Resource Partner on Biodiversity, Botanical Survey of India, Kolkata. http://www.bsienvis.nic.in

Atangana A, Khasa D, Chang S, Degrande A (2014) Definitions and classification of agroforestry systems. In: Tropical agroforestry. Springer, Dordrecht, pp 35–47

BBG (1985) Water gardening. In: Plants & gardens, vol 41. Brooklyn Botanic Garden Record, Brooklyn, NY

Brod E, Henriksen TM, Ørnsrud R, Eggen T (2023) Quality of fish sludge as fertiliser to spring cereals: nitrogen effects and environmental pollutants. Sci Total Environ 875:162541. https://doi.org/10.1016/j.scitotenv.2023.162541

Castle SE, Miller DC, Merten N et al (2022) Evidence for the impacts of agroforestry on ecosystem services and human well-being in high-income countries: a systematic map. Environ Evid 11:10. https://doi.org/10.1186/s13750-022-00260-4

Cerozi BS, Arlotta CG, Richardson ML (2022) Fish effluent as a source of water and nutrients for sustainable urban agriculture. Agriculture 12:1975. https://doi.org/10.3390/agriculture12121975

Chaturvedi OP, Handa AK, Uthappa AR, Sridhar KB, Kumar N, Chavan SB, Rizvi J (2017) Promising agroforestry tree species in India. Jointly published by the Central Agroforestry Research Institute (CAFRI), Jhansi, and the South Asia regional Programme (based in Delhi, India) of the World Agroforestry Centre. https://www.cifor-icraf.org/publications/downloads/Publications/PDFS/B17969.pdf

Cook CDK (1996) Aquatic and wetland plants of India. Oxford University Press, Oxford. ISBN:9780198548218, 0198548214

Dollinger J, Jose S (2018) Agroforestry for soil health. Agrofor Syst 92(2):213–219

Dubey R, Choudhary AK, Singh S, Ajay A, Kumar S, Kumar R, Mondal S, Singh VK (2023) Assessing the impact of air pollution on trees and crops in the eastern Gangetic Plains of India. Curr Sci 124(8):956–963. https://doi.org/10.18520/cs/v124/18/956-963

Fagerholm N, Torralba M, Burgess PJ, Plieninger T (2016) A systematic map of ecosystem services assessments around European Agroforestry. Ecol Indic 62:47–65

Fruscella L, Kotzen B, Paradelo M, Milliken S (2023) Investigating the effects of fish effluents as organic fertilisers on onion (Allium cepa) yield, soil nutrients, and soil microbiome. Sci Hortic 321:112297. https://doi.org/10.1016/j.scienta.2023.112297

Gao Y, Zhang H, Peng C, Lin Z, Li D, Lee CT, Wu W-M, Li C (2021) Enhancing nutrient recovery from fish sludge using a modified biological aerated filter with sponge media with extended filtration in aquaponics. J Clean Prod 320:128804. https://doi.org/10.1016/j.jclepro.2021.128804

Garrett HE, Buck L (1997) Agroforestry practice and policy in The United States of America. For Ecol Manage 91(1):5–15

Garrity DP (2004) Agroforestry and the achievement of the millennium development goals. Agrofor Syst 61(1):5–17

Garrity DP, Akinnifesi FK, Ajayi OC, Weldesemayat SG, Mowo JG, Kalinganire A, Larwanou M, Bayala J (2010) Evergreen agriculture: a robust approach to sustainable food security in Africa. Food Secur 2(3):197–214

GOI (2014) National agroforestry policy 2014. Government of India, Department of Agriculture & cooperation, Ministry of Agriculture, New Delhi. https://www.nitiforstates.gov.in/public-assets/Policy/policy-repo/agriculture-and-allied-services/PNC511A000509.pdf

Götsch E (2018) Planting water – Agroforestry research at Ernst Götsch in Bahia. A blog published on July 3, 2018 in reNature website: https://www.renature.co/articles/ernst-gotsch-planting-water/

Grover N (2023) 10 trees to beat air pollution. Environment, Planning Tank. https://planningtank.com/environment/10-trees-to-beat-air-pollution. Accessed 30 Apr 2023

Grow Billion Trees (2025) Agroforestry in India: bridging agriculture and sustainability. Grow Billion Trees. https://growbilliontrees.com/

Holzmueller EJ, Jose S (2012) Biomass production for biofuels using agroforestry: potential for the north central region of the United States. Agrofor Syst 85(2):305–314

Illera-Vives M, López-Fabal A, López-Mosquera ME, Ribeiro HM (2015) Mineralization dynamics in soil fertilized with seaweed-fish waste compost. J Sci Food Agric 95(13):3047–3054. https://doi.org/10.1002/jsfa.7207

Jaies I, Qayoom I, Saba F, Khan S (2023) Fish wastes as a source of fertilizers and manures. In: Maqsood S et al (eds) Fish waste to valuable products, sustainable materials and technology, pp 329–338. https://doi.org/10.1007/978-981-99-8593-7_15

Jordan N, Warner KD (2013) Towards multifunctional agricultural landscapes for the upper Midwest region of the USA. In: Wratten SD, Sandhu H, Cullen R, Costanza R (eds) Ecosystem services in agricultural and urban landscapes. Wiley, pp 139–156

Jordon MW, Willis KJ, Harvey WJ, Petrokofsky L, Petrokofsky G (2020) Implications of temperate agroforestry on sheep and cattle productivity, environmental impacts and enterprise economics. A systematic evidence map. Forests 11(12):1321

Jose S (2009) Agroforestry for ecosystem services and environmental benefits: an overview. Agrofor Syst 76(1):1–10

Jose S, Bardhan S (2012) Agroforestry for biomass production and carbon sequestration: an overview. Agrofor Syst 86(2):105–111

Jose S, Gold MA, Garrett HE (2012). The future of temperate agroforestry in the United States. In: Nair PKR, Garrity D, editors. Agroforestry-the future of global land use. Dordrecht: Springer; 2012. p. 217–245

Kambale SS, Kolte RR, Janarthanam MK (2016) An extended distribution of recently described Natesh's cape pondweed (Aponogetonaceae: Aponogeton noteshi) to the state of Karnataka, India. J Threat Taxa 8(4):8641–8744. https://doi.org/10.11609/jott.2016.8.4.8641-8744

Lall SP, Kaushik SJ (2021) Nutrition and metabolism of minerals in fish. Animals 2021(11):2711. https://doi.org/10.3390/ani11092711

Lassoie JP, Buck LE, Current D (2009) The development of agroforestry as an integrated land use management strategy. In: North American Agroforestry: an integrated science and practice, 2nd edn. American Society of Agronomy, Madison, WI, pp 1–24

Lovell ST, Dupraz C, Gold M, Jose S, Revord R, Stanek E, Wolz KJ (2018) Temperate agroforestry research: considering multifunctional woody polycultures and the design of long-term field trials. Agrofor Syst 92(5):1397–1415

Lynch A, Heng S, Pringle H (2018) Sustainability in aquaculture-agroforestry systems: case studies from Southeast Asia. J Sustain Agric 42(5):510–524

Madende M, Hayes M (2020) Fish by-product use as biostimulants: an overview of the current state of the art, including relevant legislation and regulations within the EU and USA. Molecules 25(5):1122. https://doi.org/10.3390/molecules25051122

Miller EC (2025) Plants for banks of a fish pond. Elisabeth C. Miller Library. University of Washington. https://depts.washington.edu/hortlib/pal/plants-for-banks-of-a-fish-pond/

Montambault JR, Alavalapati JRR (2005) Socioeconomic research in agroforestry: a decade in review. Agrofor Syst 65(2):151–161

Mosquera-Losada MR, Moreno G, Pardini A, McAdam JH, Papanastasis V, Burgess PJ, Lamersdorf N, Castro M, Liagre F and Rigueiro-Rodríguez A (2012). Past, present and future of agroforestry systems in Europe. In: Nair PKR, Garrity D, editors. Agroforestry—the future of global land use. Dordrecht: Springer; 2012. p. 285–312

Moyo B, Masika P, Hugo A, Muchenje V (2015) Nutritional value of Moringa (Moringa oleifera) leaves for human consumption. Food Res Int 65:151–157

Nair PKR, Garrity D (eds) (2012) Agroforestry—the future of global land use. Springer, Dordrecht. https://doi.org/10.1007/978-94-007-4676-3

Nature WorldWide (2024) Breathing fresh: top 10 air purifying trees in India. Nature WorldWide 2024. https://natureworldwide.in/breathing-fresh-top-10-air-purifying-trees-in-india/?utm_source. Accessed 26 Feb 2024

Neem Foundation (2025) Greening India with neem. Neem Foundation. https://neemfoundation.org/greening-india-with-neem/?utm_source

Razafindratsima OH, Kamoto JFM, Sills EO, Mutta DN, Song C, Kabwe G, Castle SE, Kristjanson PM, Ryan CM, Brockhaus M et al (2021) Reviewing the evidence on the roles of forests and tree- based systems in poverty dynamics. Forest Policy Econ 2021(131):102576

Roy RN, Finck A, Blair GJ, Tandon HLS (2006) Plant nutrition for food security: a guide for integrated nutrient management. In: FAO Fertilizer and Plant Nutrition Bulletin no. 16. Food and Agriculture Organization of the United Nations, Rome

Sadhale N (1996) Surapala's Vrikshyurveda (the science of plant life by Surapala). Agri-history bulletin no. 1. Asian Agri-History Foundation, Secunderabad

Sanduni Madhusanki FH, Kankanamge Epa UP et al (2023) Use of aquaculture pond sludge and fish waste to produce manure worm, Eisenia fetida (Lumbricidae), vermiwash. J Fish Environ 47(3):126–137

Santiago-Freijanes JJ, Mosquera-Losada MR, Rois-Díaz M, Ferreiro-Domínguez N, Pantera A, Aldrey JA, Rigueiro-Rodríguez A (2021) Global and European policies to foster agricultural sustainability: agroforestry. Agrofor Syst 95(5):775–790

Sileshi G, Tadesse T, Tesfaye G (2009) Agroforestry in Ethiopia: practices and benefits. FAO, Rome

Smith J, Pearce BD, Wolfe MS (2012) Reconciling productivity with protection of the environment: is temperate agroforestry the answer? Renew Agric Food Syst 28(01):80–92

Traverso V (2020) Which trees reduce air pollution best? BBC Future. https://www.bbc.com/future/article/20200504-which-trees-reduce-air-pollution-best. Accessed 4 May 2020

Tsonkova P, Böhm C, Quinkenstein A, Freese D (2012) Ecological benefits provided by alley cropping systems for production of woody biomass in the temperate region: a review. Agrofor Syst 85(1):133–152

Udawatta RP, Jose S (2012) Agroforestry strategies to sequester carbon in temperate North America. Agrofor Syst 86(2):225–242

UN SDG 2030 (2015) Transforming our world: the 2030 Agenda for Sustainable Development. 70/1. Resolution adopted by the General Assembly on 25 September 2015 [without reference to a Main Committee (A/70/L.1)]. United Nations General Assembly. Seventieth session, Agenda items 15 and 116. Accessed 21 Oct 2015

USDA (2011) USDA agroforestry strategic framework, fiscal year 2011–2016. US Department of Agriculture, Washington, DC

USDA (2013) Agroforestry USDA reports to America, fiscal years 2011–2012—comprehensive version. U.S. Department of Agriculture, Washington, DC

USDA (2019) USDA agroforestry strategic framework, fiscal year 2019–2024. U. S. Department of Agriculture, Washington, DC

Van Noordwijk M, Coe R, Sinclair F (2019) Agroforestry paradigms. In: Sustainable development through trees on farms: agroforestry in its fifth decade. World Agroforestry (ICRAF) Southeast Asia Regional Program, Bogor, Indonesia, pp 1–14

Waldron A, Garrity D, Malhi Y, Girardin C, Miller DC, Seddon N (2017) Agro- forestry can enhance food security while meeting other sustainable development goals. Trop Conserv Sci 10:1–6

Wilson M, Lovell S (2016) Agroforestry-the next step in sustainable and resilient agriculture. Sustainability 8(6):574

Wolz KJ, Lovell ST, Branham BE, Eddy WC, Keeley K, Revord RS, Wander MM, Yang WH, DeLucia EH (2018) Frontiers in alley cropping: transformative solutions for temperate agriculture. Glob Chang Biol 24(3):883–894

Yuan J, Zhang L, Xie L (2019) Effects of floating aquatic plants on water quality improvement in integrated aquaculture systems. Aquacult Res 50(7):2042–2051

Zhang H, Gao Y, Shi H, Lee CT, Hashim H, Zhang Z, Wu W-M, Li C (2020) Recovery of nutrients from fish sludge in an aquaponic system using biological aerated filters with ceramsite plus lignocellulosic material media. J Clean Prod 258:120886. https://doi.org/10.1016/j.jclepro.2020.120886

Zhou D, Wang Y, Liu J (2016) The role of sedges and aquatic plants in pond edge stabilization and water quality in aquaculture systems. Environ Sci Pollut Res 23(8):7964–7974

MIX
Papier aus verantwortungsvollen Quellen
Paper from responsible sources
FSC® C105338

If you have any concerns about our products,
you can contact us on
ProductSafety@springernature.com

In case Publisher is established outside the EU,
the EU authorized representative is:
Springer Nature Customer Service Center GmbH
Europaplatz 3, 69115 Heidelberg, Germany

Printed by Libri Plureos GmbH
in Hamburg, Germany